Multiple Regression and Causal Analysis

Multiple Regression and Causal Analysis

McKee J. McClendon
Case Western Reserve University

Prospect Heights, Illinois

To Pat and Kara

For information about this book, contact:
Waveland Press, Inc.
P.O. Box 400
Prospect Heights, Illinois 60070
(847) 634-0081
www.waveland.com

2002 reissued by Waveland Press, Inc.

ISBN 1-57766-243-1

Printed in the United States of America

7 6 5 4 3 2 1

Contents

Preface

First and foremost, the purpose of *Multiple Regression and Causal Analysis* is to describe applied regression analysis as it is practiced in the social and behavioral sciences. Another goal, however, is to integrate coverage of regression analysis and causal analysis; these two subjects are usually treated in separate volumes. Of the two, the majority of space is devoted to regression analysis (Chapters 2 through 7). The coverage of multiple regression is, however, based on the assumption that most social scientists who are using this statistical technique are either explicitly or implicitly using it to conduct nonexperimental causal analyses; that is, their goal is explanation, not simply description or prediction. Therefore, the various multiple regression topics covered in this book are presented from the perspective of causal analysis. This is accomplished in several ways. First, an overview of issues and techniques for conducting causal analyses of nonexperimental data is given in Chapter 1. Second, the causal perspective is evident in the extended discussion in Chapter 3 of how multiple regression statistically controls other independent variables in order to estimate the effect of any particular independent variable, when experimental control (i.e., randomization) is not available. Third, this perspective more subtly informs the coverage of much of the other material on multiple regression in Chapters 4 through 7.

Most importantly, however, multiple regression and causal analysis are united in the last two chapters of the book (Chapters 8 and 9), which are entirely devoted to principles and techniques of causal analysis in nonexperimental research. In these chapters, multi-equation causal models are presented as direct extensions of the single-equation regression models covered in the previous chapters. Only models that can be estimated with ordinary least-squares regression (i.e., recursive models) are covered, with the exception that students are taught to recognize nonrecursive models—those with feedback loops—and to understand why they can't be estimated with ordinary least-squares. Because of the continuity of terminology, notation, statistical method (ordinary least-squares), and causal perspective, students should perceive these chapters on causal modeling not as qualitatively new subject matter, but as more of the same old stuff. At the same time, however, the introduction of multi-equation causal models in Chapter 8 permits a distinction to be made between different types of causal effects (total, indirect, and direct effects) and allows a more precise discussion of the notion of spuriousness than was possible before. The final chapter (Chapter 9) deals with the decomposition of variances and covariances in recursive models. Along with the fact that it enhances our understanding of causal systems, this material is a useful preparation for the more advanced study of causal models with latent variables (e.g., so-called LISREL models).

As stated above, the most important goal of the book is simply to give students a working knowledge of the fundamentals of applied regression analysis as practiced in the social and behavioral sciences. Although a causal perspective is emphasized, the regression material is conventional in content and can be taught in a conventional manner by instructors not interested in stressing or teaching causal analysis. Bivariate regression and correlation are reviewed in some detail in Chapter 2. It is emphasized that these statistics can be understood and interpreted in terms of three statistical building blocks: the mean, the variance, and the covariance. In addition, the Pearson correlation is defined and interpreted as a standardized regression coefficient to maintain continuity with the regression orientation of the book.

The multiple regression equation is introduced in Chapter 3. In this chapter, we define, interpret, and discuss all the basic descriptive statistics for multiple regression: the unstandardized and standardized partial slopes, the intercept, the standard error of estimate, the partial correlation, the squared partial correlation, the semi-partial correlation, the squared semi-partial correlation, the multiple correlation, the squared multiple correlation, and the adjusted (shrunken) squared multiple correlation. In a somewhat unique approach, statistical control through multiple regression is heuristically presented as involving the residualization of each independent variable on all the other independent variables. In the case of two independent variables, this perspective allows one to define and understand the various partial regression and correlation coefficients in terms of series of bivariate regression equations. Thus, conceptually, multiple regression and correlation are presented as more complicated aspects of bivariate regression. The concept of bivariate regression was interpreted in terms of the three most elementary statistics, i.e., the mean, variance, and covariance. Thus, as new and more complicated statistics are introduced here, and in subsequent chapters, they are grounded conceptually in terms of more elementary statistics that have been covered previously.

Tests of statistical significance (inferential statistics) are not introduced until Chapter 4. The concept of the sampling distribution of a slope is introduced with a hypothetical example consisting of the slopes for all distinct samples that could be selected from a small population. The characteristics of the theoretical sampling distribution are covered next, followed by the necessary assumptions about the error term and the consequences of violating these assumptions. The t tests and F tests for the slopes of the independent variables, the F tests for subsets of variables, and the F test for the equation as a whole are then described and explained. The chapter concludes with a sizeable section on detecting and correcting for heteroskedasticity (including the use of weighted least-squares), which is arguably the violation of regression assumptions that is of most concern to social scientists.

Whereas in Chapters 2 through 4 the core statistical concepts of linear regression analysis are covered, in Chapters 5 through 7 we explain how to deal with three common problems in linear regression—nominal independent variables, nonlinear relationships, and nonadditive relationships. In Chapter 5, we look at dummy, effects, and contrast coding of nominal independent variables, as well as their identity with ANOVA. The topic of Chapter 6 is the estimation of nonlinear relationships by nominalization of quantitative variables, power polynomials, and transformations for linearity. In Chapter 7, we describe how to test and interpret simple and complex nonadditive relationships (i.e., interactions) by using products of variables. No new statistical concepts are used in these chapters. Instead, the focus is on how to use transformations of variables—such as dummy coding, powers, and products—in order to incorporate nominal variables, nonlinearity, and nonadditivity into regression equations. Although the nonlinear and nonadditive techniques covered for single-equation models in Chapters 6 and 7 could be extended to the multi-equation models covered in Chapters 8 and 9, this has not been done. To do so, I believe, would unduly increase the level of complexity and difficulty in order to cover material that is seldom utilized in social science research.

The prerequisite for a course using this book should be a fairly standard introductory course in social statistics covering inferential statistics, bivariate regression and correlation, and one-way ANOVA. Although the primary audience is probably graduate students in the social and behavioral sciences taking their second statistics course, the book would also be appropriate for undergraduates at major research universities who are required to take a two-course statistics sequence.

The book is intended to be accessible to readers without a great deal of mathematical sophistication. It does not assume any knowledge of calculus or matrix algebra; however, an introduction to matrix algebra for multiple regression is presented in Appendix 3A. Mathematical derivations are not given for the formulas and equations presented. This helps to minimize the mathematical difficulty of the material. Definitional formulas, however, are emphasized for important statistics because these formulas provide precise and insightful definitions. Computational formulas are not used because it is now the norm, and a virtual necessity in many cases, to let computers do the calculations. Some extended calculations for small hypothetical data sets are presented in Chapters 2 and 3 to illustrate concretely the various elements and terms that comprise the definitional formulas. Otherwise, examples of SPSS[1] commands are provided for use in generating the computer outputs that illustrate the regression analyses throughout book.

The style of exposition emphasizes a more verbal mode of presentation in comparison to one of mathematical elegance and parsimony. Thus, it is hoped that the book will be more accessible to social science students than the typical econometrics text. It also does not assume that the students have backgrounds steeped in experimental design and analysis of variance, as some regression texts tend to do.

I acknowledge permission from SPSS Inc. to include commands from their software package to illustrate how to do regression analyses with the computer. The Inter-University Consortium for Political and Social Research provided the General Social Surveys and the Quality of Employment Survey used throughout the book. The various data and function plots were constructed with the SYSTAT software package.[2] Path diagrams and other drawings were made with Canvas by Deneba Software.

I am indebted to many people for encouragement and for constructive criticism. In particular, I would like to thank David Bass, Richard T. Campbell, John C. Henretta, Kenneth Bollen, Edgar F. Borgatta, Herbert Costner, Scott Long, and Roger Wojtkiewicz. I am also grateful for the feedback I have received over the years from the numerous graduate students who have been exposed to this material in one form or another in my "Multivariate" course. All remaining deficiencies of the book are entirely my responsibility. I could not have completed this project without the resources provided by The University of Akron. Many thanks are owed to Pat McClendon for word processing assistance; to copyeditor Janet Tilden, from whom I learned a great deal; to production editor Kim Vander Steen; and to F. Edward Peacock, who made it all possible. And finally, my special gratitude goes to my wife and daughter; they made the difference in this project, as in all others.

1. SPSS is a registered trademark of SPSS Inc. of Chicago, Illinois, for its proprietary computer software.
2. SYSTAT is a registered trademark of SYSTAT, Inc. of Evanston, Illinois.

1 Causality and Social Science

Causal analysis is of central importance to social science research. In particular, good causal analytic techniques are vital for the success of *explanatory* research. Good *descriptive* research is also undoubtedly of great value, as it provides necessary information about social structures and social change. But while the purpose of description is to tell us *what* exists, the goal of explanatory research is to tell us *why* things exist as they do. As we improve our ability to explain why social phenomena exist, we are better able to predict the future. We are also better able to control social events, either to maintain the status quo or to make desired social changes. To answer the *why* question, that is, to explain social phenomena, it is necessary to attempt to find their causes. Thus, causal analysis and explanation are one and the same.

The use of true experimental designs is a powerful method for conducting causal research. There are, however, many types of social phenomena for which it is not ethical or practical to conduct social experiments (e.g., social conflict, health, and widowhood/divorce). Furthermore, experimental research is not without its own problems of internal and external validity (Cook and Campbell 1979); for example, laboratory conditions may sometimes be so artificial as to make generalizations to the real world extremely risky. Thus, the majority of studies in the social sciences that are done for the purpose of causal analysis use nonexperimental data. Because of the absence of experimental controls, nonexperimental causal analysis must rely more heavily on statistical adjustments (i.e., statistical control) in order to achieve the controls necessary for drawing valid causal inferences from such data. The use of multivariate techniques, particularly multiple regression, for conducting causal analyses of nonexperimental data is the subject of this book.[1]

The Imagery of Causality

There is a large body of literature in the philosophy of science dealing with the meaning and existence of causality.[2] There is no general agreement among philosophers and scientists, however, on the meaning of causality, nor on whether its existence can be logically or empirically demonstrated. Furthermore, most social scientists who conduct causal research use rather commonsense notions of the meaning of cause and effect. Therefore, a rigorous definition of causality will not be provided. Still, most conceptions of causality involve the idea that one event produces another or that certain occurrences follow from other occurrences. There is a sense of agency present in this notion, a sense of connection between events. Events don't just happen to coincide; certain phenomena are responsible for the occurrence of other phenomena. Perhaps most importantly, the essence of causality may be captured by the notion of *manipulation*. If one could intervene without changing the surrounding circumstances and make a change in the first thing, a change in the second thing would follow from the original manipulation. Even when it is not possible to manipulate anything, if we can imagine the occurrence of such a manipulation, followed by the change in the second thing, then we are thinking about a causal relationship. The importance of manipulation for understanding causality is indicated by the fact that manipulation is an important component of experimental designs: the investigator manipulates the experimental treatment to see if the predicted effect occurs.

Although social scientists' conceptualizations of causality may be rather commonsensical, this does not mean that the statistical techniques used by social scientists to make causal inferences are also commonsensical. Modern multivariate techniques are the product of a long intellectual evolution (which originated around 1700) involving the interplay of mathematical concepts and the needs of several applied sciences including astronomy, geodesy, genetics, experimental psychology, and sociology (Stigler 1986). Nor is the language used by social scientists to discuss causality a common one. The person on the street might say, "Katharine married Roger *because* he is rich."[3] A social sci-

1. Multiple regression is also a flexible and powerful statistical tool for analyzing experimental data (Cohen and Cohen 1983; Pedhazur 1982). Although analysis of variance is the classical and more typical statistical technique used in experimental research, analysis of variance and multiple regression are each special cases of the general linear model.

2. Cook and Campbell (1979) and Marini and Singer (1988) provide reviews and discussions of this literature as well as of current empirical techniques for conducting causal analysis.

3. The pervasiveness of the idea of causality in everyday language and thinking is suggested by how frequently we seem to use the word *because*. Paradoxically, however, children seem to deny the existence of causality when they simply reply "Because!" when asked "Why did you do that?"

entist, on the other hand, might say, "An increase in wealth causes an increase in the marriageability of males (and females)" or "Wealth has a positive effect on marriageability." Whereas the average person is more likely to talk about individual cases, the social scientist conducting statistical causal analyses will seldom mention individual cases, focusing instead on aggregates of cases and *variables* (e.g., wealth and marriageability). The social scientist divides the social world into variables, where the categories and values of these variables represent aggregates of individuals (or other units of analysis) with similar or identical attributes, such as the amount of their wealth in dollars, or their marriageability (e.g., *very marriageable, pretty marriageable, somewhat marriageable*, or *not too marriageable*, as rated by others or themselves).[4] Thus, this book will use a language of variables to discuss empirical causal analysis.

The variables used in causal analyses may be either ratio, interval, ordinal, or nominal variables. **Ratio** variables, such as age, have a true zero origin and a known interval between categories or values of the variable. **Interval** variables also have a known interval but do not have a true zero value. Wealth is an example of an interval variable because, although it may take on the value of zero, it is not a true zero value because wealth may have negative values for persons who are in debt. **Ordinal** variables, such as marital opportunity (very marriageable, pretty marriageable, and so on), have ordered or ranked categories, but the interval or numerical distance by which the attributes differ is unknown. And finally, the categories of **nominal** variables cannot even be ranked. Marital status (married, divorced, widowed, or never married) and race are examples of nominal variables.

Strictly speaking, regression analysis assumes that all variables are measured at either the interval or ratio level. Practically speaking, all levels of measurement can be (and often are) used in regression analysis with meaningful results, if proper care is taken. When ordinal variables are used, we must assume that the numbers we assign to the ranked categories approximately represent the true (but unknown) intervals between them. Thus, we might assign the numbers 1 to 4 to the categories of marriageability (e.g., not too marriageable = 1, somewhat marriageable = 2, pretty marriageable = 3, and very marriageable = 4), knowing that this surely doesn't represent the true distances between them but believing that it is close enough to derive meaningful results (although probably somewhat biased in an unknown direction). With nominal variables, there are other special coding conventions (e.g., dummy variable coding) which will allow them to be utilized as independent variables in regression analyses (see Chapter 5).[5]

4. Although the *units of analysis* in this example are individuals, social scientists also use multiple regression to study other units of analysis such as families, firms, states, or countries.

5. There are specialized statistical techniques, such as logit and probit analysis (Aldrich and Nelson 1984), that should be used when you want to treat nominal variables as dependent variables. Although there are special situations in which dummy dependent variables can be used in regression analysis without serious biases, this usually is to be avoided.

The most commonly used symbols for variables in causal analyses are ***X*** and ***Y***. *X* is usually referred to as the **independent** variable and *Y* is referred to as the **dependent** variable. This terminology clearly indicates that the variables are, or are believed to be, in a causal relationship. To say that *X* is the independent variable means that the level of *X* does not depend on, or is not affected by, the level of *Y*. In this context, independent does not mean that *X* is not related or correlated with *Y*; the assumption or hypothesis is that they probably are correlated. Furthermore, saying that *X* is the independent variable doesn't mean that it is independent of all other variables but only that it is independent of *Y*. Calling *Y* the dependent variable means that its values are believed to depend on *X*, that is, to be affected by *X*. Thus, *X is a cause and Y is an effect.* It also implies that *X* and *Y* were given these labels because *X* comes before *Y* in the alphabet, just as the cause precedes its effect temporally. Although not everyone who uses the *X* and *Y* symbols may mean to suggest a cause-and-effect relationship, the fact that the symbols are so closely associated with the independent–dependent terminology suggests that this will usually be the case. Thus, in this book whenever the *X* and *Y* symbols are used, it can be assumed that *X* is a cause and *Y* is an effect. You should be warned, however, that sometimes when more than one *X* is being considered (X_1, X_2, and so on), some *X*'s will be specified as being causes of other *X*'s.

Criteria for Causal Inferences

Several criteria are used for making inferences about whether *X* has an effect on *Y*. The use of the term *inference* is intended to imply that there is always some uncertainty about conclusions regarding causality. *Inferential statistics* refers to the various types of statistical tests that may be carried out to determine the probability that what we have observed in a sample accurately represents what exists in the total population from which the sample is drawn. The use of significance tests to determine the probability of *Type 1 errors* (e.g., incorrectly concluding that *X* is statistically related to *Y*) is an important part of the process of making causal inferences. This source of uncertainty is only one of several problems that are involved in making causal inferences, however, and it is probably the easiest problem with which to deal because there are well-developed statistical procedures for coping with this empirical problem. The other sources of uncertainty are both empirical and theoretical in nature, and the latter are usually the most problematic, especially in nonexperimental research. These problems will be discussed as we take up three of the common criteria that are usually listed as necessary for making causal inferences: temporal sequence, covariance, and nonspuriousness.

Temporal Sequence

It is commonplace to say that the effect follows the cause. Some writers, however, maintain that some effects can occur *simultaneously* with their causes.

Although this issue will not be addressed here, it can be said with certainty that an effect cannot precede its cause. The possibility of simultaneous effects, however, does imply that there may be varying time lags between the occurrence of a cause and the realization of its effect. This fact suggests that we need to pay careful attention to the proper time lag when measuring X and Y; if Y is measured too soon after the occurrence of X, for example, the effect of X will not be detected.

Despite the potential complexity of the time lag issue, establishing that Y follows X (or that Y does not precede X) is an important criterion for making causal inferences. This is one of the strengths of true experimental designs. In the classical experimental design, the dependent variable is measured at Time 1 (Y_1) for both the experimental and control groups, the experimental group is exposed to some treatment (X), and then after an appropriate period the dependent variable is measured again (Y_2). If it is then found that the change in the dependent variable ($Y_2 - Y_1$) is greater for the group receiving the experimental treatment X, it can be concluded that X caused a change in Y (with a probability given by an appropriate statistical test). This research design meets the temporal criteria for making causal inferences because the experimenter has the necessary control over the experimental treatment to ensure that X occurred before the observed change in Y.

In nonexperimental research, however, the researcher typically doesn't have this control over the timing of X and Y.[6] A cross-sectional survey, in which all of the variables are measured at the same time, is the most frequently used nonexperimental design. This means that it is often impossible to know whether Y achieved its observed level before or after X reached its observed level.

For some X's and Y's, however, it is possible to know the temporal sequence. For example, although we might determine (measure) each respondent's race, father's education, own education, and current income all at the time of the interview, we know with a great deal of certainty (based on knowledge external to the survey) that these variables achieved their observed levels in the following temporal sequence: since the respondent's race is extremely likely to be the same as his or her father's race, race was determined before the father completed his education; almost all fathers completed their education prior to the time that the respondents completed their education; and almost all respondents completed their education prior to the time that they achieved their current income level (especially if we limit our analysis to those who are working full time). Therefore, the following temporal sequence is valid:

6. I say typically because the researcher might be able to manipulate X and observe subsequent changes in Y but not have control over which individuals were exposed to the experimental stimulus and which were not (i.e., there was no random assignment or randomization). This case is not a true experimental design; it is usually called a *quasi*-experimental design.

Race → F's Ed → R's Ed → R's Income

This sequence clearly rules out certain causal relationships (no variable can be a cause of any variable to the left of it in the chain), and it is empirically possible that each variable may have an effect on all variables to the right of it. (There are also good theoretical reasons for expecting each of these effects to exist in this case.)

Most of the variables in a cross-sectional survey, however, cannot be ordered so unequivocally. The temporal sequence, for example, of income and various attitudes (such as alienation and conservatism) would be unknown because we would have no way of knowing which of these variables reached its current level first. Furthermore, this problem cannot easily be solved by changing the design from cross-sectional to longitudinal, such as a panel design in which observations are made on the same units of analysis at two or more points in time. If we determined income during the first wave of interviews in a panel study (Time 1) and measured alienation in the second wave of interviews a year later (Time 2), it might be tempting to claim that the temporal sequence was established because income was *measured* before alienation. However, the respondents' observed levels of alienation at Time 2 might have existed prior to the measures of income taken at Time 1. And measuring both income and alienation at Times 1 and 2 would not help either. Although this would allow us to measure the changes in income (income at Time 2 minus income at Time 1) and the changes in alienation and to examine the relationship between these changes, we still would not know which changed first because we would not know at what point between the first and second waves of measurements the changes occurred.

A final possibility with the income and alienation example is to ask respondents to make retrospective reports about important changes in these variables and the timing of these changes. Although they could probably accurately report major changes in income (which might result from promotions or job changes), research has shown that they cannot make accurate retrospective reports about how their beliefs and attitudes at some point in the past differ from what they currently report them to be (Smith 1984). Therefore, establishing the timing of attitude changes is likely to be very difficult, if not impossible. Still, it is possible to obtain information on the temporal sequence of certain objective phenomena (such as changes in jobs, marital status, and place of residence) which can then be analyzed with statistical techniques such as *event history analysis* (Allison 1984).

To summarize, it is very difficult to meet the temporal sequence criterion for most variables in cross-sectional designs, longitudinal designs do not necessarily solve this problem, and knowing the timing of measurements cannot be substituted for knowledge about the timing of changes in the variables. The difficulty in satisfying this criterion for some particular X and Y, however, does not mean that we should not make the inference that X is a cause of Y. It does mean, however, that we will have to use theory or logic to make such an inference, plus empirical support for the additional criteria to be discussed below. Good theoretical arguments are often accepted in this regard, although the inference will certainly be more uncertain than if the temporal sequence could be empirically established. The point to emphasize is that social scientists do proceed with causal analyses even when the temporal sequence cannot be empirically determined. There is still important information that can be extracted about the causal processes that may be operating, information that may be very important for theoretical developments, even if it cannot be said to constitute "proof" of the causal relationships. Equally important, such analyses may cast doubt on existing theoretical formulations by failing to find any relationship between X and Y. Although the failure to find confirming evidence is usually not as satisfying as findings that support theories, it is also an important part of the social science enterprise.

Covariance

If X is a cause of Y, then there should be some covariance or correlation between the two variables—that is, they should be correlated. For example, if high values of X cause high values of Y, medium values of X cause medium values of Y, and low values of X cause low values of Y, then the effect of X on Y will have created a positive covariance between X and Y. This means that variance in Y is coordinated with variance in X, i.e., they *covary*.

The covariance of X and Y is something that we can empirically observe or measure. We cannot, however, directly observe the causal effect that we may believe created the covariance. This is what makes causality such a metaphysical concept. The covariance may have been created by the hypothesized effect of X on Y, but it might also have resulted from an effect of Y on X. It might even have been produced by a third variable that is a cause of both X and Y, even though neither X nor Y is a cause of the other. If, for example, increases in some third variable directly cause increases in X and also directly cause increases in Y, there will be a positive covariance between X and Y because of their common cause. Because of this possibility, it is frequently noted that correlations (or covariances) do not prove causality.

Although it is true that covariance is not sufficient to infer the existence of causality, it is equally true that we cannot infer causality without the existence of covariance. Thus, if a theory predicts that X is a cause of Y and we find no reliable evidence of covariance between X and Y, then there is no support for the theory. If the absence of covariance is replicated, perhaps in different circumstances and/or with different measures of X and Y, then there will be considerable doubt about the validity of the theory. Note, however, that we do not usually say that the theory is rejected; it is not appropriate in statistics to conclude that the null hypothesis (X and Y are not related) is true just because our empirical observations are not sufficiently reliable to reject it.

There is yet another reason to be cautious about our conclusions when we find no covariance between X and Y. It is possible that the predicted relationship is being hidden or masked by another variable(s) that is correlated with X and is also a cause of Y. If we examine only the bivariate (zero-order) relationship between X and Y, that is, if we fail to control this troublesome variable statistically, then we will mistakenly conclude that there is no association between X and Y. This possibility is referred to as statistical *suppression.* The types of relationships between X, Y, and another variable that produce suppression will be covered later. The point here is that the covariance criterion is not a simple one since additional variables may have to be taken into account. Although suppression appears to be a relatively infrequent phenomenon in social research, it does occur frequently enough to warrant attention.

Nonspurious Covariance

Even if the temporal sequence and covariance criteria are met, there is still another threat to valid causal inference. This threat is the possibility that the covariance between X and Y may be **spurious**. It was noted in the previous section that the covariance between X and Y may be caused entirely by a third variable that is a common cause of X and Y. Consider the following set of variables again:

Race → F's Ed → R's Ed → R's Income

Since the father's education precedes the respondent's education, if we find a positive covariance between these two variables we may want to infer that this is due to a causal effect. There is, however, another possibility. Since race precedes both of these variables, it is a potential common cause that may produce a spurious covariance between the two education variables. If, for example, race caused black fathers to have lower educations than white fathers, and race also caused black respondents to have lower educations than white respondents, these effects would create a positive covariance or correlation between the father's education and the respondent's education. Respondents with low education would tend to have fathers with low education and respondents with high education would tend to have fathers with high education—a positive covariance—because the low-education groups on both variables would be disproportionately black and the high-education groups would be disproportionately white. If the father's education itself did not have an effect on the respondent's education, then the covariance between the two education variables would be produced entirely by the effects of the race variable; that is, it would be a spurious covariance.

In order to test for a spurious covariance between X and Y, we have to "control" for the variable that we believe may be a source of spuriousness. This means controlling for variables that are prior to X and Y in temporal sequence, i.e., variables that are *antecedent* to X and Y, or controlling for variables that on the basis of theory are believed to be causes of X and Y. In the above example, we should control for race. This could be done by dividing the respondents into black and white groups and examining the covariance between X and Y (the two education variables) in each group separately. If the covariance was zero, or not statistically significant, in each group, we would conclude that the covariance that we observed for both races combined was spurious.

The above example was not very realistic since it required us to imagine that the covariance between the fathers' and the respondents' educations would disappear entirely after race was controlled. This is not likely to happen for these two variables, and in general, it does not usually occur. A more frequent result of controlling for potential sources of spuriousness is that the original bivariate association is reduced, but not eliminated, after controlling for other variables. This points up the fact that spuriousness is a matter of degree rather than being an all-or-nothing question. In causal analysis we are not only interested in making inferences about whether or not a causal effect exists, we are also interested in estimating the size and strength of that effect.[7] Thus, it is important to control for variables that are sources of partial spuriousness. Failing to control for such variables means that we will be getting upwardly biased estimates (overestimates) of the sizes of effects.

In addition to controlling for sources of spuriousness by physically separating cases into two or more groups (such as blacks and whites), procedures such as multiple regression can be used to accomplish **statistical control**. Statistical control through multiple regression is particularly advantageous when the control variable has several categories, or takes on many values, and when it is desirable to control for several variables simultaneously. Multiple regression is probably the most frequently used method for achieving statistical control in nonexperimental research.

It is instructive to note the ways in which nonexperimental and experimental research differ in how they control for spuriousness. In contrast to statistical control in nonexperimental research, experimental designs achieve the desired control over other variables by random assignment of respondents to the experimental and control groups—that is, through *randomization*. Randomization is used to make the two groups equivalent, or nearly so, on *all* variables/characteristics prior to administering the experimental treatment (X). If the two groups are equivalent, then any difference observed on the dependent variable after the experimental treatment can be attributed to the effect of the treatment. The important point here is that the experimenter doesn't have to anticipate theoretically and measure other variables that may be sources of

7. The form of a causal relationship is also important to determine: i.e., is it a linear, nonlinear, or nonadditive relationship?

spuriousness. Thus, experimental research is somewhat atheoretical because the experimenter only has to focus on the independent variable (the treatment) and the dependent variable. Randomization takes care of all other variables.[8] Nonexperimental research puts a heavier burden on the causal analyst because it is necessary to anticipate and measure all variables that may be sources of spuriousness; if they aren't theoretically specified in advance they probably won't be measured, and if they aren't measured they can't be controlled. Thus, nonexperimental research requires much more planning in order to achieve valid causal inferences.

Causal Models

Multiple Causation

The discussions of covariance and nonspurious covariance indicated that variables in addition to X and Y must be taken into account in any causal analysis. This implies that multiple causes will almost always be involved in such analyses. Thus, we will never focus on just one X; we must simultaneously analyze multiple X's (X_1, X_2, X_3, and so on). This implies that we should not say that X causes Y or that X is *the* cause of Y. It is more appropriate to say that X is *one* cause of Y. The existence of multiple causation also means that we may want to know which X is the more important cause of Y. Thus, we will need to examine statistical measures that allow us to compare the sizes and strengths of effects. Furthermore, since we will never be able to specify and measure all of the causes of Y, we will have to make some assumptions about those causes that have been omitted from the analyses. Thus, all of our statistical models will include terms for both measured and *unmeasured* variables. This suggests that we will often want to know how well our measured variables do in explaining Y as compared to the unmeasured variables.

The presence of multiple causation requires us to estimate the effects of all of the X's on Y simultaneously. We cannot estimate the effects of each X by analyzing just one XY pair at a time. This is because the independent variables will almost always be correlated with one another to varying degrees in nonexperimental research.[9] The presence of intercorrelated X's means that if each XY pair is analyzed separately from the others, biased estimates (overestimates or underestimates) will undoubtedly occur. As discussed above, spurious XY relationships may be confused with causal effects, and valid causal effects may be underestimated, or even missed entirely, because of suppression. Multiple regression is an effective technique for conducting the needed simultaneous

8. The case being described is the simplest experimental design. The design can be made more complex by using several experimental treatments (multifactorial designs) and including additional control variables (covariates) which are not experimentally manipulated.

9. Another important advantage of experimental research designs is that they can be set up in such a way that multiple experimental treatments (X's) will be uncorrelated with one another.

analysis since it statistically controls for the correlations between the independent variables when estimating the effect of each X. It is not infallible, however, because there are cases in which the independent variables are so highly correlated with one another that multiple regression cannot reliably distinguish between their separate effects. Thus, the existence of intercorrelated causes is a major problem for making causal inferences in nonexperimental research.

Causal Mechanisms

When we specify a causal relationship between X and Y or make an inference about the effect of X on Y, it is important in building a convincing case for the specification/inference to be able to say why the effect is believed to occur. This involves specifying a *causal mechanism* by which X has its effect on Y. We might ask, for example, how or why a father's education might affect his child's education. One mechanism might be that more educated fathers socialize their children to place a greater value on education. Another might be that more educated fathers have more cognitive skills and knowledge with which to help their children succeed in school. Thus, being able to specify causal mechanisms helps both the analyst and other social scientists to assess the plausibility of certain causal specifications. Furthermore, if better mechanisms can be specified for an effect of X on Y than for an effect of Y on X, then the latter possibility may be ruled out. This illustrates the important role that theory plays in drawing causal inferences from empirical research.

More importantly, however, specifying causal mechanisms suggests additional variables that should be measured and included either in the present study or in future studies. In the above educational example, the first mechanism suggests a new variable that might be called "educational aspiration." In addition to the original causal relationship between the father's and the child's education, there are now two additional specifications: the father's education is a cause of the child's educational aspiration, and the child's educational aspiration is a cause of the child's educational attainment. Educational aspiration might be measured during the senior year of high school. Using a longitudinal design, the original high school seniors could be reinterviewed five or ten years later to determine their educational attainments. The fathers' educations could be determined during either the first or second interview. In this research design, educational aspiration is an example of what is called an **intervening** variable. If we find support for both of the new hypotheses, it would be said that educational aspiration **mediates** the effect of the father's education on the child's education. If after controlling for the child's aspiration we no longer found any relationship between the father's education and the child's education, it would be concluded that aspiration totally mediated the effect of the father's education. Finding that the original relationship washes out after controlling for educational aspiration, however, would not lead to the conclusion that the original relationship was spurious. Aspiration is an intervening variable and as such is not a cause of the father's education. Spuriousness is created by a third variable that is a cause of both X and Y. Thus, we would conclude that the causal effect of the father's education on the child's education is *interpreted* by the child's educational aspiration.

The introduction of intervening variables suggests one way that causal analysis can be made more complex. We might add several intervening variables to interpret further the original relationship. This process is sometimes referred to as *elaboration*. Elaboration may also involve adding antecedent variables (causes of both X and Y), which was previously discussed as a method of checking for spuriousness. The outcome of including intervening variables and antecedent variables is referred to as a **causal model**. The term *causal model* usually refers to a system of variables that contains more than one dependent variable and in which a single variable may be both a dependent variable and an independent variable. The educational example that we have been discussing, for example, is a causal model in which the child's educational aspiration is a dependent variable with respect to the father's education and an independent variable with respect to the child's own education.

Conditional Effects

The types of effects that were discussed in the section on multiple causality implied that when there are multiple causes of Y, each of the X's has its effect independently of the others. This does not mean that the X's are uncorrelated with one another. It also has nothing to do with the fact that all of them may be called independent variables. Independence of effects means that a given level of X_1 has the same effect on Y for all levels of X_2. That is, the effect of X_1 is independent of the level of X_2. If, for example, educational aspiration has the same effect on educational attainment for children of fathers with high education as it does for children of fathers with low education, the effect of aspirations is independent of the father's education. Independent effects are also called *additive* effects. The term *additive* implies that the effects of all the X's can be added together to determine the level of Y.

Conditional effects, however, indicate that the effect of X_1 depends on the level of X_2. X_1 may have a greater effect when X_2 is at a high level than when X_2 is at a low level. Thus, the effects of X_1 are conditional upon, or are conditioned by, the level of X_2. For example, mental health researchers examine the effect of stressful life events (X_1) on depression (Y). The death of a relative or the loss of a job may cause an increase in depression. However, the effect of the loss may be less when the person has a high level of social support (X_2) than when the person has little or no support. Thus, social support conditions (buffers) the effect of stressful events. Conditional effects are called nonadditive effects. The effects simply can't be added together because they are not independent of one another. In statistical terminology this is called an *interaction* effect.

When there are several independent variables, each of which can take on numerous values, conditional effects can become very complex and difficult to detect. It can be a very tedious task to explore all of the possible types of interactions. Thus, theory is important for predicting which of the numerous possibilities should be examined. Despite the difficulties involved, discovering conditional effects where they exist is just as important as determining the intervening causal mechanisms through which a cause has its effect; both of these elaborations clarify our understanding of causal relationships.

Causal Diagrams

The existence of multiple causation and the process of elaboration by adding intervening and antecedent variables means that causal models will often be relatively complex. These models may be described verbally, as was done in the examples above. Total reliance on verbal descriptions, however, will often be lacking in clarity, precision, and the economy of expression so often valued by scientists. The models can also be described by writing a series of mathematical equations (or regression equations), one for each dependent variable. Although it is valuable to write these equations because they specify the parameters/effects of the causal model that are to be estimated by multiple regression, many social scientists are not comfortable (perhaps because of limited mathematical skills) with total reliance on equations. This discomfort might be called symbol shock. Instead, causal diagrams are often used to achieve an intermediate level of precision and economy in describing the causal model. Even though they are somewhat lacking in precision as compared to the use of equations, they provide a picture of the causal model that is very helpful for quickly grasping both the total structure and the more detailed aspects of the causal processes described by the model. A picture is worth a thousand words, or a thousand mathematical symbols, as the case may be.

Although a causal diagram can be used to describe a causal model with only a single XY pair, the pervasiveness of multiple causation suggests that such a model will seldom be worth investigating. Therefore, a simple causal model with two independent variables and one dependent variable will be diagrammed first (Figure 1.1). For obvious reasons, causal diagrams such as this are often called *path diagrams*.

Straight lines with single-headed arrows between variables specify causal relationships. The arrows, of course, point from the causes to the effects and thus represent causal paths. Thus, this model specifies that X_1 and X_2 are both causes of Y.

The diagram also contains a path running from e to Y. The term e represents all variables other than X_1 and X_2 that affect Y. Thus, the diagram specifies that X_1 and X_2 are not the only causes of Y. The inclusion of e, which may be thought of as an error term, means either that there is no theory to specify what the

FIGURE 1.1 Path Diagram with One Dependent Variable

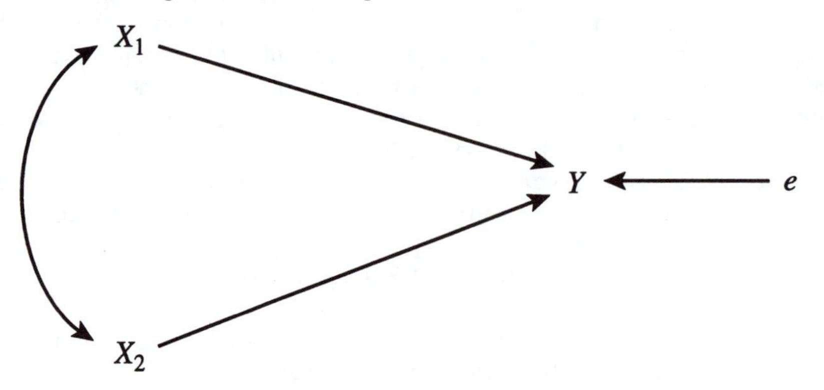

other causes of Y are or that in some particular study we do not have any measures of them. In empirical analyses, therefore, e stands for all *unmeasured* variables. Practically speaking, it will never be possible to specify all the causes of Y, nor to measure them. Thus, causal models should always have an error term. If it is left out, it is always implicitly assumed to exist.

There is one further interpretation of e that should be mentioned. Measures of theoretical variables are never free of measurement error. Random measurement error cannot, by definition, be explained by the measured variables, nor by the unmeasured theoretical variables not included in the model. Therefore, in any empirical study, e also includes random measurement error in Y.

The curved double-headed arrow between X_1 and X_2 means that these two variables are correlated; that is, they covary. It does not mean that each has a causal effect on the other (i.e., reciprocal causation). Double-headed arrows are never used to represent causal effects. If we felt that each variable affected the other, two single-headed arrows pointing in opposite directions would be placed between the two variables. Since the double-headed arrow does not represent a causal effect, it is not called a path. The fact that X_1 and X_2 do not have any paths leading into them, however, does not mean that they have no causes (they are not uncaused causes). It only means that this theory or this study is not concerned with them. In fact, not only do we assume that all variables have causes, the specified correlation between the *X*'s implies that they share some common causes. Since the diagram says that the *X*'s do not cause each other, the only way that they could be correlated is if they shared a common cause or causes.

One final caveat about the absence of a causal path between X_1 and X_2 is in order. We should not always take the absence of such a path at face value. Sometimes the researcher may feel that there is a causal relationship between X_1 and X_2 but simply does not care to investigate it in that particular study. In

FIGURE 1.2 Path Diagram with Two Dependent Variables

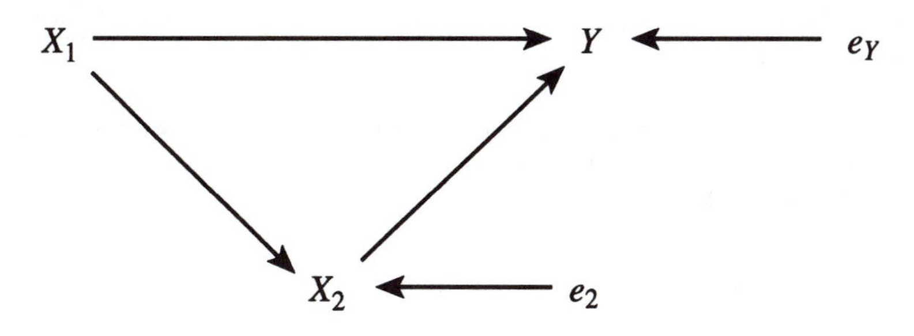

such cases, the path diagram may be specified to reflect only the paths that will be investigated.

It is also important to note that there are no paths or double-headed arrows connecting e with the *X*'s. This means that the unmeasured variables that also cause *Y* are specified to be uncorrelated with the measured *X*'s. Correlations between e and the *X*'s can be specified and included in the model if they are believed to exist. Since the variables represented by e are unmeasured, however, we can't empirically estimate their correlations with X_1 and X_2. If we really believe the unmeasured variables are correlated with the measured variables, we should measure them and include them in the model. Otherwise, we won't be able to control for these variables that are correlated with the specified *X*'s, and consequently, we will get biased estimates of the effects of the *X*'s. Since we can't control what has not been measured, however, path diagrams typically specify that e is uncorrelated with the *X*'s.

Let us now elaborate the causal model specified in Figure 1.1. Imagine that some new information has now come to our attention about the temporal sequence of X_1 and X_2 which would allow us to respecify the model as shown in Figure 1.2.

This causal model specifies that there is a causal relationship between X_1 and X_2, i.e., X_1 is a cause of X_2. X_1 and X_2 are still specified to be causes of Y, however. There are now two dependent variables (X_2 and Y) instead of just one. Since X_2 is a dependent variable relative to X_1 but an independent variable relative to Y, X_2 is an *intervening* variable in this model. Since the conventional *XY* notation breaks down for an intervening variable, it has been labeled as an *X* variable to suggest that Figure 1.2 is just an alternative causal model for the three variables shown in Figure 1.1.

There are also now two error terms, e_Y and e_2, which must be distinguished by their subscripts (the subscripts represent the dependent variable that is affected by these unmeasured variables). Note that the error term for *Y* (now called e_Y) is still uncorrelated with X_1 and X_2. The error term for X_2 (e_2) is also uncorrelated with X_1. Finally, e_Y and e_2 are uncorrelated with one another. The latter fact means that the unmeasured causes of Y are entirely different from those of X_2. Although it may not be clear at this point, this specification is consistent with the first model (Figure 1.1).

Perhaps the most important point to be noted from the second model is that there are now two types of causal effects that X_1 may have on Y. First, there is the direct path between the two variables. Second, there is an indirect path that passes through X_2:

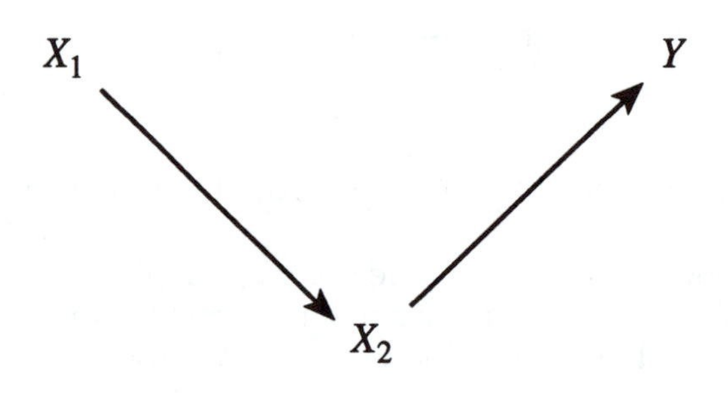

The indirect path means that X_1 first causes a change in X_2, and the change in X_2 subsequently causes a change in Y. The two types of effects that can be estimated with the second model are called, logically, **direct** effects and **indirect** effects. Indirect effects are effects that are mediated by intervening variables. When we discover indirect effects, it helps us to interpret causal effects.

Knowledge about indirect effects is one of the most important types of information to be gained from causal analyses. Remember that we said earlier that specifying causal mechanisms was important for helping us to understand why X affects Y. The discovery of intervening variables that mediate the effect of X on Y is precisely what is meant by a causal mechanism.

Since intervening variables in a causal model are both dependent variables and independent variables, it is valuable to introduce two new terms to help us to distinguish between different types of variables that are included in such models. First, **exogenous** variables are those that develop or originate from without. Thus, the model does not specify any causes for exogenous variables. In the second model above (Figure 1.2), X_1 and the two error terms are exogenous variables. **Endogenous** variables, however, are those that develop or originate from within. That is, the model specifies the causes of endogenous variables, which include both the measured and unmeasured causes. In Figure 1.2, Y and X_2 are both endogenous variables. Thus, intervening variables are one type of endogenous variable, but variables such as Y that are dependent variables relative to all other variables in the model are also endogenous variables. It may be helpful to think of exogenous variables as *external* variables and endogenous variables as *internal* variables.

Discussion

This book is about using multiple regression to conduct causal analyses of nonexperimental data. In this chapter we have outlined and discussed some of the important issues and problems that are involved in such analyses. In order to carry out these causal analyses, the researcher must formulate a causal model. This involves specifying which variable is the dependent variable and which variables are the independent variables or, in more complex models, which variables are endogenous and which are exogenous. As has been pointed out, this is often a difficult decision because the temporal ordering of variables is often unknown. It may also be difficult to know which variables should be included in the model. If important variables are overlooked, the statistical associations that are estimated for the variables in the model may be biased (e.g., spurious). This comes down to two issues: for each pair of variables, have we correctly specified which is the dependent variable, and have we correctly specified which other variables should be included as causes of the dependent variable? The answers to these questions must usually be provided by theory or other information external to the data.

Once we have specified a causal model, we can then use multiple regression to quantify the various relationships or paths specified by the model. We can also use multiple regression to help us make inferences about these relationships in the population from which our sample data were selected. This book is primarily about estimating the relationships specified by the model, once the model has been constructed. It, however, has very little to do with constructing causal models; model construction should be covered by theory books or theory construction books.

It is very important to emphasize that using multiple regression to quantify a causal model for nonexperimental data is not the same as *testing* the model in the sense that an experiment tests a causal hypothesis. If an experiment is carried out properly, it has the necessary internal validity to allow the investigator to conclude that *X is a cause of Y*, with a certain probability. But when we use nonexperimental data to quantify the relationships between *X* and *Y* in a causal model, we will often not be able to make the same claim validly. This is because the time order of *X* and *Y* may be unknown and because we cannot be as certain as in an experiment that we have controlled all sources of spuriousness in the *XY* relationship. Thus, strictly speaking, we are not testing whether there is a causal effect of *X* on *Y*, but instead, we are estimating the magnitude of the causal effect *if X* is truly a cause of *Y*. This is not an unimportant thing to do, especially if it is not possible to conduct experiments, but it must be recognized (if, perhaps, not fully understood yet) that it is not the same thing as conducting an *empirical* test of a hypothesized causal relationship.

On the other hand, there are social scientists and statisticians who believe there is little or no value in estimating causal models for nonexperimental data.

An important recent article by Freedman (1991) provides an example of this viewpoint.[10] Freedman uses examples to argue that nonexperimental studies have difficulty distinguishing the independent variables from dependent variables and also do not include sufficiently valid control variables in the regression equations to rule out spuriousness. Furthermore, he argues that the assumptions about error terms in regression models (see Chapter 4) are often not valid in social science applications and that most investigators never attempt to check the validity of these assumptions.

Freedman's (1991) ideas should receive our careful attention. The use of regression analysis in situations where assumptions are seriously violated is not appropriate and can lead to erroneous causal inferences. His criticisms of causal models that are incorrectly specified with respect to the measured variables, however, are not inconsistent with the view stressed in this chapter—that is, that the construction of causal models for nonexperimental data must be done with care and theoretical insight. All other things being equal, social scientists should make more use of experimental designs whenever possible. But Blalock's (1991) question, "Are There Really Any *Constructive* Alternatives to Causal Modeling?", is also valid in the large number of research contexts in which experiments are not possible.

Some texts on multiple regression techniques do not say much about principles of causality and causal analysis until after they have covered the statistical techniques in some detail. Most do not have much to say at all about causal analysis. It is my feeling that this creates considerable confusion concerning appropriate and inappropriate uses of regression techniques. Thus, I have presented material on causal analysis in some detail in the first chapter in the hope that it will help you to make more sense of the statistical techniques that will be covered. We will now turn to the statistical techniques that may be used to estimate causal effects in nonexperimental research. After you have immersed yourself in the statistical techniques described in Chapters 2 through 7, the material that has been presented in this chapter should in turn become more meaningful. After covering the various regression techniques that are relevant for causal analysis, we will return to some of the topics introduced in this chapter in Chapters 8 and 9.

References

Aldrich, John H., and Forrest D. Nelson. 1984. *Linear Probability, Logit, and Probit Models*. Beverly Hills: Sage Publications.

Allison, Paul D. 1984. *Event History Analysis: Regression for Longitudinal Event Data*. Beverly Hills: Sage Publications.

---

10. Berk (1991), Blalock (1991), and Mason (1991) have written largely sympathetic responses to Freedman that are worth studying.

Berk, Richard A. 1991. "Toward a Methodology for Mere Mortals." Pp. 315–324 in *Sociological Methodology 1991*, edited by Peter V. Marsden. Oxford: Basil Blackwell.

Blalock, Hubert M. 1991. "Are There Really Any *Constructive* Alternatives to Causal Modeling?" Pp. 325–335 in *Sociological Methodology 1991*, edited by Peter V. Marsden. Oxford: Basil Blackwell.

Cohen, Jacob, and Patricia Cohen. 1983. *Applied Multiple Regression/Correlation Analysis for the Behavioral Sciences* (2nd edition). Hillsdale, New Jersey: Lawrence Erlbaum Associates.

Cook, Thomas D., and Donald T. Campbell. 1979. *Quasi-Experimentation: Design & Analysis Issues for Field Settings*. Chicago: Rand McNally.

Freedman, David. 1991. "Statistical Models and Shoe Leather." Pp. 291–313 in *Sociological Methodology 1991*, edited by Peter V. Marsden. Oxford: Basil Blackwell.

Marini, Margaret, and Burton Singer. 1988. "Causality in the Social Sciences." Pp. 347–410 in *Sociological Methodology 1988*, edited by Clifford C. Clogg. Washington, D.C.: American Sociological Association.

Mason, William M. 1991. "Freedman Is Right as Far as He Goes, but There Is More, and It's Worse. Statisticians Could Help." Pp. 337–351 in *Sociological Methodology 1991*, edited by Peter V. Marsden. Oxford: Basil Blackwell.

Pedhazur, Elazar J. 1982. *Multiple Regression in Behavioral Research: Explanation and Prediction* (2nd edition). New York: Holt, Rinehart and Winston.

Smith, Tom W. 1984. "Recalling Attitudes: An Analysis of Retrospective Questions on the 1982 GSS." *Public Opinion Quarterly*, 48:639–649.

Stigler, Stephen M. 1986. *The History of Statistics: The Measurement of Uncertainty before 1900*. Cambridge, Massachusetts: Belknap Press of Harvard University Press.

2 Bivariate Regression and Correlation

Bivariate regression and correlation refers to the case in which only one X and one Y are being analyzed at a time. Much of the material covered here should be a review for the reader. Some new material will undoubtedly be covered, however, and we will emphasize certain points and formulas that may not have been stressed in your first course in statistics.

We begin with three statistics—the mean, the variance, and the covariance—which may be considered building blocks for the bivariate regression and correlation statistics covered in this chapter and for the various multiple regression and correlation statistics covered in Chapter 3 as well. We then turn to linear bivariate relationships—the simplest type of causal model—and describe how to find the slope and intercept of the line that best fits a sample of observations of X and Y. The least-squared-residuals criterion for the best-fitting line is illustrated and contrasted with another plausible criterion, the least-absolute-residuals. The formulas for the least-squares slope and intercept are described and interpreted. Although these formulas define the best-fitting line, two additional statistics—the coefficient of determination and the standardized slope—are presented to measure how closely the line fits the observed values of X and Y. The equivalence between the standardized slope and the well-known Pearson correlation is illustrated. Finally, we provide an introduction to regression diagnostics that includes simple measures of two types of outliers. These measures help to locate observations that may unduly influence the slope and intercept of the regression line.

A Note on Formulas. Mathematical derivations of formulas will be entirely omitted. This does not mean, however, that the formulas themselves are unimportant. Formulas provide precise definitions of statistical concepts, and they must be learned and understood in order to maximize intelligent use of regression techniques. If you merely attempt to memorize formulas without making the effort to understand them, you will certainly run into difficulties. Once you have developed an understanding of a formula, it will be much easier to recall it because you should then be able to recreate it by utilizing the logic that underlies the development of the formula.

Formulas will not be used for computational expediency, however. Although *computational* formulas exist for many of the more elementary regression statistics, hand and/or pocket-calculator solutions will not be emphasized. The availability of the computer has freed us from this drudgery and thus rendered many of the computational formulas obsolete. Although computational formulas expedite hand calculations, they are often obtuse and do not enhance our understanding of the meaning and logic of the statistic. *Definitional* formulas, on the other hand, provide a sharper view of the meaning of the statistic and thus must be learned and understood. Although the distinction between the two types may be said to be in the "eye of the beholder," the difference should become clearer as we proceed. With only a few exceptions, this book will present only those formulas that are considered to be definitional formulas.

Statistical Building Blocks

Mean

The first question that is usually asked about a set of observations on any variable is "what is a typical person or case like?" There are three types of averages that most frequently are used to answer this question: the **mean**, the **median**, and the **mode**. These averages are called measures of central tendency. Since *central tendency* implies a focus on the middle of a distribution, perhaps this explains why all of these terms begin with the letter *m*. Measures of central tendency are one example of *univariate* statistics, which are used to describe one variable at a time. In a course focusing on causal analysis and multiple regression, however, we will not be very concerned with univariate statistics. Our interest will be more with bivariate statistics (two variables) and especially with multivariate statistics (more than two variables). The discussion of spuriousness and causal mechanisms in Chapter 1 emphasized the need to examine three or more variables at a time.

Of all the characteristics of a univariate distribution, central tendency is probably the one with which we will be least concerned. Causal analysis is concerned more with differences between people or cases than with what is average or most common. Causal analysis attempts to explain why differences exist. Yet that does not mean that we have no use for measures of central tendency. In fact, one measure in particular, the *mean*, is a fundamental building block from which more complex multivariate statistics are constructed. Although Sir Francis Galton emphasized the median when he invented regression analysis in the nineteenth century (Stigler 1986), it was soon supplanted by the mathematical superiority of the mean.

The mean of a distribution is the *arithmetic average* of its scores:

$$\overline{X} = \frac{\sum X}{n} \tag{2.1}$$

This is a definitional formula. There are grouped-score formulas for the mean that facilitate hand computations when the data are in the form of a frequency distribution. The mean is analogous to a *center of gravity* in the sense that it balances the scores that are above and below it, as we will see shortly. Not only is the mean a building block for other statistics, it will be seen that many more complex statistics are arithmetic averages of different kinds of scores. The mean is the average of the raw scores (X's).

Variance

There are different ways to measure the differences, or dispersion, among a set of scores (e.g., the range or the interquartile range). The statistics that are used by regression and correlation analysis are the **variance** and the **standard deviation**. These measures use the mean as a reference point from which to calculate the dispersion of the distribution. The definitional formula for the standard deviation is

$$s = \sqrt{\frac{\sum (X - \overline{X})^2}{n}} \tag{2.2}$$

The variance is defined as the expression under the square root sign (radical) in the formula for the standard deviation. Since squaring the standard deviation removes the radical, the symbol that will be used for the variance is s^2. Thus, the definitional formula for the variance is

$$s^2 = \frac{\sum (X - \overline{X})^2}{n} \tag{2.3}$$

The important term in the formulas for the variance and standard deviation is $X - \overline{X}$, which is called a **deviation score**. This indicates where the standard deviation gets its name. There are computational formulas for the variance and standard deviation that contain only raw scores (X), but these raw-score formulas do not indicate as clearly why these statistics serve as measures of dispersion or variance.

Let us turn now to an example to help us examine some of the important characteristics of these statistics. Table 2.1 contains three scores on X and Y for three hypothetical persons. Although such a small number of cases is totally

TABLE 2.1 Example Data, Means, and Standard Deviations

Case	X	Y	$X - \overline{X}$	$Y - \overline{Y}$	$(X - \overline{X})^2$	$(Y - \overline{Y})^2$	$(X - \overline{X})(Y - \overline{Y})$
1	1	2	−2	−1	4	1	2
2	3	2	0	−1	0	1	0
3	5	5	2	2	4	4	4
Σ	9	9	0	0	8	6	6

$$\overline{X} = \frac{\sum X}{n} = \frac{9}{3} = 3 \qquad \overline{Y} = \frac{\sum Y}{n} = \frac{9}{3} = 3$$

$$s_X^2 = \frac{\sum (X - \overline{X})^2}{n} = \frac{8}{3} = 2.67 \qquad s_Y^2 = \frac{\sum (Y - \overline{Y})^2}{n} = \frac{6}{3} = 2$$

$$s_X = \sqrt{\frac{\sum (X - \overline{X})^2}{n}} = \sqrt{\frac{8}{3}} = \sqrt{2.67} = 1.63 \qquad s_Y = \sqrt{\frac{\sum (Y - \overline{Y})^2}{n}} = \sqrt{\frac{6}{3}} = \sqrt{2} = 1.41$$

unrealistic for conducting any kind of analysis, it will make it easy to follow the calculations as we examine the characteristics and meaning of these and other statistics. The calculations are presented only to enhance understanding by making the statistical concepts as concrete as possible. Imagine that X is the number of siblings of each person and Y is the number of his or her own children. The temporal sequence of these variables would suggest that X may be a cause of Y, and you also should be able to think of causal mechanisms to account for this effect.

The mean of each variable is 3. The deviations from the mean (deviation scores) may be positive or negative. Thus, the deviation scores indicate how far each person is above or below the mean. Since these scores represent departures from the average, they are logical candidates for use in developing a measure of dispersion. One natural inclination would be to add them up to measure the total departure or deviation from the mean. The sums of the two deviation score columns are zero, however. You can see that for each variable the negative deviations totally balance out the positive deviations. This is why the mean is somewhat like a center of gravity. Since the sum of the deviations is zero, the mean of the deviation-score distribution is zero. Although the mathematical proof will not be given, the mean of any deviation-score distribution will always be zero.

To correct for the zero sum, our first thought might be simply to sum the absolute values (ignoring the negative signs) of the deviation scores to get a measure of total deviations. We would then see that the total absolute deviation for both X and Y equals 4 ($\Sigma |Y - \overline{Y}| = \Sigma |X - \overline{X}| = 4$). Although this measure is sometimes used ($\Sigma |X - \overline{X}|/n$ is called the *average deviation*), it does not behave well mathematically for purposes of deriving other statistics that we will be using.

The solution to the zero-sum problem that is used in calculating the variance is to square the deviation scores, e.g., $(X - \overline{X})^2$, thus removing the negative signs. The sum of squared deviation scores, e.g., $\Sigma (X - \overline{X})^2$, is called the **sum of squares**. An important characteristic of the sum of squares is that it is smaller than the sum of squared deviations around any other value of X. For example, $\Sigma (X - \overline{X})^2 < \Sigma (X - \text{Median})^2$, assuming $\overline{X} \neq \text{Median}$. This characteristic is called a **least-squares** characteristic. If we wanted to choose a single value to represent a distribution of scores, the mean would be the best choice if the criterion were to minimize the sum of squared differences between each score and the representative value. Thus, the mean is a *least-squares* statistic.

Table 2.1 shows that the sum of squares for X (which equals 8) is larger than the sum of squares for Y (which equals 6). Thus, X has greater dispersion (variation) around the mean than Y, according to this measure. The reason for this is that squaring deviation scores gives a disproportionate weight to the largest or most extreme deviations.

Look closely at the deviation scores for X and Y. There are two people below the mean on Y and only one below the mean on X. Yet the sum of the negative deviations on Y is equal to the single negative deviation on X because the two below-average persons on Y are not as far below average as the single below-average person on X. Squaring the deviations, however, eliminates this equality. The squared deviation for the single below-average person on X is two greater than the sum of the squared deviations for the two below-average persons on Y. This demonstrates, by example, the fact that scores at the extreme ends of a distribution receive more weight when squared deviations are used to measure dispersion than they otherwise would. Thus, the variance and the standard deviation are sensitive to extreme deviation scores.

The formulas indicate that we divide by n after summing the squared deviations.[1] This means that the variance is a certain type of arithmetic average. Whereas the mean is the arithmetic average of the raw scores, the variance is the arithmetic average of the squared deviation scores. After dividing by n we find that the variance of Y equals 2 and the variance of X equals 2.67. These values are inflated relative to the scales of X and Y as a result of squaring the deviations. Taking the square roots of the variances to get the standard deviations brings these values back closer to the scales of the raw scores. Although the standard deviation of X is greater than that of Y, the difference between the standard deviations is not as great as the difference between their vari-

1. Many texts use $n - 1$ in the denominator of the formula for the *sample* standard deviation. This is correct if you want to use the sample to *estimate* the population standard deviation. If n is used to compute the sample standard deviation, the statistic will be a biased estimate (an underestimate) of the population parameter. This slight bias is corrected by dividing by $n - 1$.

ances. The ratio of the standard deviations ($s_Y/s_X = 1.63/1.41 = 1.16$) is also not as great as the ratio of the variances ($s_Y^2/s_X^2 = 2.67/2 = 1.34$). Thus, if you were comparing the dispersion of two variables (and they would have to be measured on the same scale in order to do so), you should use the variances if you were biased in such a way that you wanted to stress the difference. An unbiased social scientist, however, would present both. Neither measure is better than the other, of course, since the standard deviation is an exact function (i.e., the square root) of the variance. Regression and correlation statistics make extensive use of both.

Now that we have examined the characteristics of the variance and standard deviation, it may be instructive to look at a **computational formula**. A computational formula for the variance is

$$s^2 = \frac{\sum X^2 - \frac{\left(\sum X\right)^2}{n}}{n}$$

You can see that there are no deviation scores in this formula (the formula does not even contain the mean). Thus, this formula does not reveal that the variance is a measure of dispersion around the mean. Furthermore, the average person would probably not even be able to recognize from this formula that the variance is a measure of any kind of dispersion. Although you might occasionally want to use this formula because it saves you from having to compute deviation scores, you certainly would not want to memorize it. It is important to know the definitional formulas, however, because they convey the essential characteristics of the measures.

Covariance

In order to estimate causal effects, we must move beyond univariate statistics to bivariate statistics that measure the degree to which X and Y are empirically coordinated with one another. The most elementary measure of statistical coordination used in regression analysis is the **covariance**. This statistic can be viewed as a building block for all other bivariate and multivariate regression/correlation statistics. The definitional formula for the covariance is

$$s_{XY} = \text{Cov}(X, Y) = \frac{\sum (X - \overline{X})(Y - \overline{Y})}{n} \qquad \textbf{(2.4)}$$

The covariance is a measure of the extent to which deviations in X are coordinated with deviations in Y (hence the name *covariance*). The essential element in this formula is $(X - \overline{X})(Y - \overline{Y})$, the product of the deviation scores of X and Y, which is called the **cross-product**. Thus, $\Sigma (X - \overline{X})(Y - \overline{Y})$ is called the **sum of cross-products**.

Since the variance involves the squared deviation scores, the formula for the variance of X can be written as

$$s_x^2 = \frac{\sum (X - \overline{X})(X - \overline{X})}{n}$$

If $Y - \overline{Y}$ is substituted for one of the $X - \overline{X}$ terms, you have the formula for the covariance of X and Y. Thus, it is sometimes said that the variance of X is the covariance of X with itself. Although this analogy has its limitations, it does suggest why the covariance is a measure of the degree to which X and Y covary.

Both Cov(X,Y) and s_{XY} were included in Equation 2.4 because both are frequently used in the literature. The analogy between the variance and the covariance suggests why s_{XY} is often used. This symbol, however, is somewhat ambiguous because s is the symbol for the standard deviation, not the variance. It would be misleading to use "s^2_{XY}", though, because the covariance is not the square of the standard deviation or any other statistic. This may account for why Cov(X,Y) is also often used.

In order to understand how this statistic measures covariance, it is important to remember that deviation scores tell us the direction and distance of a raw score from the mean. Positive scores mean that the case is above the mean and negative scores mean that the case is below the mean. Thus, if a case is above the mean on both X and Y, the cross-product will be positive. If it is below the mean on both variables, the cross-product will also be positive. If a case is above the mean on one variable but below the mean on the other, however, the cross-product will be negative.

Let us now look at the examples in Table 2.2 to see how different arrangements of scores on X and Y affect the cross-products and thus the measure of covariance. The first example shows the calculations for the original scores from Table 2.1. The first case is below the mean on both X and Y, and thus the cross-product is positive. The third case is above the mean on both variables, and thus it also has a positive cross-product. The middle case is at the mean on X (a deviation score of 0) and consequently has a zero value for its cross-product. Since there are no negative cross-products, the sum of cross-products must be positive. When the sum is divided by n, we get a positive covariance of 2. Focusing only on the sign of the covariance, this positive value means that there is a tendency for cases to have scores on the same side of the mean for both X and Y; if a case is above the mean on X it tends to be above the mean on Y, and if it is below the mean on X it tends to be below the mean on Y. In terms of the variables hypothetically represented by X and Y, if a person has a below-average number of siblings, that person will tend to have a below-average number of children, and a person with an above-average number of siblings would be likely to have an above-average number of children.

TABLE 2.2 Examples of Different Possible Covariances for the Given X and Y Scores

Relationship	X	Y	$(X - \bar{X})(Y - \bar{Y})$		
1. Positive	1	2	$(-2)(-1)$	=	+2
	3	2	$(0)(-1)$	=	0
	5	5	$(+2)(+2)$	=	+4
					6
					$s_{XY} = 6/3 = 2$
2. Zero	1	2	$(-2)(-1)$	=	+2
	3	5	$(0)(+2)$	=	0
	5	2	$(+2)(-1)$	=	−2
					0
					$s_{XY} = 0/3 = 0$
3. Negative	1	5	$(-2)(+2)$	=	−4
	3	2	$(0)(-1)$	=	0
	5	2	$(+2)(-1)$	=	−2
					−6
					$s_{XY} = -6/3 = -2$
4. Positive	10,000	2	$(-20{,}000)(-1)$	=	20,000
	30,000	2	$(0)(-1)$	=	0
	50,000	5	$(+20{,}000)(+2)$	=	40,000
	$\bar{X}$ = 30,000				60,000
	s = 16,329.93				$s_{XY} = 60{,}000/3 = 20{,}000$

The second example interchanges the scores on Y of the second and third cases. This, of course, does not affect the mean of Y and thus does not affect the values of the deviation scores. It does, however, affect how some of them are paired up with deviation scores on X. The cross-product scores for the first two cases remain the same. The cross-product for the third case changes from $+4$ in the first example to -2 in the second example. When we add the cross-products, the negative product for the third case exactly offsets the positive product for the first case, producing a zero sum. The covariance is thus zero; the two variables do not covary. There is no overall tendency for cases to have scores on the same side of the mean for both variables or for cases to have scores on X and Y that are on opposite sides of the mean. We would say that there is no statistical coordination between these variables.

The third example interchanges the original scores on Y of the first and third cases. Now each of these two cases has a score on Y that is on the opposite side of the mean from its score on X. Thus, their cross-product values are both negative. The sum of the products is –6, and the covariance is -2. The negative covariance means that scores on Y tend to be on the opposite side of the mean from scores on X; persons with an above-average number of siblings tend to have a below-average number of children, while those with a below-average number of siblings tend to have an above-average number of children.

These three examples show how the cross-products of deviation scores are used to produce a measure of the coordination of X and Y. A zero value of the covariance means that there is no coordination (no relationship), and the sign of a nonzero covariance indicates the **direction** of the relationship. It is difficult, however, to tell the **strength** of the relationship from the value of the covariance. Is the observed value of 2 a large or a small value?

The fourth example in Table 2.2 shows why this question is impossible to answer until we take into account additional information. Here, each value of X has been multiplied by 10,000 to create a new X variable, which might be thought of as income. The mean is now 30,000 and the standard deviation is 16,329.93. Although the mean and standard deviation have increased, the values of the new X are in exactly the same proportion to one another as are the values of the original X. The rescaling of X, however, dramatically increases the value of the covariance to 20,000. The point of this exercise was to demonstrate that the magnitude of the covariance is very sensitive to the sizes of the standard deviations of X and Y. The larger the standard deviations, all other things being equal, the greater will be the covariance. Thus, the strength of the XY relationship can't be determined by the value of the covariance alone; the values of the standard deviations must also be taken into account. We will return later to the correct way to adjust, or standardize, the covariance to remove the influences of the standard deviations. At this point, however, it should be reiterated that the covariance indicates whether there is *any* relationship between X and Y, and if so, it also gives the direction of the relationship; it does not, however, serve as a measure of the strength of the relationship.

Finally, the covariance statistic does not make any assumptions about which is the independent variable and which is the dependent variable. It is a **symmetrical** measure because the value of the covariance will be the same whether X is a cause of Y or Y is a cause of X. Since we are interested in causal analysis, we will now turn to a statistic that is built on the covariance in such a way that it can be used to describe the effect of X on Y— the regression coefficient.

Bivariate Regression

Linear Relationships

The simplest type of causal effect is a linear effect. The formula for a straight line is

$$Y = a + bX \tag{2.5}$$

FIGURE 2.1 Linear Relationship Between X and Y

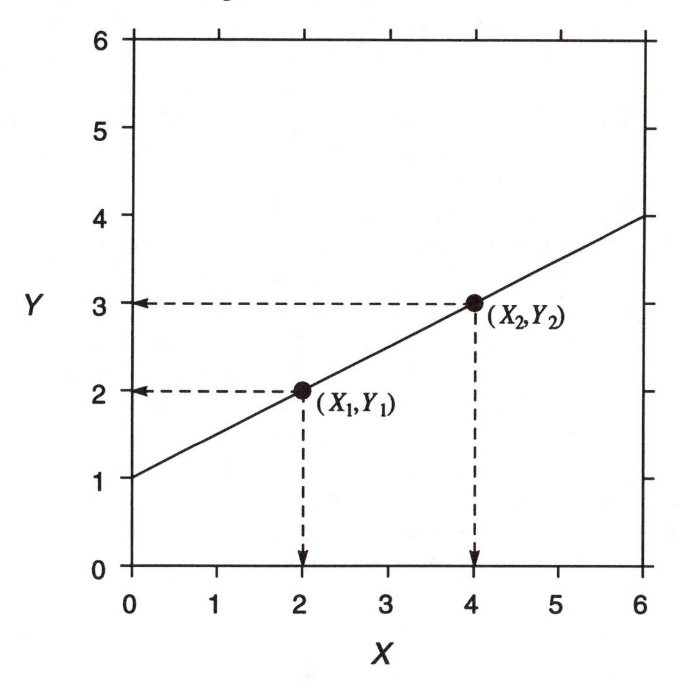

In this equation a is a constant that equals the value of Y when $X = 0$. Substituting 0 for X reduces the equation to $Y = a$. This constant is often called the **Y-intercept** because when a line is plotted on a graph, a is the value of Y at the point where the line crosses the Y-axis. The constant b is the **slope** of the line. *The slope represents the change in Y per unit increase in X*. If X were a cause of Y, then a one-unit increase in X would cause Y to change by b units. Thus, b would represent the *effect* of X on Y. This effect, represented by the slope of the line, can be either positive or negative. Thus, b could equal either an increase in Y or a decrease in Y caused by a unit increase in X, depending on its sign.

Figure 2.1 shows a graph of a positive linear relationship between X and Y, where $a = 1$ and $b = .5$. The line can be plotted by first taking any two values of X and computing their corresponding values of Y from the formula. The two pairs of XY values form two pairs of coordinates (X, Y) through which the line must pass. The coordinates in the graph in Figure 2.1 were determined by arbitrarily choosing $X_1 = 2$ and $X_2 = 4$ and then computing $Y_1 = 1 + .5(2) = 2$ and $Y_2 = 1 + .5(4) = 3$. Thus, the two pairs of coordinates are (2, 2) and (4, 3).

If, on the other hand, we were given the straight line drawn in the graph but we didn't know the formula for the line, we could determine it from the graph. The line crosses the Y-axis at $Y = 1$, so $a = 1$. This value represents the value of Y when $X = 0$ because the Y-axis is drawn through the zero value (the origin) of X, as is the usual custom. A word of caution, however, is in order. Sometimes the Y-axis is not drawn through $X = 0$. If, for example, X takes on very large positive values and there are no X values near zero, the Y-axis might be drawn through some positive value of X that is closer on the X-axis to where the observed scores are located. If this is the case, then the constant a is not equal to the Y-intercept, as plotted. What would you do?

The slope b could be determined by first selecting any two points on the line. One of these points could be the point at which the line crosses the Y-axis, which is (0, 1) in Figure 2.1. For this example, however, we will use the points in the figure represented by the coordinates (X_1, Y_1) and (X_2, Y_2). From each of these points drop a vertical line down to the X-axis to determine the X values and run a horizontal line to the Y-axis to determine Y values. The value of the slope would then be computed by subtracting the smaller X value from the larger X value to get the increase in X and then dividing the corresponding change in Y by this increase in X, as follows:

$$b = \frac{Y_2 - Y_1}{X_2 - X_1} = \frac{3 - 2}{4 - 2} = \frac{1}{2} = .5$$

When the change in Y is divided by the increase in X, we get the change in Y per unit increase in X. If we subtract the values of (X_2, Y_2) from those of (X_1, Y_1), we get the same value for the slope,

$$b = \frac{Y_1 - Y_2}{X_1 - X_2} = \frac{2 - 3}{2 - 4} = \frac{-1}{-2} = .5$$

In the latter case a decrease in X produced a decrease in Y, but we still got a positive slope. Thus, the slope $b = .5$ represents the change in Y per unit *increase* in X. If X decreases by one unit, we must change the sign of the slope to get the associated change in Y. In causal analysis, .5 is the effect of X on Y.

The Simplest Causal Model

The equation for a linear causal model with one independent variable is

$$Y = \alpha + \beta X + \varepsilon \qquad \textbf{(2.6)}$$

Greek letters are used for the intercept, slope, and error term to indicate that these are unknown and unobservable population parameters. It is often not possible to observe or measure X and Y for every member of the population to which the causal model pertains. Thus, the population parameters must be estimated with a sample of elements from the population for which we can measure X and Y. Bivariate regression analysis is a statistical technique that is often appropriate for estimating the parameters of Equation 2.6 with an appropriately selected sample.

In the ideal case, we would select the sample in such a way that every element in the population would have an equal probability of being selected, as in a simple random sample.[2] The various inferential statistics that will be presented in Chapter 4 assume that we have selected a simple random sample. The error term ε in the causal model indicates that the values of Y are not exactly determined by X. Before the regression techniques described in this book can be used, certain assumptions must be made about ε. These assumptions will be covered in Chapter 4. We now turn to the method that is used to calculate a linear regression equation from a sample of observations of X and Y.

The Best-Fitting Line

In any observed data set the XY scores will never fall exactly along a straight line. Because of the presence of multiple causation, the values of Y will vary for any given level of X as a result of the effects of other variables. Even if X were the only cause of Y, some writers maintain that Y would not be a perfect function of X since there would always be some random variation in Y due to the influence of many small, chance events. Whether or not they are correct about the presence of randomness in the social world, there will undoubtedly always be too many small effects for us to measure them all. Thus, our causal models will always need to contain an error term (such as ε in Equation 2.6) for the apparent randomness caused by these numerous small effects. Measurement error, which often appears to be random, will also prevent us from ever observing a perfect linear relationship.

Figure 2.2 shows the scatterplot for the three XY cases from Table 2.1. The data points obviously don't fall exactly along a straight line. This may result from any of the above factors or be attributed to random sampling error. Alternatively, it may be the result of a true **nonlinear** effect of X. Methods for measuring and testing nonlinear effects will be covered in Chapter 6. It is also clear from the scatterplot that Y never decreases as X increases but that it sometimes increases. That is, Y appears to be at least approximately a positive function of X. But what is the positive linear function that best approximates the XY relationship?

Figure 2.3 shows an educated guess about a line that may fit the data reasonably well. "Fitting" the data well involves drawing a line that comes as close as possible to all of the data points. The educated guess was not based on any rigorous mathematical criteria; it was simply an "eyeball" guess, but it

2. Sampling methods are not covered in this book. Sudman (1976) provides a good introduction to the most commonly used sampling techniques.

FIGURE 2.2 Scatterplot of the Observed *XY* Scores

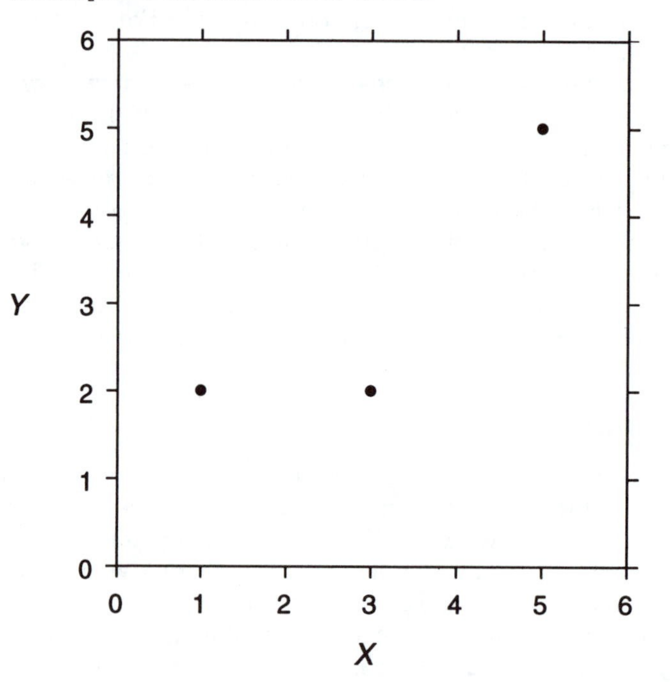

has some nice features. It passes exactly through the highest coordinate and splits the first two in half; that is, it comes equally close to the first two cases. You might fit the data better by drawing the line exactly through the first and third points, but then there would be a relatively big error for the second case. Furthermore, you would be totally ignoring the second case; that is, you would not be allowing it to influence the placement of your line. A reasonable criterion, and one that is used by all mathematical solutions, is to give at least some weight to each pair of observations in determining the best line.

The equation for the guess of the best-fitting line is $\hat{Y} = 0 + 1X$. This equation has a zero Y-intercept and a unit slope. The Y with a circumflex is commonly called "Y-hat." Whenever you see a hat on a variable or statistic it means that it is a predicted or estimated value. The predicted value of Y for the first case (for which $X = 1$) is $\hat{Y} = 0 + 1X = 0 + 1(1) = 1$. The difference between Y and the predicted value of Y (i.e., $Y - \hat{Y}$) is the error of prediction. This error is called a **residual** value. The residual for the first case is $Y - \hat{Y} = 2 - 1 = 1$. In Figure 2.3 the residuals are the vertical distance between each point and the line, as is indicated for the second case.

A logical way of evaluating the predictive power of the equation is to compute the sum of all the residuals. Since some of the residuals are always neg-

FIGURE 2.3 An Educated Guess of the Best-Fitting Line

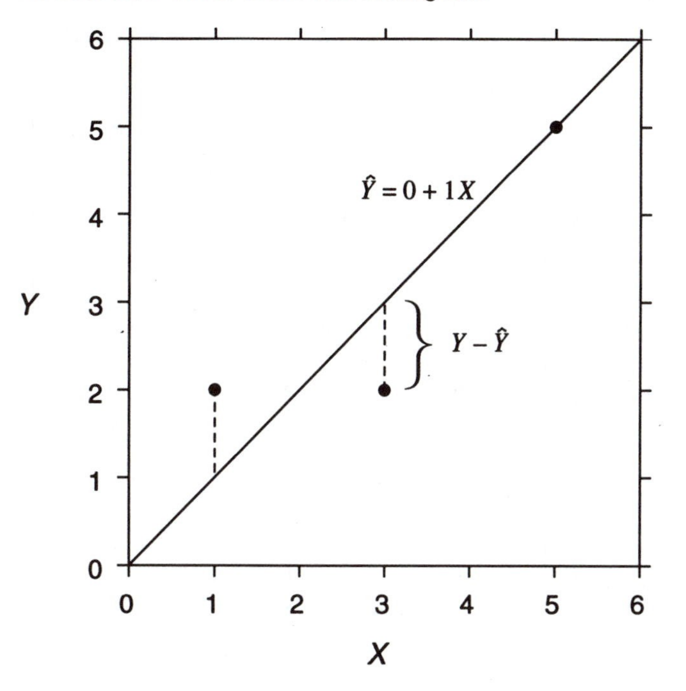

ative, we will ignore the signs and calculate the sum of the absolute values $|Y - \hat{Y}|$ to get the total amount of error. The results of these calculations are shown in Table 2.3. The sum of the absolute values of the residuals equals 2. A natural question would be whether there is another line that could reduce this total amount of error: i.e., is there a line that would make even better predictions? This question suggests that minimizing the sum of the absolute values of the residuals might be a good criterion for selecting the line that fits these data best. This criterion, however, is not the one that is used in regression analysis.

TABLE 2.3 Residuals for the Educated-Guess Line

X	Y	$\hat{Y}$	$Y - \hat{Y}$	$\lvert Y - \hat{Y}\rvert$	$(Y - \hat{Y})^2$
1	2	1	1	1	1
3	2	3	−1	1	1
5	5	5	0	0	0
				2	2

FIGURE 2.4 The Least-Squares Regression Line

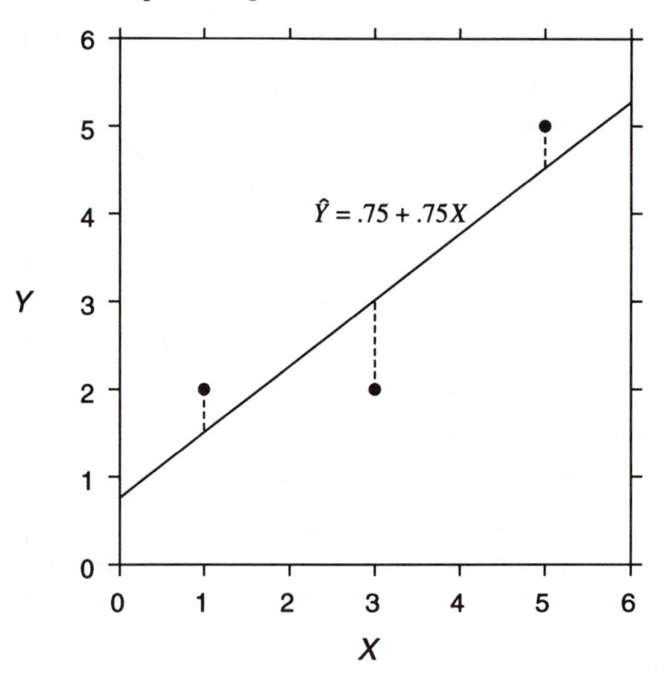

The best-fitting line according to the criterion used in regression analysis is the line that minimizes the sum of the squared residuals, $\Sigma (Y - \hat{Y})^2$. There are a number of important statistical reasons for selecting the line that minimizes the squared residuals rather than the absolute values of the residuals. These reasons will be discussed in Chapter 4. For obvious reasons, the line that minimizes the squared residuals is called the **least-squares line**.

Before covering the formulas that define the slope and intercept of the least-squares line, let us compare the least-squares line for these data with the "educated-guess" line that we have been examining. The least-squares line is plotted in Figure 2.4. Its equation is $\hat{Y} = .75 + .75X$. This line does not pass through any of the data points in Figure 2.4.

The residuals for the least-squares line are given in Table 2.4. If we ignore the signs of the residuals, the sum of the absolute values is 2. This is the same sum that we got for the "educated-guess" line. Thus, the two lines predict Y equally well if we use the absolute values of the residuals as the criteron for choosing a best-fitting line. When we compare the sums of squared residuals, however, the least-squares line has a value of 1.5, as compared to a value of 2 for the "educated-guess" line (Table 2.3). Thus, $\hat{Y} = .75 + .75X$ fits the data better than $\hat{Y} = 0 + 1X$, according to the least-squares criterion. Furthermore,

TABLE 2.4 Residuals for the Least-Squares Line

X	Y	$\hat{Y}$	$Y - \hat{Y}$	$\|Y - \hat{Y}\|$	$(Y - \hat{Y})^2$
1	2	1.5	.5	.5	.25
3	2	3.0	−1.0	1.0	1.0
5	5	4.5	.5	.5	.25
				2	1.5

mathematical theory assures us that it fits better, in terms of the sum of squared residuals, than any other line would fit. Consequently, we use the least-squares line, or more specifically, the least-squares slope, as our best estimate of the linear effect of X on Y.

You are being asked to take it on faith at this point that the least-squares line is best. It is important to realize, however, that there is no way to demonstrate on the basis of a single sample of data, such as this example data set, that a least-squares solution is better than one that minimizes the sum of the absolute values of the residuals. Thus, the least-squares line and the "educated guess" line may look equally good, and some might even prefer the latter line. The reason that the least-squares line is preferable, however, is that it is less susceptible to sampling error when it is used to estimate the linear effect of X on Y in the population; the least-squares line will have a greater chance of being close to the true population value than the results of other criteria for fitting a line to the scatterplot. We will return to this issue when we cover sampling distributions and inferential statistics in Chapter 4.

The Regression Coefficients

The least-squares equation was given above without discussing how its slope and intercept are defined. The formula for the slope of the line that minimizes the sum of the squared residuals is

$$b_{YX} = \frac{\sum (X - \overline{X})(Y - \overline{Y})}{\sum (X - \overline{X})^2} \tag{2.7}$$

The first subscript of the slope b indicates which variable is the dependent variable. It is important to note this because the regression slope is an **asymmetric** measure of association. If X were treated as the dependent variable, the value of the slope would differ, as we shall see. The numerator of the slope is the sum of the cross-products of the deviation scores, i.e., the term that is used to define the covariance. The denominator of b is the sum of the squared deviation scores of X, which is the independent variable for this equation. When the values of these terms in the example data (Table 2.1) are entered into Equation 2.7, the slope is

$$b_{YX} = \frac{\sum (X - \overline{X})(Y - \overline{Y})}{\sum (X - \overline{X})^2} = \frac{6}{8} = .75$$

Thus, when X increases by one unit, Y increases by .75 units, *on the average*. In terms of causal analysis, the linear effect of X on Y is .75.

If both the numerator and the denominator of the least-squares slope are divided by n, we get

$$b_{YX} = \frac{\sum (X - \overline{X})(Y - \overline{Y})}{\sum (X - \overline{X})^2} = \frac{\dfrac{\sum (X - \overline{X})(Y - \overline{Y})}{n}}{\dfrac{\sum (X - \overline{X})^2}{n}} = \frac{s_{XY}}{s_X^2} \quad \textbf{(2.8)}$$

This formula indicates that the regression slope is the covariance of X and Y divided by the variance of X. In other words, *the slope equals the covariance of the independent and dependent variables divided by the variance of the independent variable*. Thus, the covariance and variance—which are two of the statistical building blocks—are used to construct a measure of the effect of X on Y. Intuitively, this may make sense when you remember that the slope is the change in Y that follows from a one-unit increase in X. Since all of the points do not fall on a straight line, however, we cannot just divide the change in Y between any two points by the change in X between the same two points to get the slope. Instead, we divide the amount by which X and Y vary or change together (the covariance) by the amount that X itself varies or changes.

After the slope b has been determined, the Y-intercept can then be determined by the following formula:

$$a_{YX} = \overline{Y} - b_{YX}\overline{X} \quad \textbf{(2.9)}$$

This formula indicates that the mean of X is first multiplied by the slope and the resulting product is subtracted from the mean of Y to get the intercept. Our example data (in Table 2.1) give

$$a_{YX} = \overline{Y} - b_{YX}\overline{X} = 3 - .75(3) = .75$$

The reason that Equation 2.9 gives the intercept is that the least-squares line always passes through the pair of coordinates corresponding to the means of X and Y, i.e., $(\overline{X},\overline{Y})$. That is, when X is at its mean value, the best prediction of Y is always $\overline{Y}$. In our example data, the mean of X equals 3. If $X = 3$ is plugged into the regression equation, we have

$$\hat{Y} = .75 + .75X = .75 + .75(3) = 3 = \overline{Y}$$

Thus, the following statement always holds true:

$$\overline{Y} = a_{YX} + b_{YX}\overline{X}$$

FIGURE 2.5 The Least-Squares Regression Line When the Slope Equals Zero

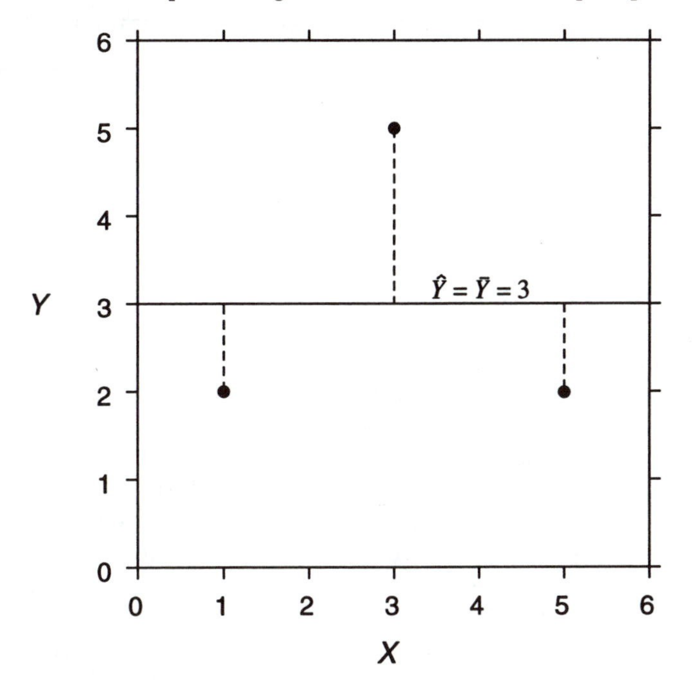

You should be able to see that Equation 2.9 for the intercept is simply an algebraic rearrangement of the above formula.

It should also be noted that if there is no relationship between X and Y (i.e., $b = 0$), then

$$\overline{Y} = a_{YX} + 0\overline{X} = a_{YX}$$

That is, the intercept equals the mean of Y. This indicates that when the slope is zero, the best prediction of Y for all values of X is the mean of Y. For instance, the second example in Table 2.2 has a covariance of zero, and thus the slope is also zero. Figure 2.5 shows the scatterplot and least-squares line for this set of XY scores.

For every value of Y, $Y - \hat{Y} = Y - \overline{Y}$; that is, each Y residual is equal to its deviation from the mean. Since this is the case, the sum of squared residuals will also equal the sum of the squared deviation scores.[3]

$$\sum (Y - \hat{Y})^2 = \sum (Y - \overline{Y})^2$$

3. This shows that the mean of Y, or the mean of any variable, is a *least-squares* statistic. The sum of squared deviations around the mean will be smaller than the sum of squared deviations around any other value.

TABLE 2.5 Effect of Doubling Y on the Slope of the Line

Y	$Y - \overline{Y}$	$(Y - \overline{Y})^2$	$(X - \overline{X})(Y - \overline{Y})$
4	−2	4	4
4	−2	4	0
10	4	16	8
18	0	24	12
$\overline{Y} = 6$	$s_Y = \sqrt{\frac{24}{3}} = \sqrt{8} = 2.83$		$b_{YX} = \frac{12}{8} = 1.50$

When the covariance was discussed, we pointed out that its magnitude was directly affected by the sizes of the standard deviations of X and Y; the greater the standard deviations are, the greater will be the covariance, all other things being equal. This means that there is no set upper limit on the value of the covariance which would hold for all XY pairs. There is also no fixed upper limit for the slope, for it is also affected by the standard deviations. In the case of the slope, however, it is not the absolute values of the standard deviations that affect the value of the slope (as it is with the covariance). Instead, it is the magnitude of the standard deviation of Y relative to the standard deviation of X, i.e., s_Y/s_X, that is directly related to the values of the slope. The greater the standard deviation of Y relative to the standard deviation of X, the greater will be the slope. It may make intuitive sense that the greater the variation of Y relative to that of X, the greater will be the change in Y per unit change in X.

Table 2.5 shows the effect of multiplying Y in our example data by 2. Doubling Y increased its standard deviation, and since the standard deviation of X was not changed, the multiplication increased the ratio s_Y/s_X. As a result of the change in the standard deviation, the sum of the cross-products also increased, and thus, the slope increased from .75 for the original variables to 1.50. You should convince yourself, however, that if X had also been multiplied by 2, the slope would not have changed because the relative sizes of the standard deviations would not have changed. Because the slope has no fixed upper limit due to its dependency in part on the relative magnitude of the standard deviations, it is often referred to as an **unstandardized** slope.

Interpreting the Regression Coefficients

The intercept a has been defined as the predicted value of Y when $X = 0$. In a causal analysis, this definition implies that a is the amount of Y that is caused by a zero value of X. There are several factors, however, that must be taken into account in order to make a meaningful interpretation of a. First, if X is a ratio variable with a true zero value, then the intercept may represent the value of Y that is caused by the *absence* of whatever X is measuring. In our hypothetical example, X equals the number of siblings, and thus $X = 0$ means the absence of any brothers and sisters. It makes sense to think of zero siblings (an only child) as causing such persons to have only .75 children (about one) on the average; since they had no siblings, they may also want to have only one child. Or, the absence of any earnings (zero income) over some period of time may cause a high level of alienation. However, if X is not measured on a ratio scale, we can't make quite the same interpretation because $X = 0$ doesn't mean the absence of what X is measuring, i.e., zero on the scale is not a true zero. If, for example, we ask three agree–disagree questions to measure alienation and give each person an alienation score equal to the number of questions with which he or she agrees (0–3 points), then some people will undoubtedly have a score of zero. This doesn't mean that they have absolutely no alienation, however. Thus, although we still might say that the intercept is the amount of Y caused by a zero value of X (approximately), we can't interpret this value as the lowest value of Y (or the highest value, depending on the sign of the effect) that could be caused by X, because zero doesn't truly represent the absence of X. Therefore, we must pay attention to the measurement scale of X when interpreting the intercept.

The second factor to which we must pay attention when interpreting the intercept is whether there are any cases that take on the value $X = 0$. If X equals age, for example, and you were investigating its effect on earnings of employed persons, there wouldn't be any persons in the study who were zero years old or even approximately zero years old. Thus, it wouldn't be justifiable, or even meaningful in this example, to extrapolate predictions beyond the range of the observed values of X.

The final factor is related to the second. There may not be enough cases at $X = 0$ to use the intercept as a prediction of Y. For instance, if X equals the number of years of schooling, there will usually be less than one percent of the individuals with a score of zero. Since there are few cases at this value of X, they won't have much weight in determining the slope of the regression line, and thus predictions may be very bad at this end of the scale. In sum, you should pay attention to the distribution of cases on X and the measurement scale before you attempt to interpret the intercept.

We must also use a to determine the level of Y that is caused by nonzero values of X. Social scientists, however, are seldom interested in estimating the value of Y that is determined by a specific value of X. Perhaps this is because we don't have enough true ratio scales for individual values to have any intrinsic meaning or interest. Or perhaps it is because the causal theories are at such a low level of development that we are primarily interested in whether X has *any* effect on Y, and if so, what is the direction of that effect. Whatever the reason, researchers are primarily interested in the slope b of the regression equation. We focus on how much and in what direction Y changes as a result of a change in X, i.e., the slope of the line. There is, practically speaking, little interest in the intercept, either in its own right or for the purpose of predicting individual levels of Y caused by a given value of X. The slope, on the other hand, represents the *effect* of X on Y.

The slope has been interpreted as the change in Y per unit increase in X. In emphasizing *change*, this interpretation is consistent with a dynamic imagery of cause-and-effect relationships. This may follow from the criterion that the effect must follow the cause, where the emphasis on time may imply that Y is changed by changes in X. It may also be due to the close association in our minds between the experimental method and causal analysis. In the experimental method the experimenter actually manipulates (changes) the cause and also frequently measures changes in Y.

There are two problems, however, with this dynamic interpretation of the regression slope in nonexperimental research. First, we may often have only cross-sectional data. If so, then it is a theoretical assumption to interpret the regression slope as the change in Y per unit increase in X. We have not measured any changes in either variable. Therefore, to represent what has been empirically observed more accurately, we should say that the slope is the *difference* in Y per positive unit *difference* in X. This interpretation, however, is offered more in the spirit of a caveat to make you aware of the empirical limitations of cross-sectional data. Many will still interpret the slope in terms of change when this is what causal theory implies.

Second, not all variables that are causes of other variables produce their effects through changes. Race and gender don't change, yet many things are affected by these variables. And although the number of siblings (X) that a child has may change as his or her parents have more children, these changes are all typically completed before the child begins having children (Y). Finally, although education changes as a person progresses through higher and higher levels of schooling, they typically aren't working at jobs where their earnings are affected by these changes until after they have completed their education. Thus, in all of these cases it is best to interpret the regression slope as the difference in Y for a positive-unit difference in X.

Regression of *X* on *Y*

Although we typically designate one variable as dependent (Y) and the other as independent (X) when using regression to estimate the effect of the latter on the former, it is also possible to calculate a regression equation that uses the variable labeled Y as a predictor of the variable labeled X. Two researchers, for example, might disagree about the direction of causality. Thus, it is instructive to examine how the regression coefficients differ in the two cases.

The least-squares formulas for X as the dependent variable and Y as the independent variable are

$$b_{XY} = \frac{\sum (X - \overline{X})(Y - \overline{Y})}{\sum (Y - \overline{Y})^2} \qquad \textbf{(2.10)}$$

$$a_{XY} = \overline{X} - b_{XY}\overline{Y} \qquad \textbf{(2.11)}$$

The structure of these formulas is identical to the original formulas for the slope and the intercept. With one exception, you get the above formulas by substituting Y for X and X for Y in Equations 2.7 and 2.9. The exception is the numerator of the formula for the slope, which remains unchanged. The meaning of the slope is still the same: it is the sum of cross-products (or the covariance) divided by the sum of squared deviations (or the variance) of the independent variable, which is now Y. The intercept is still the mean of the dependent variable (which is now X) minus the product of the slope and the mean of the independent variable (which is now Y). All of the subscripts are reversed, so that the left-hand variable is the dependent variable X.

For the example data in Table 2.1, the least-squares equation for X predicted by Y is

$$b_{XY} = \frac{\sum (X - \overline{X})(Y - \overline{Y})}{\sum (Y - \overline{Y})^2} = \frac{6}{6} = 1.00$$

$$a_{XY} = \overline{X} - b_{XY}\overline{Y} = 3 - 1(3) = 0$$

$$\hat{X} = 0 + 1Y$$

The slope and intercept differ from the values found when Y was treated as the dependent variable. The "effect" of Y on X is greater than the "effect" of X on Y in terms of the change in the dependent variable per unit increase in the independent variable. Because the coefficients differ depending on which is designated as the dependent variable, the intercept and slope are **asymmetric** statistics. The slopes differ because the variances of X and Y differ. That is, a unit change in Y is associated with a larger change in X (b_{XY}) because the variance in X (2.67) is greater than the variance in Y (2.00). If the variances of X and Y were equal, the slopes would be equal.

The fact that we have two different regression lines, one when Y is treated as the dependent variable and another when X is treated as dependent, does not mean that they can both be used as measures of the effects of each variable on the other. They can't both be correct. We have to choose one variable as dependent and the other as independent and then use the appropriate formula.[4] The two regression formulas have been introduced for the sole purpose of showing that you will get different slopes and intercepts depending on which variable is chosen as the dependent variable.

---

4. The case where X and Y may each affect the other (reciprocal effects) will be discussed in Chapter 8.

Measures of Strength of Relationship

Coefficient of Determination

It has been pointed out that neither the covariance nor the regression slope has a fixed upper (or lower) limit. They cannot be used as measures of the strength of the relationship. We know that the least-squares regression equation gives the best predictions of Y, but we don't yet have a measure of how good those predictions are. One logical candidate for use in constructing such a measure is the sum of the squared residuals $\Sigma (Y - \hat{Y})^2$, the quantity that is minimized to determine the regression equation. The sum $\Sigma (Y - \hat{Y})^2$ is called the **residual sum of squares**. These residuals are errors of prediction. Notice in Table 2.4 that the sum of the errors (unsquared) equals zero and, thus, that the mean of the residuals is equal to zero. This is always the case. Therefore, the squared errors will be used to construct our measure.

In order to understand the development of the next statistics, it is helpful to think of the residuals or errors as just another variable, a new variable that has been constructed from raw scores and predicted scores. As such, it has a mean and variance, just as any variable does. To compute the variance of the residuals, which we will call e for the moment in order to simplify notation, we would sum the squared deviations of e from its mean:

$$s_e^2 = \frac{\sum (e - \bar{e})^2}{n} = \frac{\sum (e - 0)^2}{n} = \frac{\sum e^2}{n} = \frac{\sum (Y - \hat{Y})^2}{n}$$

This formula shows that the variance of the residuals (e) is simply the sum of squared errors, divided by n. So when we are working with the sum of squared errors, it represents the variation of the residuals.

The variation of the residuals is often referred to as *unexplained variation*. Since a residual is an error of prediction, it represents that part of Y that can't be explained by X. Now think of the total variation of Y as consisting of two components, the unexplained variation plus the explained variation. If we subtract the unexplained variation in Y from the total variation, we would have the *explained variation*. A logical measure of how well our regression equation explains Y would be to divide the explained variation by the total variation. This proportion is called the **coefficient of determination** and is defined by the following formula:

$$\text{Coefficient of Determination} = \frac{\sum (Y - \bar{Y})^2 - \sum (Y - \hat{Y})^2}{\sum (Y - \bar{Y})^2} \qquad \textbf{(2.12)}$$

The numerator represents the explained variation because the sum of squared *residuals* (unexplained variation) is subtracted from the sum of squared *deviations* (the total variation). Because this explained variation is divided by the total variation, *the coefficient of determination equals the proportion of the variation (variance) in Y that is explained by X.* The calculations for the example data are shown in Table 2.6, where the statistic is labeled as $r^2_{\hat{Y}X}$, for reasons to be given later. The value of the coefficient is .75, which means that .75 (or 75 percent) of the variation in Y is explained by X.

Notice also in Table 2.6 that the coefficient of determination for X (when X is the dependent variable) is given as $r^2_{\hat{X}Y}$. This formula is the same as the coefficient of determination for Y, except that X is substituted for Y in Equation 2.12. The coefficient of determination for X also equals .75. Although the amount of explained variation in X ($8 - 2 = 6$) is greater than the explained variation in Y ($6 - 1.5 = 4.5$), the total variation in X (which equals 8) is also greater than the total variation in Y (which equals 6). Thus, when the explained variation in X is expressed as a proportion of the total variation in X, we find that this proportion (.75) equals the explained variation in Y expressed as a proportion of the total variation in Y(.75). Thus, the coefficient of determination is a **symmetrical** measure of the strength of association. It doesn't make any difference which variable is considered to be dependent; each variable "explains" .75 of the variation of the other. This, of course, means that X cannot do any better or worse at explaining Y than Y can do at explaining X.

The coefficient of determination can also be written in a slightly different form:

$$\text{Coefficient of Determination} = \frac{\sum (\hat{Y} - \overline{Y})^2}{\sum (Y - \overline{Y})^2} \qquad \textbf{(2.13)}$$

TABLE 2.6 Example Data for the Coefficients of Determination

$\hat{Y}$	$Y - \hat{Y}$	$(Y - \hat{Y})^2$	$\hat{Y} - \overline{Y}$	$(\hat{Y} - \overline{Y})^2$	$\hat{X}$	$X - \hat{X}$	$(X - \hat{X})^2$
1.5	.5	.25	−1.5	2.25	2	−1	1
3.0	−1.0	1.00	0.0	0.00	2	1	1
4.5	.5	.25	1.5	2.25	5	0	0
9.0	0.0	1.5	0.0	4.50	9	0	2

$$r^2_{\hat{Y}X} = \frac{\sum (Y - \overline{Y})^2 - \sum (Y - \hat{Y})^2}{\sum (Y - \overline{Y})^2} = \frac{6 - 1.5}{6} = .75 \qquad r^2_{\hat{X}Y} = \frac{\sum (X - \overline{X})^2 - \sum (X - \hat{X})^2}{\sum (X - \overline{X})^2} = \frac{8 - 2}{8} = .75$$

$$r^2_{\hat{Y}X} = \frac{\sum (\hat{Y} - \overline{Y})^2}{\sum (Y - \overline{Y})^2} = \frac{4.5}{6} = .75$$

$$s_{Y-\hat{Y}} = \sqrt{\frac{\sum (Y - \hat{Y})^2}{n}} = \sqrt{\frac{1.5}{3}} = .707 \qquad \hat{s}_{Y-\hat{Y}} = \sqrt{\frac{\sum (Y - \overline{Y})^2}{n - 2}} = \sqrt{\frac{1.5}{1}} = 1.225$$

In this formula, $\Sigma(\hat{Y} - \overline{Y})^2$ is a direct measure of explained variation. It represents the sum of the squared departures of Y from its mean that are accounted for by X. Since $\hat{Y}$ is exactly determined by the regression equation (i.e., $\hat{Y}$ is an exact linear function of X), $\Sigma(\hat{Y} - \overline{Y})^2$ is called the *regression sum of squares*. Therefore, the deviation of $\hat{Y}$ from the mean $(\hat{Y} - \overline{Y})$ represents the part of Y that is explained or caused by X.

As shown in Table 2.6, the second equation for the coefficient of determination (2.13) gives exactly the same value as the first (2.12). This is because both Equations 2.12 and 2.13 equal the explained variation divided by the total variation. In Equation 2.12 the explained variation is expressed as

$$\sum (Y - \overline{Y})^2 - \sum (Y - \hat{Y})^2 = 6 - 1.5 = 4.5$$

whereas in Equation 2.13 the explained variation is expressed as

$$\sum (\hat{Y} - \overline{Y})^2 = 4.5$$

As this example illustrates, both are equivalent expressions. The above expressions show that the explained variation plus the unexplained variation equals the total variation:

$$\sum (\hat{Y} - \overline{Y})^2 + \sum (Y - \hat{Y})^2 = \sum (Y - \overline{Y})^2$$
$$4.5 + 1.5 = 6.0$$

Although Equations 2.12 and 2.13 are equivalent, the first seems to be used more frequently, possibly because it contains the squared residuals, the quantity that is minimized by the least-squares solution.

The range of values of the coefficient of determination is fixed. When Y is perfectly determined by X, the residuals $Y - \hat{Y}$ will be exactly zero for each case, and thus the sum of the squared residuals will be zero. Equation 2.12 will then reduce to

$$\text{Coefficient of Determination} = \frac{\sum (Y - \overline{Y})^2 - 0}{\sum (Y - \overline{Y})^2} = 1.00$$

Unity is thus the maximum value of the coefficient, indicating perfect prediction. When there is no relationship between X and Y, i.e., the slope is zero, the best prediction for each case will be the mean of Y. Equation 2.12 will then reduce to

$$\text{Coefficient of Determination} = \frac{\sum (Y - \overline{Y})^2 - \sum (Y - \overline{Y})^2}{\sum (Y - \overline{Y})^2} = \frac{0}{\sum (Y - \overline{Y})^2} = .00$$

Thus, the minimum value of the coefficient is zero. Obviously, the higher the value of the coefficient, the stronger the relationship between X and Y.

It was emphasized above that the sum of the squared residuals divided by n is the variance of the residuals. Thus, the following is the standard deviation of the residuals:

$$s_{Y-\hat{Y}} = \sqrt{\frac{\sum (Y - \hat{Y})^2}{n}} \tag{2.14}$$

This is the definition of the standard deviation of the sample residuals. But, as with any standard deviation, dividing by n produces an underestimate of the population standard deviation (see footnote 1). We divide by $n - 1$ to correct for this bias when we want to estimate the standard deviation of Y. When we want to estimate the population standard deviation of the residuals $Y - \hat{Y}$, we must divide by $n - 2$ because we have one less degree of freedom as a result of using one independent variable to estimate $\hat{Y}$:

$$\hat{s}_{Y-\hat{Y}} = \sqrt{\frac{\sum (Y - \hat{Y})^2}{n - 2}} \tag{2.15}$$

The hat over s indicates that it is the estimate of the population parameter. This statistic is called the **standard error of estimate**, to reflect the fact that it is a standard deviation of residuals or errors. The values given by Equations 2.14 and 2.15 are calculated in Table 2.6.

Standardized Regression Coefficient

A second approach to developing a measure of the strength of the XY relationship is based directly on the regression equation itself. This approach takes into account the fact that the magnitude of the unstandardized regression slope is directly related to the ratio of the standard deviation of Y to the standard deviation of X, as was discussed in some detail earlier; the greater s_Y/s_X is, the greater will be b_{YX}. It will be seen that if we can transform the X and Y variables so as to make them have equal standard deviations, this influence can be eliminated.

This transformation involves computing **standardized scores** for X and Y. The formulas for standardized scores are

$$z_X = \frac{X - \overline{X}}{s_X} \tag{2.16}$$

$$z_Y = \frac{Y - \overline{Y}}{s_Y} \tag{2.17}$$

A standardized score is a deviation score divided by the standard deviation. Each case has a standardized score. The deviation score indicates the distance between the score and the mean. When divided by the standard deviation, the result indicates how many standard deviations a score is from the mean of its distribution. The standardized scores for X and Y are given in Table 2.7. For the first case on X, the standardized score is

$$z_X = \frac{X - \overline{X}}{s_X} = \frac{1 - 3}{1.63} = -1.23$$

This means that the person with one sibling is 1.23 standard deviations below the mean number of siblings.

Notice that the sum of z scores for both X and Y is zero. Thus, the mean of the standard scores for any variable is always zero. This is because the mean of the deviation scores is zero, and since we are dividing the deviation scores by a constant (the standard deviation), the mean will still be zero. Table 2.7 also gives squared z scores for each variable. Since the mean of z is 0, a squared z score is a squared deviation score. Remember that z is just another variable that we have created, and as such, it has deviation scores and a standard deviation. The standard deviation of the z scores of X is

$$s_{z_x} = \sqrt{\frac{\sum (z - \bar{z})^2}{n}} = \sqrt{\frac{\sum (z - 0)^2}{n}} = \sqrt{\frac{\sum z^2}{n}} = \sqrt{\frac{3.02}{3}} = \sqrt{1} = 1.00 \quad \textbf{(2.18)}$$

This result occurs because the sum of the squared z scores for X is 3.02, which, except for a slight rounding error, is equal to n. The sum of squared z scores for Y is also equal to n, except for another slight rounding error. In general, $\sum z^2 = n$. Thus, the standard deviation of z will always have n in the numerator and n in the denominator, and consequently it will always have a value of unity.

The z scores are simply a linear transformation of the raw scores. For example,

$$z_X = \frac{X - \overline{X}}{s_X} = \frac{1}{s_X}X - \frac{\overline{X}}{s_X} = \frac{1}{1.63}X - \frac{3}{1.63} = .612X - 1.837$$

TABLE 2.7 Standardized Scores for the Example Data

z_X	z_Y	z_X^2	z_Y^2	z_Xz_Y
−1.23	−.71	1.51	.50	.87
.00	−.71	.00	.50	.00
1.23	1.41	1.51	1.99	1.73
.00	.00	3.02	2.99	2.61

$$r_{YX} = B_{z_Xz_Y} = \frac{\sum z_Xz_Y}{n} = \frac{2.61}{3} = .87$$

Figure 2.6 shows the linear relationship between X and z_X in our example data.

The figure shows that a unit difference on z corresponds to a difference on X of $s = 1.63$, the standard deviation of X. Thus, a change in X of 1.63 (one standard deviation in X) corresponds to a change in z of 1 (the standard deviation of z).

We have transformed X and Y into standardized scores which have a standard deviation equal to unity for both X and Y. If we use these standardized scores in a regression analysis, we will then have eliminated the influence of s_Y/s_X on the regression slope because the ratio of the standard deviations will be unity.

Now, if we take the least-squares formula for the regression slope and simply substitute z_X for X and z_Y for Y, we will have the **standardized** regression slope,

$$B_{z_Y z_X} = \frac{\sum (z_X - \bar{z}_X)(z_Y - \bar{z}_Y)}{\sum (z_X - \bar{z}_X)^2} = \frac{\sum z_X z_Y}{\sum z_X^2} = \frac{\sum z_X z_Y}{n} \tag{2.19}$$

Because the means of the z scores are 0 and the sums of the squared z scores are n, the standardized slope reduces to a very simple formula, the sum of the

FIGURE 2.6 The Linear Transformation of X into z

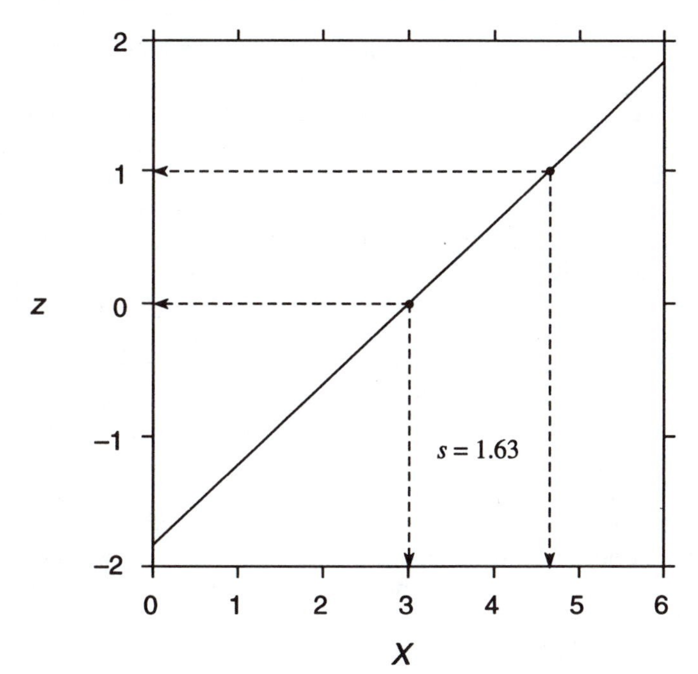

cross-products of the standardized scores divided by n. Remembering that the means are zero, you should be able to see that this formula represents the **covariance of the standardized scores**. Thus, we have wound up computing a standardized regression coefficient that is a standardized covariance.

If we also plug the z scores into the formula for the intercept, we get

$$A_{z_Y z_X} = \bar{z}_Y - B_{z_Y z_X} \bar{z}_X = 0 - B_{z_Y z_X} 0 = 0 \quad \textbf{(2.20)}$$

Thus, the intercept equals 0. Notice that capital letters have been used for the standardized intercept and slope to distinguish them from the unstandardized values. The formula for the standardized regression equation is

$$\hat{z}_Y = B_{YX} z_X \quad \textbf{(2.21)}$$

Since the variables are designated by z scores in the equation, it is not necessary to put z's in the subscript for the slope. The maximum value of B is 1 and the minimum value is -1 (remember, slopes can be negative). Thus, it serves as a measure of the strength of the effect of X on Y; the closer the value of B is to 1 or -1, the stronger the relationship. As computed in Table 2.7, the standardized slope equals .87. This is a strong relationship, as it is close to unity.

The standardized slope also has a very clear interpretation based on the fact that it is a slope. B indicates the change in the standard scores of Y per unit increase in the standard scores of X. In our example, an increase of 1 in z_X causes an increase of .87 in z_Y. Remember that a z score indicates how many standard deviations a raw score is from the mean. Thus, a change in z_X of 1 means that X changes by one standard deviation, and the associated change in z_X of .87 means that Y changes by .87 standard deviation. When we are dealing with z scores, the scale of measurement is standard deviations. Therefore, we can say that if X increases by one standard deviation, Y increases by .87 standard deviation.

You undoubtedly noticed that the formula for the standardized slope in Table 2.7 was specified as being equal to r_{YX}, the symbol for the **Pearson product-moment correlation coefficient**. Equation 2.19 for the standardized regression coefficient is also often given as a formula for the correlation coefficient. The two are one and the same. The importance of Equation 2.19 for the correlation is indicated by the term "product-moment," which precedes it above. In mathematics, taking the "moment" of a distribution of scores is another way of saying divide by n to get the arithmetic average. Thus, product-moment correlation means the arithmetic average of the products of the z scores for X and Y (Equation 2.19). The identity of the standardized slope and the Pearson correlation means that the correlation can be given a very explicit interpretation in terms of a slope. Thus, the correlation not only tells us the strength of association between X and Y, it also indicates the change in Y associated with a change in X.

One other formula that is frequently given for the correlation is

$$r = \frac{s_{XY}}{s_Y s_X} \quad \textbf{(2.22)}$$

In this formula, the covariance is divided by the products of the standard deviations in order to remove the influence that they have on the covariance. Thus, this formula also allows us to interpret the **correlation as a standardized covariance**.

If we square the correlation coefficient or the standardized slope, we get $r^2 = B^2 = (.87)^2 = .76$. This is equal to the value of the coefficient of determination that was given in Table 2.6, except for a slight rounding error. Thus, we see that the coefficient of determination is equal to the square of the correlation, and that is why it was given that symbol in Table 2.6. It is important to emphasize, however, that the coefficient of determination has its own formula and interpretation. Although it is equal to r^2, it is not clear from the definition of r as a standardized slope, or as defined in any other way, that squaring it will allow us to interpret its value in the way that we have defined and interpreted the coefficient of determination. It is acceptable to refer to the coefficient of determination as r^2; this is a common practice. However, although the variance is simply the square of the standard deviation, the coefficient of determination is interpreted differently from the correlation coefficient, even though we can calculate its value by squaring the correlation.

Finally, there are two useful formulas that allow you to switch back and forth between the correlation coefficient (or B) and the unstandardized regression coefficient:

$$b_{YX} = \frac{s_Y}{s_X} r \quad \textbf{(2.23)}$$

$$r = \frac{s_X}{s_Y} b_{YX} \quad \textbf{(2.24)}$$

Equation 2.23 shows that the unstandardized regression slope is a function of the strength of the relationship between X and Y, as indicated by r, and the ratio of the standard deviations of Y and X. The latter influence was discussed several times previously. Now we see both factors contained in a single formula such that the greater the correlation between X and Y, and the greater the standard deviation of Y relative to that of X, the greater will be the change in Y caused by a unit increase in X. Equation 2.24, on the other hand, shows how to remove the influence of the standard deviations on the slope in order to get a measure of the strength of relationship, the correlation (or B).

Outliers and Influential Cases

Outliers are cases that are unusual or distinct in some respect. Sometimes outliers may have "undue" influence on the regression coefficients. This may be the result of an error in the outlier's score on either X or Y, such as a coding error or an erroneous interview response. Or the case may be an outlier because of some other variable or variables that are not included in the causal model. In either instance, it is important to determine whether the outliers are highly influential cases and, if so, to attempt to determine the reason for this so that appropriate modifications to either the data or the model may be made.

When we look again at the scatterplot for the XY scores in our example data (Figure 2.7a), Case 3 looks somewhat like an outlier because of its distance in the plot from the other two cases. But when we draw the regression line on the plot and show the distance of each case from the regression line (residual scores) (Figure 2.7b), Case 3 does not look any more unusual than Case 1 and Case 2 looks the most like an outlier. Thus, Figure 2.7 indicates that whether we consider a case to be an outlier depends on our frame of reference.

It is useful in terms of regression analysis to use two frames of reference. First, we may consider a case an outlier if it has an unusual score on X, that is, if it lies a great distance from the mean of X. Such cases may be called **X outliers.** A standardized score (z score) may be used as an indicator of this type of outlier. Table 2.8 gives the values of z^2 in order to eliminate negative signs.[5] According to this criterion, Cases 1 and 3 are both equally "unusual," whereas Case 2 is at the mean of X and thus is just an average case. X outliers are often said to have *leverage* (Fox 1991; Bollen and Jackman 1990). In order to influence the regression coefficients greatly a case must have great leverage. But leverage is not sufficient for influence; leverage is best considered to be an indicator of *potential* influence.

The second criterion for being an outlier is to be unusual relative to the regression line. That is, if a case has an unusually large residual, it will be an outlier. This type will be called a **regression outlier.** If a case is a regression outlier, its value of Y is unusual relative to its value of X. The residuals in Table 2.8 as well as Figure 2.7b indicate that Case 2 is the biggest regression outlier. Standardized residuals are often helpful in judging the magnitude of the

TABLE 2.8 Outlier and Influence Data for the XY Scores

Case i	z^2_x	$Y - \hat{Y}$	$e/\hat{s}_e$	$b(-i)$	*DFBETA*
1	1.5	.5	.41	1.50	−.75
2	.0	−1.0	−.82	.75	.00
3	1.5	.5	.41	.00	.75

5. The squared z scores are a linear function of h ($z^2 = nh - 1$), where h is the "hat" value, which is frequently used as an indicator of X outliers (Fox 1991; Bollen and Jackman 1990). In the bivariate case, z^2 is also equal to **Mahalanobis's distance**, when $n - 1$ is used to compute the sample variance.

FIGURE 2.7 (a & b) Scatterplots for the *XY* Scores

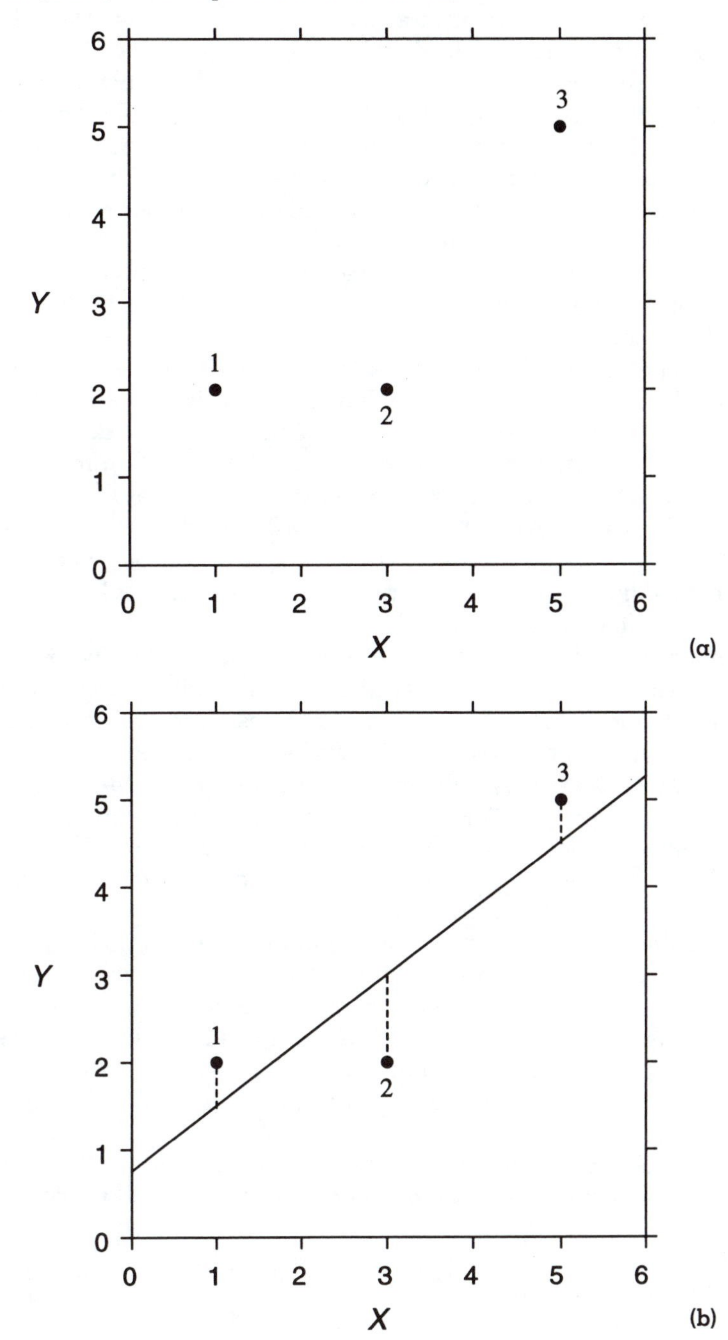

regression outliers. Standardized residuals may be computed by dividing the residuals by the standard error of estimate: $(Y - \hat{Y})/\hat{s}_{Y-\hat{Y}} = e/\hat{s}_e$.[6] The standardized residuals in Table 2.8 are not large.

A logical way to measure the extent to which an outlier actually influences the regression coefficients is to delete the case and recompute the regression equation. We will focus on how much influence an outlier has on the regression slope. The slope with the ith case deleted will be noted as $b_{(-i)}$. The difference between the slope based on all n cases and the slope with the ith case deleted will be the indicator of influence: $d_{(-i)} = b - b_{(-i)}$. There will be one $d_{(-i)}$ for each of the n cases. These d's are often called DFBETA. More complex measures of influence on the slope are described and discussed by Bollen and Jackman (1990) and Fox (1991).

Table 2.8 shows the $b_{(-i)}$'s and $d_{(-i)}$'s for each of the three cases in our example data. In this limited data set, when one case is deleted there are only two cases left. The regression line after deletion will pass exactly through the remaining pair of points in the scatterplot, as shown in Figure 2.8. You will remember that the slope for all three cases equals .75. When Case 1 is deleted, the slope increases to 1.5 (see $\hat{Y}_{(-1)}$ in Figure 2.8) and $d_{(-1)} = -.75$. When Case 3 is deleted, the slope decreases to .0, or $d_{(-3)} = +.75$. But when Case 2 is deleted, there is no change in the slope, i.e., $d_{(-2)} = 0$. Thus, Cases 1 and 3 are most influential, and equally so.

It is important to consider how the indicators of the two types of outliers fared in predicting how much influence each case would have on the regression slope. Although Case 2 is the largest regression outlier, it does not have any influence on the regression line. This is because it is equal to the mean of X. Cases with scores at or near the mean of X contribute little or nothing to the slope of the regression line. This can be seen by looking again at the equation for the least-squares slope, $b_{YX} = \Sigma (X - \overline{X})(Y - \overline{Y})/\Sigma (X - \overline{X})^2$. If X equals the mean of X, then the deviation scores for X in the numerator and denominator will both be zero for that case. Thus, the case with a mean value of X will contribute nothing to the covariation in the numerator and nothing to the variation in the denominator. Consequently, deleting the case will not delete anything from the sums that are used to compute the slope; therefore, the slope will not change after deletion. Also, cases that are quite near the mean will have little effect on the slope. Even though such cases may be relatively large regression outliers, as was Case 2, they are not influential with respect to the least-squares equation.

In sum, influence is a function of being at least somewhat unusual with respect to both X and the regression equation. Having the characteristics of an

6. Better methods of standardizing residuals, such as the *studentized residual*, are described in Fox (1991) and Bollen and Jackman (1990). The different standardizations, however, usually give approximately equal values in large samples (Fox 1991, p. 27).

FIGURE 2.8 Least-Squares Regression Lines with the *i*th Case Deleted

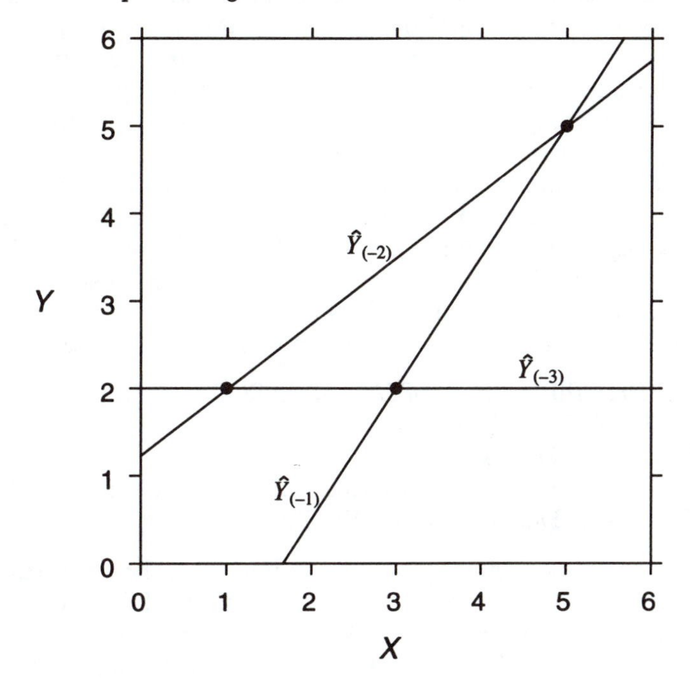

outlier with respect to one frame of reference but not with respect to the other will lead to little or no influence on the regression coefficients.

When an influential outlier has been detected, it might be tempting simply to delete the case from the data set. It would be more appropriate to investigate carefully why the case is an outlier. If an incorrect score on X or Y is found to be the problem, then the score should be corrected if possible and the case retained in the regression analysis. If there does not appear to be a data error, then the characteristics of the case should be carefully examined. Perhaps something else peculiar about the case will be noticed, such as another variable on which the case is also unusual and which also affects Y. If this third variable is included in the regression equation (see Chapter 3), then the new variable may help correct the bad prediction of Y that was based on X alone and consequently eliminate the case as an influential regression outlier.

An Example: Siblings and Fertility

A real example of the use of bivariate regression and correlation for studying the effect of the number of a person's siblings on the number of his or her own children will be presented to provide a review of the statistical concepts presented in this chapter. The data were collected in face-to-face interviews as part of the 1986 General Social Survey (GSS), a nationwide survey of approximately 1,500 adults in the United States conducted annually by the National Opinion Research Center (Davis and Smith 1993). The analysis will be confined to ever-married women (never-married women are excluded) who were 35 years of age or older at the time of the interview. The age restriction was for the purpose of ensuring that most of the women would have completed their childbearing. The independent variable X is SIBS, the number of brothers and sisters of the respondent (including stepbrothers and stepsisters). The dependent variable Y is CHILDS, the number of children that the respondent has had. The regression and correlation statistics for the data that we will examine may be computed by SPSS (e.g., Norusis 1990) with the following commands.

```
REGRESSION DESCRIPTIVES = DEFAULTS XPROD/
  VARIABLES = CHILDS SIBS/
  DEP = CHILDS/ ENTER SIBS/
  RESIDUALS = OUTLIERS (MAHAL,RESID,COOK) ID (SIBS)/
  CASEWISE = ALL DEPENDENT PRED DFBETA
```

It is interesting to note the means of these variables shown in Table 2.9. The mean SIBS is considerably higher than the mean CHILDS. If you add 1 to the mean SIBS to take into account the fact that the respondent herself is not counted in the number of SIBS, we see that the average number of children in the families in which these women grew up is almost twice as great as the number of children that these women have had. This reflects a tremendous change in fertility in American society.

We, however, are primarily interested in the relationship between SIBS and CHILDS. These two variables meet the temporal criteria for making causal inferences because in the great majority of cases all of the respondent's brothers and sisters would have been born before the respondent began to have her own children. It was expected, of course, that respondents who were reared in larger families (large numbers of siblings) would have larger families of their own because of intergenerational transmission of fertility norms.

Table 2.9 shows the formulas and computations for various regression and correlation statistics that have been covered in this chapter. The table gives the means, sums of squares, and sums of cross-products that are required for the calculations. The elements in the Sums of Squares and Cross-Products Matrix are the sum of squares of SIBS ($\Sigma (X - \overline{X})^2 = 4940.673$), the sum of squares of CHILDS ($\Sigma (Y - \overline{Y})^2 = 1845.720$), and the sum of cross-products of SIBS and CHILDS ($\Sigma (X - \overline{X})(Y - \overline{Y}) = 551.181$). Carefully examine all of the formulas and computations to make sure you understand them.

The bivariate regression slope indicates that for each additional sibling, the predicted number of respondent's children increases by .112. This value is positive and thus in the expected direction. Notice that it is much smaller than the

TABLE 2.9 Regression of Number of Children on Number of Siblings

	Mean	Std Dev
SIBS	4.286	3.166
CHILDS	2.736	1.935
n	493	

Sums of Squares and Cross-Products Matrix

	SIBS	CHILDS
SIBS	4940.673	551.181
CHILDS	551.181	1845.720

$$b_{YX} = \frac{\sum (X - \overline{X})(Y - \overline{Y})}{\sum (X - \overline{X})^2} = \frac{551.181}{4940.673} = .112$$

$$a_{YX} = \overline{Y} - b_{YX}\overline{X} = 2.736 - .112\,(4.286) = 2.258$$

$$\hat{Y} = 2.258 + .112\,X$$

	Sum of Squares
Regression	61.48959
Residual	1784.23049
Total	1845.72008

$$\hat{s}_{Y-\hat{Y}} = \sqrt{\frac{\sum (Y - \hat{Y})^2}{n - 2}} = \sqrt{\frac{1784.230492}{491}} = \sqrt{3.63387} = 1.906$$

$$r^2_{YX} = \frac{\sum (Y - \overline{Y})^2 - \sum (Y - \hat{Y})^2}{\sum (Y - \overline{Y})^2} = \frac{1845.720 - 1784.23049}{1845.720} = \frac{61.48959}{1845.720} = .033$$

$$r_{YX} = B_{YX} = \frac{s_{XY}}{s_X s_Y} = \frac{551.181/493}{(3.166)(1.935)} = .182$$

.75 value for the slope in the hypothetical three-case example. The influence of size of family of origin in the real world is much less than in the artificial data used for illustrative purposes in this chapter. (It is valid to compare the two unstandardized slopes because both are based on the same scales, i.e., number of siblings and number of children.) The intercept indicates that respondents with no siblings (i.e., only children) are nevertheless predicted to have 2.258 children of their own.

The standard error of estimate shows that the standard deviation of the residuals equals 1.906, that is, the cumulative effects of other variables that have not been included in the equation have a standard deviation of 1.906 children. This is almost as great as the standard deviation of CHILDS itself. The coefficient of determination shows that SIBS explains .032 (or 3.2%) of the variance in CHILDS, a small proportion of the variance. (Imagine what kind of a society we would have if our parents' fertility strongly determined our fertility.) The standardized regression slope or correlation coefficient indicates that an increase of 1 standard deviation in the number of siblings will cause an increase of .182 standard deviation in the number of children.

Figure 2.9 shows the scatterplot for children and siblings. These are discrete variables (i.e., noncontinuous) for which fractional values are not possible, a not uncommon case in the social sciences. In large samples such as the GSS ($n = 493$ in this analysis) this often means that a number of cases pile up at many of the points on the plot. In Figure 2.9 it is not possible to see how many cases may be at each data point. Thus, we cannot see very well where the data are dense and where they are sparse, which hinders our perception of the relationship between siblings and children.

One method of unpiling the data is to add to each pair of *XY* scores a small random number. In this case, a random number between –.5 and +.5 from a uniform distribution was added to each value of *X* and each value of *Y*. This required the generation of $2 \times 493 = 986$ random numbers, two for each case. This procedure is called *jittering* the data. It is equivalent to adding a small random error to each score on each variable. The jittered variables are not used to estimate the regression coefficients. They are used only to convert the discrete variables into pseudo-continuous variables in order to present a better picture of the relationship between *X* and *Y*. In addition, unfilled circles are used for plotting symbols in place of the solid black symbols used in Figure 2.9. The result is shown in Figure 2.10. It can now be seen that the greatest concentration of cases lies approximately in the area bounded by 1–6 siblings and 1–4 children.

Figures 2.9 and 2.10 also show three circled data points that have been labeled with case numbers. The *X* and *Y* values for these three cases can easily be read from Figure 2.9. Case 228 is the largest *X* outlier. It stands out more distinctly than any other case in the plot. This woman has nineteen siblings (remember that siblings includes stepbrothers and stepsisters) and six children of her own. The z score for 19 siblings equals 4.64 (i.e., 4.64 standard deviations above the mean) and $z^2 = 21.56$. This case, however, is not a particularly large regression outlier. The largest residual belongs to Case 475, a woman with only one sibling but 8 or more children of her own (the top category for the CHILDS variable corresponds to 8 or more children). The residual for this case equals 5.63; that is, she has 5.63 more children than expected based upon the number of her siblings.

Neither the largest *X* outlier nor the largest regression outlier is the most influential case. This honor goes to Case 59, a woman with sixteen siblings and

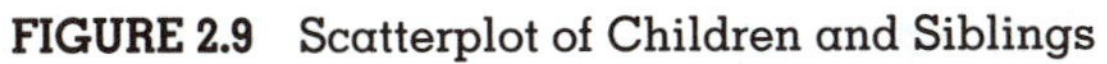

FIGURE 2.9 Scatterplot of Children and Siblings

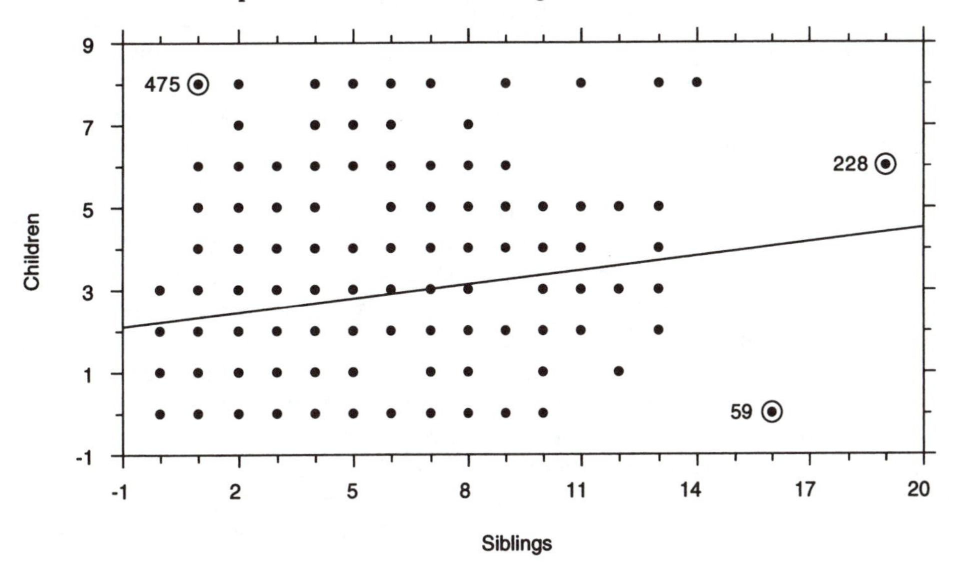

FIGURE 2.10 Jittered Scatterplot of Children and Siblings

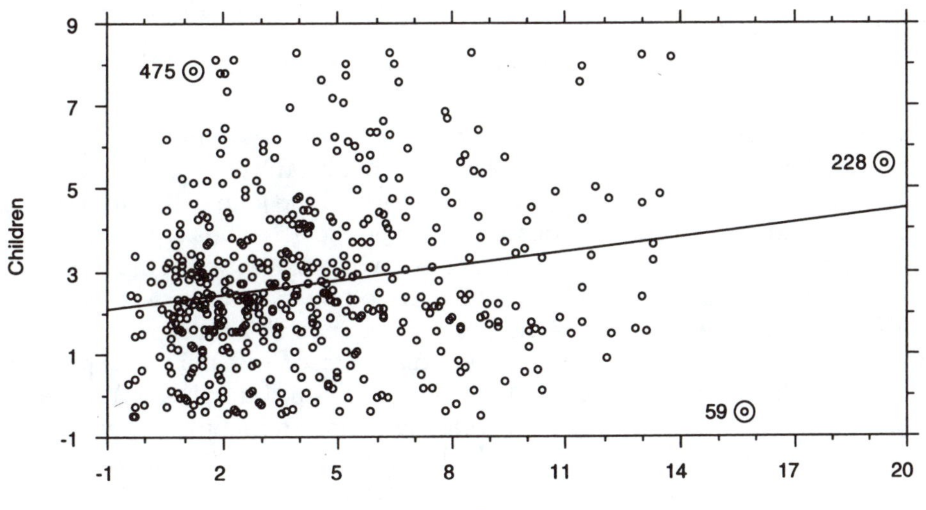

zero children of her own. DFBETA = –.0099 for Case 59; that is, if this case is deleted, the slope will rise from $b = .112$ to $b = .122$. Although this case is clearly the most influential, it still does not have a great deal of influence. This is a good illustration of the fact that no matter how unusual a case may be, one case alone will probably not have much influence in a large sample. Nevertheless, it is of value to take a closer look at the characteristics of this case. Since it was not practical to check the data records to see if any errors might have been made on either SIBS or CHILDREN, her characteristics on some other variables were examined. This woman was a 66-year-old divorcee at the time of the interview. She had reported that she had married when she was 23 years of age. Unfortunately, the GSS does not determine how long respondents were married before divorcing. She also indicated that she considered five children to be the ideal number of children for a couple to have. Thus, this woman married young enough to have children and indicated that she considers having children to be desirable, but she had none. This suggests that she might have divorced relatively soon after marriage. Although it is only speculation, it appears advisable to control for number of years married (if it were known) in order to get a more valid estimation of the effect of number of siblings on number of children.

In conclusion, the effect of siblings appears real but relatively small. In a world of multiple causation, what other important variable(s) (other than years married) should be included in a causal model for number of children? Also, do you think that the relationship between SIBS and CHILDS will turn out to be *spurious* when these additional variables are controlled, or do you think that our failure to control these other variables might actually be *suppressing* the bivariate relationship that we have observed?

Summary

This chapter has defined and interpreted the least-squares regression equation for estimating the bivariate linear relationship between Y and X. This equation is defined by the slope and intercept that minimize the sum of squared differences or residuals between the observed and predicted values of Y. In general, the slope of a least-squares line equals the covariance between the independent and dependent variables (X and Y) divided by the variance of the independent variable (X). This slope indicates the change in Y that corresponds to a one-unit increase in X. The intercept equals the mean of Y minus the product of the slope and the mean of X; it indicates the predicted value of Y when X equals zero. The slope and intercept of the regression equation are asymmetric; that is, their values will be different if X is being predicted by Y. The justification for the least-squares solution is that it minimizes sampling error, a phenomenon which will be examined in more detail in Chapter 4.

Although we can be assured that the formulas for the slope and intercept give us the best-fitting line (according to the least-squares criterion), it is an unstandardized solution and thus does not indicate how well the line fits the data (i.e., the strength of the relationship). The standardized slope and the coefficient of determination are closely related measures of the strength of the relationship between X and Y; they are also equivalent to the Pearson correlation (r) and the squared Pearson correlation (r^2), respectively. The standardized slope (and r) equals the change in the standardized score of Y (z_Y) per unit increase in the standardized score of X (z_X), and the coefficient of determination (and r^2) equals the proportion of the variance in Y that can be predicted or explained by X.

An introduction to regression diagnostics emphasized that cases with the greatest influence on the least-squares slope tend to be both X outliers (e.g., as indicated by a relatively large value of z^2) and regression outliers (e.g., as indicated by a relatively large residual on Y). Influence can be measured by how much the slope changes when a case is dropped (DFBETA). Close inspection of influential cases may reveal errors in the data or suggest additional causal variables that might account for the large residuals of influential cases.

Although the bivariate regression slope may be used as an estimate of the linear effect of X on Y in the simplest causal model, the need to incorporate additional variables into the analysis looms large in nonexperimental research (see the discussion of nonspurious covariance and multiple causation in Chapter 1). In the next chapter, we will discuss how multiple regression controls for additional variables in order to estimate the nonspurious effects of two or more independent variables.

References

Bollen, Kenneth A., and Robert W. Jackman. 1990. "Regression Diagnostics: An Expository Treatment of Outliers and Influential Cases." In *Modern Methods of Data Analysis*, edited by John Fox and J. Scott Long. Newbury Park: Sage.

Davis, James A., and Tom W. Smith. 1993. *General Social Survey 1972–1993: Cumulative Codebook*. Chicago: National Opinion Research Center.

Fox, John. 1991. *Regression Diagnostics*. Newbury Park: Sage.

Norusis, Marija J. 1990. *SPSS Base System User's Guide*. Chicago: SPSS Inc.

Sudman, Seymour. 1976. *Applied Sampling*. New York: Academic Press.

3 Multiple Regression

Multiple regression is used to estimate the effects of variables in causal models with two or more independent variables. Since most variables have more than one cause, it is necessary to use multiple regression in causal analyses, assuming that measures of more than one cause are available. Multiple regression is particularly appropriate when the causes (independent variables) are intercorrelated, which again is usually the case.

A multiple regression equation contains a single dependent variable and two or more independent variables. A number of new statistics have been created to describe different aspects of the relationships between the dependent variable and the independent variables. The primary objective of this chapter is to develop a clear understanding of the unstandardized partial regression coefficient or slope that describes the linear relationship between the dependent variable and one of the independent variables *with all of the other independent variables held constant.* This partial regression coefficient is perhaps the most important statistic in this book because it provides an estimate of the effect of one independent variable while allowing us to control for all of the other independent variables. Another important statistic to be described in this chapter is the standardized partial regression slope, which provides a means of comparing the slopes of independent variables that are not measured on the same scale. Unlike the bivariate case, however, the standardized slopes in multiple regression do not have fixed upper and lower limits. Therefore, the partial correlation and the semipartial correlation are introduced to describe two different aspects of the strength of the relationship between the dependent variable and one of the independent variables. We will also discuss the interpretations of the squares of each of these correlations, interpretations that are analogous to the squared Pearson correlation (i.e., the coefficient of determination). Finally, the multiple correlation and the squared multiple correlation will be shown to provide measures of the strength of the relationship between the dependent variable and the entire set of independent variables.

The partial regression slopes and correlations may either be weaker than their bivariate counterparts or they may be stronger or have an opposite sign from their bivariate analogs. These two cases are referred to as redundancy and suppression. In the final section of the chapter we will look at the different patterns of relationships between the variables in the regression equation that are associated with redundancy and suppression.

The multiple regression statistics mentioned above will be presented as an extension of bivariate regression and correlation. If you have a good understanding of the bivariate case, you possess all the conceptual tools necessary for understanding multiple regression. This is not to say that the multiple regression case is not more complex than the bivariate case. All of the new statistics, however, can be understood as consisting of a series of bivariate regressions or correlations that are used first to transform the variables in order to eliminate certain correlations among them and then to describe the relationships between these transformed variables. From this perspective, it is possible to develop a clear understanding of how multiple regression statistically controls for correlations among the independent variables in order to get unbiased estimates of the effect of each variable. Achieving statistical control for other X's that are causes of Y is of fundamental importance in nonexperimental research.

Introducing multiple regression as a series of bivariate regressions and correlations is somewhat unorthodox. It also involves some fairly complex mathematical notation that you will have to learn. Once you have become comfortable with the notation, however, you should be able to see the basically simple concepts that are involved in multiple regression statistics. A good understanding of these fundamentals will make you a more intelligent user of multiple regression.

We will begin with causal models involving only two independent variables and one dependent variable (single-equation models). Most of the statistical concepts and issues that arise in multiple regression analysis can be presented in terms of models with only two X's. The statistics for equations with three or more X's are rather straightforward extensions of the equation with two independent variables.

A Model with Two Independent Variables

The equation for a causal model with two independent variables is

$$Y = \alpha + \beta_1 X_1 + \beta_2 X_2 + \varepsilon \tag{3.1}$$

This equation represents the causal model diagrammed in Figure 1.1. It specifies that Y is caused by two variables, X_1 and X_2. These variables have linear and additive effects on Y, which are represented by β_1 and β_2. Each β represents the change in Y produced by a unit increase (or difference) in the X by which it is multiplied, with the other X held constant. The constant α represents the value of Y when X_1, X_2, and ε are equal to zero. The error term ε represents the cumulative effect of all causes other than X_1 and X_2. These additional causes are usually conceptualized as being so numerous and small that they behave in the aggregate as a random variable. Models containing such random error terms are called **stochastic** models.

The use of the Greek symbols β, α, and ε indicates that Equation 3.1 is a theoretical model for the true effects that exist in some *population* that is under investigation. Although the β's are sometimes referred to as population regression coefficients, you should not think of them as regression coefficients. Regression coefficients are computed from empirical observations, usually taken from a sample of the population, in order to estimate the true effects of the variables. Although the computed regression coefficients may be the best estimates of the coefficients in the theoretical model, there are several reasons for making a clear distinction between regression coefficients and true coefficients.

First, we usually have observations for only a sample of the population. Therefore, the regression coefficient may differ from the true coefficients due to sampling error. Tests for sampling error will be taken up in a later chapter. Second, the regression coefficients may be in error even if we have data for the entire population. Since there is always measurement error in our empirical observations, the regression coefficients will always be at least somewhat biased (for a discussion of the consequences of measurement error for regression statistics, see Berry and Feldman 1985). Third, there is a philosophical difference between the regression coefficient and the β coefficient in the theoretical model that stands for the effect of an independent variable. If we have observations for the entire population, if all of our variables are measured perfectly, and if other assumptions that we make about the error term are correct (these will be covered later), the regression coefficients will *equal* the true effects of the independent variables. The fact that they are equal, however, does not mean that they are conceptually the same. The effects of the independent variables (the β's) are phenomena that are "out there" in the sense that they exist before and independently of any investigation of them. They are part of the "real world." The regression coefficients are mathematical tools that have been invented to help us investigate these phenomena. If we are very fortunate, our regression coefficients may be equal to the true effects of the variables, but they are not the same thing. This third point, perhaps, is rather metaphysical and may not be agreed upon by all methodologists. However, it seems best to make a clear distinction between the statistical tools that we are using (the regression coefficients) and the phenomena that we are investigating (primarily, the β's).

The multiple regression equation that is used to estimate the theoretical model represented by Equation 3.1 is

$$\hat{Y} = a_{Y \cdot 12} + b_{Y1 \cdot 2}X_1 + b_{Y2 \cdot 1}X_2 \quad \textbf{(3.2)}$$

Notice that the equation contains only Roman letters, contains no error term, and has Y-hat as the dependent variable. These are all tip-offs that this is a regression equation.

There are some relatively complex subscripts for a and the two b's in Equation 3.2 that need to be clearly understood. Subscripts are used to clarify the meaning of the statistic or variable to which they are attached. This "dot notation" may be used in two different ways.

In the first type of dot notation, the dot may be translated as "controlling" or "holding constant." This is the case when the subscript is attached to a measure of *partial association*, such as $b_{Y1 \cdot 2}$, the **partial regression slope**. The two symbols to the left of the dot indicate the dependent and independent variables, with the symbol for the dependent variable coming first.[1] The symbol(s) to the right of the dot indicates the variable(s) to be "controlled" or "held constant." Notice that the subscripts of the b's do not contain the full symbols for the independent variables (which would be X_1 and X_2) but use only the number contained in the subscript of each X; thus, 1 refers to X_1 and 2 refers to X_2. According to these rules, $b_{Y1 \cdot 2}$ is the slope for the dependent variable Y regressed on the independent variable X_1, with X_2 as the variable that is controlled. Alternatively, it is the change in Y associated with a unit increase in X_1 when X_2 is held constant. An important aspect of this chapter is to clarify what is meant by "controlling" or "holding constant" a variable in multiple regression analysis.

In the second meaning of the dot notation, the dot may be translated as "predicted by." The subscript in $a_{Y \cdot 12}$ provides an illustration. In this usage, there is only one variable to the left of the dot; this is the dependent variable. The single variable to the left of the dot is a tip-off that this statistic is not a measure of partial association; measures of partial association always have symbols for two variables to the left of the dot. The two symbols to the right of the dot in $a_{Y \cdot 12}$ indicate which variables are the independent variables or predictors. Thus, $a_{Y \cdot 12}$ indicates that a is a constant in an equation with Y as the dependent variable and with X_1 and X_2 as the independent variables or predictors. In this case, $a_{Y \cdot 12}$ is the predicted value of Y when the independent variables are equal

1. In the case of symmetric measures of association, such as r, where the value of the measure is the same no matter which variable is thought of as dependent, the order of the two symbols is arbitrary.

FIGURE 3.1 A Ballantine for Two Uncorrelated Independent Variables

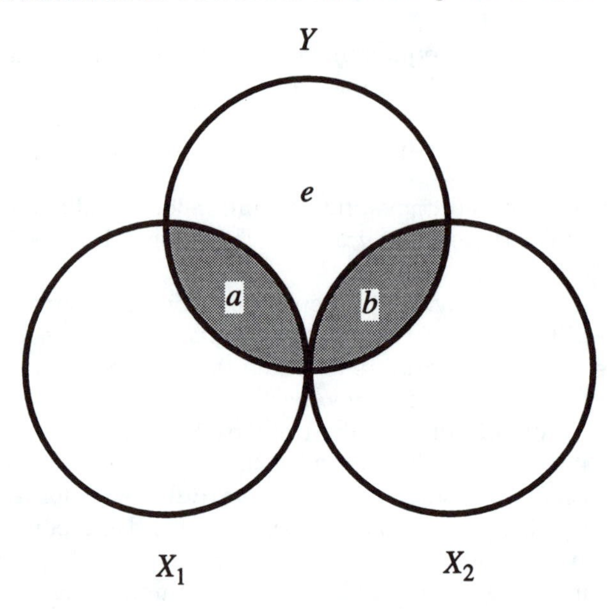

to zero. We will see that this "predicted by" dot notation can also be used with other types of statistics.

Equation 3.2 included full subscript notation primarily as a means to explain the notation. Actually, it is not necessary to include the full notation for the intercept and slopes when the full equation is written, because the equation itself indicates which are the independent and dependent variables and which variables are being controlled. Thus, the following equation contains all the information that is necessary:

$$\hat{Y} = a + b_1X_1 + b_2X_2 \quad \textbf{(3.3)}$$

The mere presence of X_2 in the equation means that we can interpret the slope for X_1 as a partial slope with X_2 held constant. The full subscript notation is necessary, however, when the symbol for the statistic is standing alone in a sentence or table or when it is used in another equation where its meaning would not be clear without the full notation.

The slopes in Equations 3.2 and 3.3 have been interpreted as partial slopes that indicate the change in Y per unit increase in one X, with the other X's held constant. An examination of Figures 3.1 and 3.2 may help to clarify why it is necessary to hold constant other X's and how we might go about doing it. These figures are sometimes called *ballantines*.

FIGURE 3.2 A Ballantine for Two Correlated Independent Variables

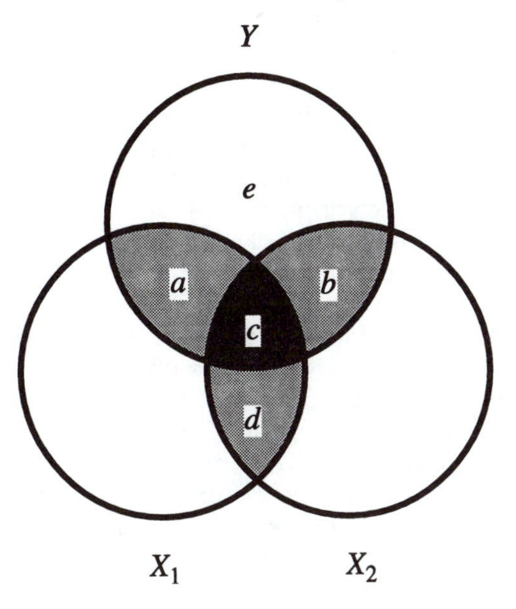

Each circle may be thought of as representing the variance of the variable, where the variable has been standardized to have a unit variance (a z score). The overlap between two circles indicates the proportion of variance in each variable that is shared with the other. Since the proportion of variance in a variable that is shared or explained by another is defined as the coefficient of determination, or r^2, the shaded areas *a* and *b* in Figure 3.1 indicate the coefficient of determination for Y with X_1 and for Y with X_2, respectively. The fact that X_1 and X_2 do not overlap means that they are not correlated. As a consequence, area *a* does not overlap with area *b*. This means that the variance that X_1 and Y share is separate and distinct from the variance that X_2 and Y share. In the case of two uncorrelated independent variables, therefore, we could compute two separate bivariate regression equations to determine the effect of each variable on Y. Although there are some other statistical advantages to using multiple regression when the independent variables are uncorrelated (which will be discussed later), it would not be absolutely necessary to do so in this case.

Figure 3.2 shows a ballantine for two correlated causes of Y. The shared variance between X_1 and Y is represented by $a + c$, and the shared variance between X_2 and Y is represented by $b + c$. Thus, *c* is a new area created by the correlation between X_1 and X_2; *c* represents the proportion of the variance in Y that is shared jointly with X_1 and X_2.

To which variable should we assign this jointly shared variance c? If we computed two separate bivariate regression equations, we would assign this variance to both independent variables, and it would clearly be incorrect to count it twice. Therefore, bivariate regression would be wrong.

If we remember that an additive effect is the effect of one cause that is *independent* of the effect of another, we would not want to assign area c to either variable. Thus, area a is the variance of Y that is caused by X_1 independently of X_2, and area b is the variance of Y that is caused by X_2 independently of X_1. This indicates that running two bivariate regressions would overestimate the independent effect of each variable on Y.

The ballantine in Figure 3.2 also suggests what we might do to estimate areas a and b. If we removed from X_1 all of the variance that it shares with X_2 $(c + d)$, this would remove area c and the only remaining shared variance between X_1 and Y would be a, the variance in Y caused by the independent effect of X_1. Likewise, if we removed from X_2 the variance it shares with X_1, only area b would be left to assign to X_2. The general solution, therefore, would be to remove from all of the independent variables the variance that they share with one another, leaving only independent or unique variance in each X with which to estimate their effects on Y. The method for removing this shared variance will be taken up in the next section.

Finally, there is still the question of what to do with area c; which variable gets the credit for c? The first apparent solution might be to just give it to both independent variables jointly and call it a joint effect. The notion of a joint effect, however, doesn't appear to have any validity in the social scientific community. Although there has been considerable debate over how to distribute this overlapping variance among the independent variables, the solution that appears most valid today requires the specification of a causal model with some independent variables as causes of others, as in Figure 1.2. If, for example, it were valid to specify X_1 as a cause of X_2, then we could say that the area $c + d$ that X_1 and X_2 share is variance in X_2 that is explained by X_1. Since c is also variance that X_2 shares with Y, we could thus allocate c to X_1. Consequently, $a + c$ would be the total variance in Y explained by X_1. This procedure is illustrated later, at the end of the section on the semipartial correlation.

Multiple Regression Statistics

The Example Data

A small, hypothetical data set will be used to illustrate all of the multiple regression and correlation statistics that will be covered in this chapter. The data provide an example of two variables that affect salaries in the academic marketplace: number of years of experience since receiving the Ph.D., and number of publications. Although this is a simplified example, it involves two variables (experience and publications) that appear to have widespread importance. These variables will be designated as

$$X_1 = \text{Years of Experience}$$
$$X_2 = \text{Number of Publications}$$
$$Y = \text{Annual Salary in Thousands of Dollars}$$

Furthermore, we will imagine that we know the true effects of these variables. For many academic units there are salary criteria that specify the monetary value of different variables. Therefore, there may be explicit "causal laws" operating in many of these small social systems, and these effects might be ascertained simply by asking the appropriate person to tell you what variables are used to determine salary and what weight (i.e., effect) is given to each variable. In the social system for which this hypothetical data set has been created, the equation for determining salary is

$$Y = 20 + 1X_1 + 2X_2 + \varepsilon$$

This equation is analogous to Equation 3.1, which stands for the true population values of the causal system. In this system, a professor with no experience and no publications will receive a salary of $20,000 (the intercept); each additional year of experience causes a salary increase of $1,000; and each additional publication causes a salary increase of $2,000. The error term ε might represent the discretionary power of the official determining salary to make minor adjustments depending on his or her subjective assessment of additional factors that should be taken into account for each professor. Since these adjustments are subjective, they may appear to be random or unpredictable.

The values for X_1 and X_2 are given in Table 3.1. These are exogenous variables that serve as inputs to the salary equation. The values of ε are also givens. The administrator making the salary decision simply counts the number of years of experience and the number of publications for each professor, multiplies them by the weights (effects), adds the constant to the sum of these products, and then adds or subtracts one or two thousand dollars in an apparently unpredictable fashion for each person. The results are as follows:

$$\begin{aligned} Y &= 20 + 1X_1 + 2X_2 + \varepsilon \\ &= 20 + 1(0) + 2(2) + 2 = 26 \\ &= 20 + 1(2) + 2(1) - 1 = 23 \\ &= 20 + 1(5) + 2(8) - 2 = 39 \\ &= 20 + 1(8) + 2(5) - 1 = 37 \\ &= 20 + 1(10) + 2(9) + 2 = 50 \end{aligned}$$

TABLE 3.1 Data and Bivariate Regression Equations for the Academic Salary Example

X_1	X_2	ε	Y	$X_1 - \bar{X}_1$	$(X_1 - \bar{X}_1)^2$	$X_2 - \bar{X}_2$	$(X_2 - \bar{X}_2)^2$
0	2	+2	26	−5	25	−3	9
2	1	−1	23	−3	9	−4	16
5	8	−2	39	0	0	3	9
8	5	−1	37	3	9	0	0
10	9	+2	50	5	25	4	16
25	25	0	175	0	68	0	50
$\bar{X}_1 = 5$	$\bar{X}_2 = 5$		$\bar{Y} = 35$				

$$s_1 = \sqrt{\frac{68}{5}} = 3.6878 \qquad s_2 = \sqrt{\frac{50}{5}} = 3.1623 \qquad s_Y = \sqrt{\frac{470}{5}} = 9.6954$$

$Y - \bar{Y}$	$(Y - \bar{Y})^2$	$(X_1 - \bar{X}_1)(Y - \bar{Y})$	$(X_2 - \bar{X}_2)(Y - \bar{Y})$	$(X_1 - \bar{X}_1)(X_2 - \bar{X}_2)$
−9	81	45	27	15
−12	144	36	48	12
4	16	0	12	0
2	4	6	0	0
15	225	75	60	20
0	470	162	147	47

$$b_{Y1} = \frac{\sum(X_1 - \bar{X}_1)(Y - \bar{Y})}{\sum(X_1 - \bar{X}_1)^2} = \frac{162}{68} = 2.382 \qquad b_{Y2} = \frac{\sum(X_2 - \bar{X}_2)(Y - \bar{Y})}{\sum(X_2 - \bar{X}_2)^2} = \frac{147}{50} = 2.940$$

$$a_{Y1} = \bar{Y} - b_{Y1}\bar{X}_1 = 35 - 2.382(5) = 23.09 \qquad a_{Y2} = \bar{Y} - b_{Y2}\bar{X}_2 = 35 - 2.94(5) = 20.30$$

$$\hat{Y} = 23.09 + 2.382X_1 \qquad \hat{Y} = 20.3 + 2.94X_2$$

The above values of Y were entered in Table 3.1 after they were computed. It must be emphasized that these values of Y were created by applying the known causal laws to the given distributions of scores for X_1 and X_2 and then adding a random error score.

Now that we have seen how the Y scores were generated, let us imagine that we don't know the causal equation. Since we are outside investigators, no one may be willing to tell us the true causal parameters. It may be possible, however, for us to find out from public records the salary, years of experience, and number of publications of each professor. If it were not, we could conduct a survey of the professors and ask them directly for the information on these three variables. The values of these variables are observable. The values of the causal parameters and the error term, however, are not directly observable. Thus, we must somehow estimate them.

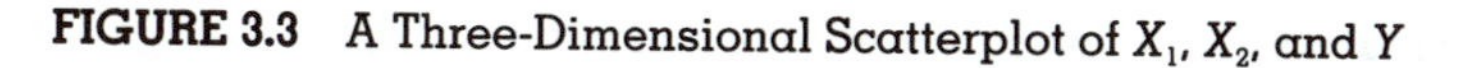

FIGURE 3.3 A Three-Dimensional Scatterplot of X_1, X_2, and Y

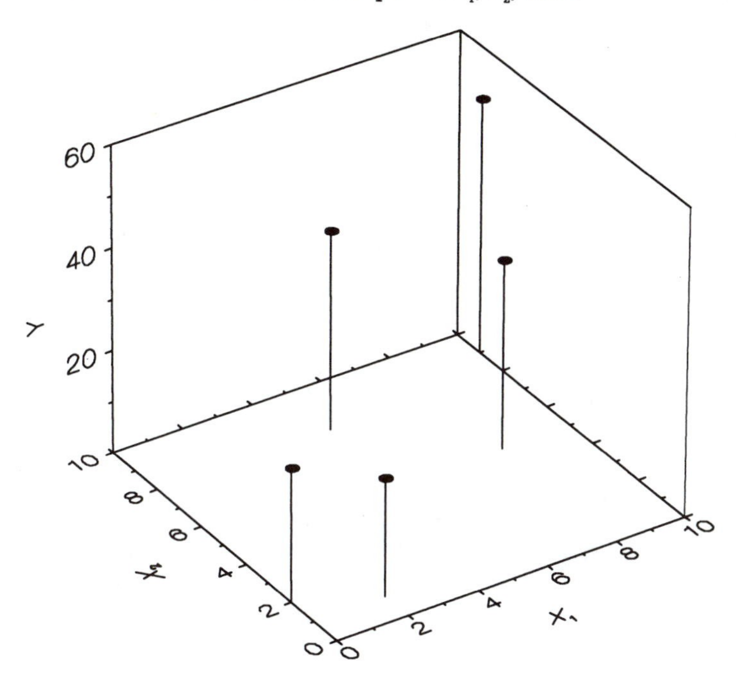

Before conducting any statistical analyses, however, we may construct a three-dimensional scatterplot to help us visualize the relationship between the three variables (Figure 3.3). Each point represents an (X_1, X_2, Y) coordinate in the Cartesian space. As the points move from low values of X_1 and X_2 at the front of the plot to high values at the rear of the plot, there is a clear increase in Y. That is, there appears to be a positive relationship between X_1 and Y and between X_2 and Y (remember that as investigators we have not seen the actual values of the parameters for X_1 and X_2). At this point, however, we cannot visualize the values of the slopes of Y on X_1 and Y on X_2 that will provide the best estimates of these parameters.

The three-dimensional plot in Figure 3.3 is formed by three two-dimensional planes, the X_1Y plane, the X_2Y plane, and the X_1X_2 plane. The projections of the data points onto the X_1Y plane and the X_2Y plane are shown in Figures 3.4a and b, respectively. These are simply two-dimensional plots of the relationship between Y and each independent variable. The bivariate least-squares line is plotted on each. At first glance these lines might appear to provide good estimates of the effects of each independent variable. Thus, we will examine more carefully the characteristics of each of these two lines.

FIGURE 3.4 (a & b) Bivariate Views of the Three-Dimensional Plot

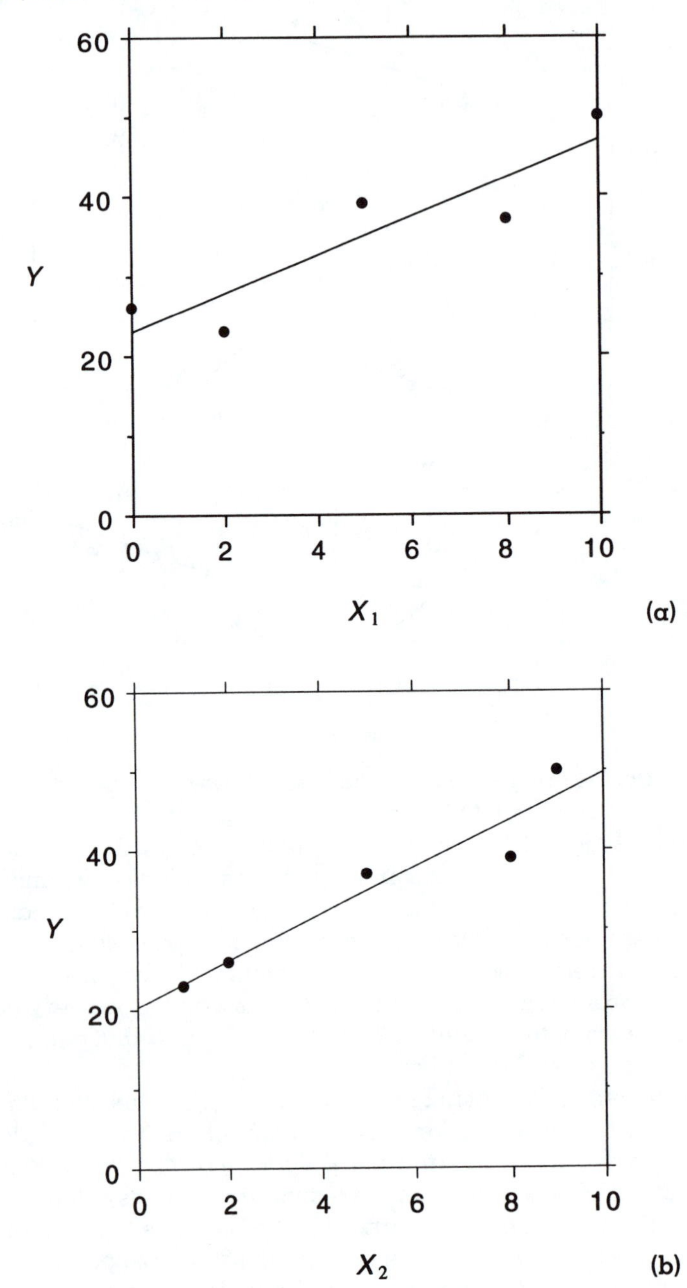

Bivariate Regression Equations

The bivariate regressions for salary and experience and for salary and publications are first computed. All of the necessary statistics for computing these bivariate regressions are given in Table 3.1. Using the usual formulas for the slope and intercept, the results are

$$\hat{Y} = 23.09 + 2.382X_1$$

$$\hat{Y} = 20.3 + 2.94X_2$$

The slopes imply that each additional year of experience increases salary by $2,382 (each year of experience is worth $2,382) and each additional publication increases salary by $2,940 (each publication is worth $2,940). These bivariate slopes are overestimates of the true slopes for experience and publications, which are $1,000 and $2,000. Since the researcher doesn't know the true values, however, these values might unwittingly be accepted as good estimates of the effects of experience and publications. A careful examination of these results and the data, however, would provide two clues that something is wrong. First, how would the results of the two regressions be combined to get salary predictions from an equation that includes both experience and publications? An equation that includes the bivariate slopes would be

$$\hat{Y} = a + 2.382X_1 + 2.94X_2$$

How would we choose a value for a? If we summed the two bivariate intercepts, we would get 43.39, which would be the predicted salary of a person with no publications and no experience. This prediction would be far too high. If we took the mean of the two intercepts, the value would be 21.695. Although this value would be a reasonable prediction for a person with no experience and no publications, the equation would predict a salary of 71.975 for the person with 10 years of experience and 9 publications. This prediction would be far above that person's actual salary of 50. You might suggest that the predicted value of Y should be the mean of Y when each X is at its mean, just as in the bivariate case. If so, then the intercept could be determined as follows:

$$\hat{a} = \overline{Y} - 2.382\overline{X}_1 - 2.94\overline{X}_2 = 35 - 2.382(5) - 2.94(5) = 8.39$$

The use of this intercept would also predict far too low a salary for those persons with few or no years of experience and publications, and it would even predict too high a salary for the person with 10 years of experience and 9 publications (58.67). No matter what value might be chosen for a, the resulting

FIGURE 3.5 Relationship Between X_1 and X_2

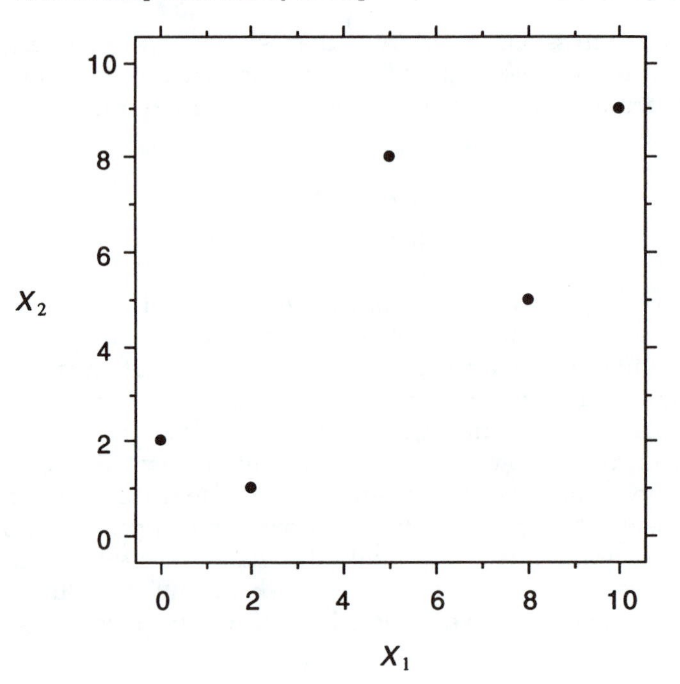

equation would give very poor predictions. The predicted range of salary for the observed values of the X's would be far too great because the slopes are too steep. That is, the poor predictive ability is not the fault of the intercept, it is the fault of the upwardly biased slopes.

The second clue that something is wrong with the bivariate approach is suggested by a careful examination of the scores for X_1 and X_2. The slope or effect of an independent variable should indicate the change in Y produced by a unit increase in that variable alone, that is, when the other independent variable doesn't change. Figure 3.5, however, indicates that when X_1 increases, X_2 also tends to increase. That is, persons with more experience also tend to have more publications (the variables are positively correlated). If both of these variables affect Y, as we suspect they do, then the change in Y that is *associated* with an increase in X_1 in the bivariate case is partly caused by the increase in X_1 and partly caused by the corresponding increase in X_2. The bivariate slope, however, attributes all of this increase in Y to the increase in X_1, thereby overestimating the effect of X_1. Thus, we need a statistical technique for holding constant X_2 in order to determine the extent to which changes in Y result from a change in X_1 alone.

TABLE 3.2 Multiple Regression Equation for the Academic Salary Example

$$b_{12} = \frac{\sum (X_1 - \overline{X}_1)(X_2 - \overline{X}_2)}{\sum (X_2 - \overline{X}_2)^2} = \frac{47}{50} = .940 \qquad b_{21} = \frac{\sum (X_1 - \overline{X}_1)(X_2 - \overline{X}_2)}{\sum (X_1 - \overline{X}_1)^2} = \frac{47}{68} = .6912$$

$$a_{12} = \overline{X}_1 - b_{12}\overline{X}_2 = 5 - .94(5) = .30 \qquad a_{21} = \overline{X}_2 - b_{21}\overline{X}_1 = 5 - .6912(5) = 1.5441$$

$$\hat{X}_1 = .30 + .94X_2 \qquad \hat{X}_2 = 1.5441 + .6912X_1$$

$\hat{X}_{1\cdot 2}$	$X_1 - \hat{X}_{1\cdot 2}$	$(X_1 - \hat{X}_{1\cdot 2})(X_2 - \overline{X}_2)$	$(X_1 - \hat{X}_{1\cdot 2})^2$	$(X_1 - \hat{X}_{1\cdot 2})(Y - \overline{Y})$
2.18	−2.18	6.54	4.7524	19.62
1.24	.76	−3.04	.5776	−9.12
7.82	−2.82	−8.46	7.9524	−11.28
5.00	3.00	0.00	9.0000	6.00
8.76	1.24	4.96	1.5376	18.60
	0.00	0.00	23.8200	23.82

$$b_{Y1\cdot 2} = \frac{\sum (X_1 - \hat{X}_{1\cdot 2})(Y - \overline{Y})}{\sum (X_1 - \hat{X}_{1\cdot 2})^2} = \frac{23.82}{23.82} = 1.00$$

$\hat{X}_{2\cdot 1}$	$X_2 - \hat{X}_{2\cdot 1}$	$(X_2 - \hat{X}_{2\cdot 1})(X_1 - \overline{X}_1)$	$(X_2 - \hat{X}_{2\cdot 1})^2$	$(X_2 - \hat{X}_{2\cdot 1})(Y - \overline{Y})$
1.5441	.4559	−2.2795	.2078	−4.1031
2.9265	−1.9265	5.7795	3.7114	23.1180
5.0000	3.0000	0.0000	9.0000	12.0000
7.0735	−2.0735	−6.2205	4.2994	−4.1470
8.4559	.5441	2.7205	.2960	8.1615
	0.0000	0.0000	17.5146	35.0294

$$b_{Y2\cdot 1} = \frac{\sum (X_2 - \hat{X}_{2\cdot 1})(Y - \overline{Y})}{\sum (X_2 - \hat{X}_{2\cdot 1})^2} = \frac{35.0294}{17.5146} = 2.00$$

$$a_{Y\cdot 12} = \overline{Y} - b_{Y1\cdot 2}\overline{X}_1 - b_{Y2\cdot 1}\overline{X}_2 = 35 - 1.00(5) - 2.00(5) = 20$$

$$\hat{Y} = 20 + 1X_1 + 2X_2$$

Multiple Regression Equation

The solution to the problem of how to hold constant one variable in order to examine the effect of another was suggested in the discussion of the ballantines. To determine the effect of X_1, for example, it is first necessary to remove from X_1 the variance that it shares with X_2. The first step is to compute the bivariate regression equation for X_2 as a predictor of X_1 (see Table 3.2). Entering the appropriate quantities from Table 3.1 into the formulas at the top of Table 3.2 gives

$$\hat{X}_1 = .30 + .94X_2$$

The slope indicates that for each additional publication there is an increase in experience of .94 years. The calculation of this equation is not meant to imply that the number of publications is believed to be a *cause* of experience; such a causal specification would not make much sense. This is merely the first step in the procedure for obtaining statistical control.

The values of X_2 from Table 3.1 are then entered into the above equation to get predicted X_1 scores for each case. These predictions are given in Table 3.2 as $\hat{X}_{1\cdot2}$. The number 1 to the left of the dot indicates that X_1 is the predicted variable and the number 2 to the right of the dot indicates that X_2 is the predictor. Remember that when there is only one symbol to the left of the dot, the dot means "predicted by."

The next step is to compute the residuals $X_1 - \hat{X}_{1\cdot2}$. The residual values are shown in Table 3.2. Note that they sum to zero like any set of residual scores; therefore, the mean of $X_1 - \hat{X}_{1\cdot2}$ is zero. The residual scores should be thought of as an important new variable that has been created in order to control or hold constant X_2. These residuals represent the amount of X_1 that cannot be predicted by X_2. We may say that X_1 has been residualized on X_2 or that X_2 has been partialled out of X_1. As a result, the residuals of X_1 are uncorrelated with X_2, as residuals are always uncorrelated with the predictor variable. This is shown in the third column in Table 3.2 where it can be seen that the sum of the products of the residuals of X_1 times the deviation scores of X_2 is zero; that is, the covariation between $(X_1 - \hat{X}_{1\cdot2})$ and $(X_2 - \overline{X}_2)$ is zero. [Note that it is not necessary to subtract the mean of the residuals from $(X_1 - \hat{X}_{1\cdot2})$ to compute the covariation because the mean of the residuals is zero.] The absence of any correlation between the X_1 residuals and X_2 means that the variance of these residuals is that portion of the total variance of X_1 that X_1 does not share with X_2. This is what the ballantine suggested we look for; the unique variance of X_1 relative to X_2. Column four of Table 3.2 shows that the sum of the squared residuals (i.e., the variation of the residuals) equals 23.82, which is considerably less than the sum of squared deviation scores of X_1 (which equals 68) shown in Table 3.1. Thus, in partialling X_2 out of X_1 we have created a new residual variable with considerably less variance than the original variable.

Figure 3.6 shows a scatterplot with the new residualized variable on the horizontal axis and X_2 on the vertical axis. There is no discernible relationship between the residuals and X_2. As shown in the scatterplot, the best prediction of X_2 for all values of the residuals is the mean of X_2 (i.e., 5). Thus, when the residuals change, there is no linear change in X_2; that is, X_2 is held "constant." X_2 is not literally held constant; there is change in X_2 as the residuals of X_1 change. But there is no tendency for X_2 either to increase or decrease as the residuals of X_1 increase. Therefore, since X_2 and the X_1 residuals are uncorre-

FIGURE 3.6 Scatterplot of X_1 Residuals and X_2

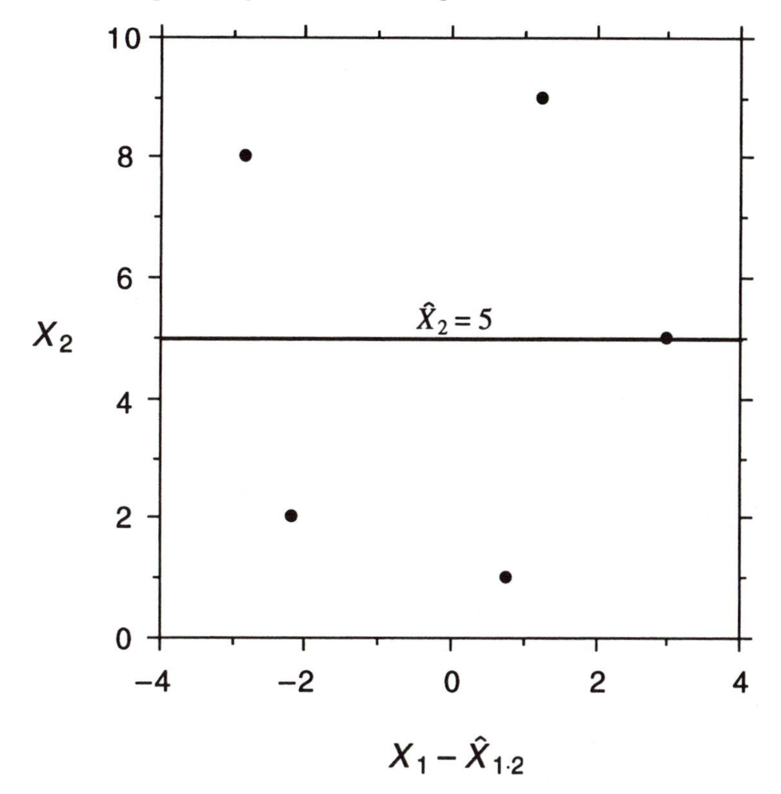

lated, there is no way to confuse changes in Y that might be related to the residuals of X_1 with changes in Y that are related to X_2.

Notice in Table 3.2 that the individuals are not ranked the same on the residuals of experience as they are on the raw scores. The fourth person has the highest residual (3.00) but is only second highest in actual number of years of experience. The positive residual for this case means that this person has more experience than is expected based on the number of his or her publications. That is, given the modest number of publications for this person, we would expect fewer years of experience than this person has. The person with the smallest residual on experience (-2.82) is right in the middle on actual years of experience. The negative residual results because this person has less experience than would be expected on the basis of his or her relatively high number of publications. In sum, the residuals indicate years of experience relative to the number of publications. Thus, we will be using differences in relative experience to estimate the effect of X_1 on Y with X_2 held constant.

FIGURE 3.7 Scatterplot for Y and the Residuals of X_1

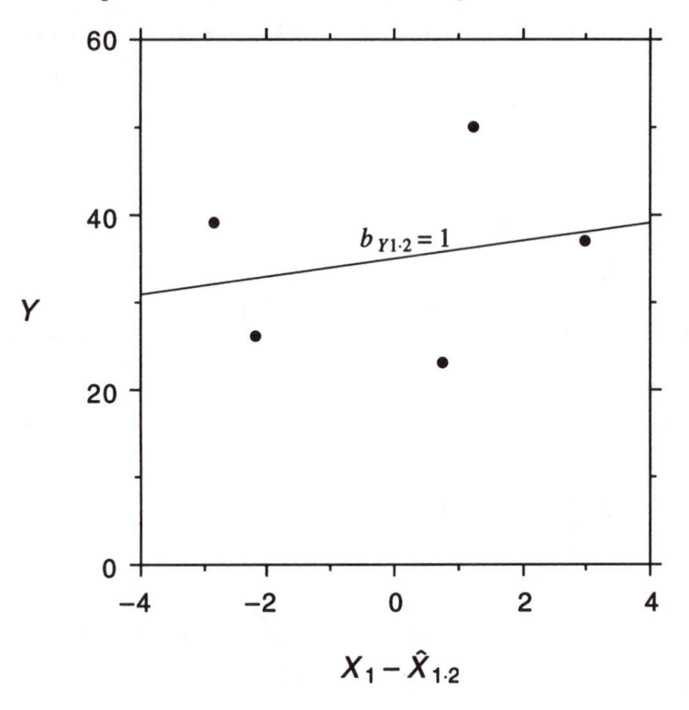

In order to do this we can regress Y on the residuals of X_1 to find out how much Y changes per unit increase in X_1, with X_2 held constant. The formula for the slope of Y regressed on the residuals of X_1 is

$$b_{Y1\cdot 2} = \frac{\sum (X_1 - \hat{X}_{1\cdot 2})(Y - \overline{Y})}{\sum (X_1 - \hat{X}_{1\cdot 2})^2} \qquad \textbf{(3.4)}$$

The subscript for b indicates that this is a partial measure of association because there are two symbols to the left of the dot (Y and 1). Thus, the dot means "holding constant" or "partialling out" variable 2. The numerator of the equation is the covariation between Y and the new independent variable, the residuals of X_1. The denominator is the variation of the independent variable, i.e., the residuals of X_1. Thus, this formula is the usual formula for the slope between two variables, as applied to a residualized X. Because a residualized X is used, however, it is called a partial slope rather than a bivariate slope. This partial regression line is shown on the scattergram for Y and the residuals of X_1 (Figure 3.7).

As indicated in Table 3.2, when the values of the covariance and variance are entered into the formula, the partial slope is found to equal unity. This is the value for the effect of X_1 in the original causal equation that was used to create the values of Y. Thus, the method of residualizing to hold constant publications has allowed us to discover (or recover) the true effect of experience on salary in this particular social system.

An analogous residualizing procedure is used to find the effect of publications on salary. First, X_2 is regressed on X_1 to get the following bivariate regression equation (see top of Table 3.2):

$$\hat{X}_2 = 1.5441 + .6912X_1$$

The predicted values $\hat{X}_{2 \cdot 1}$ are computed from the above equation, and then the residuals $X_2 - \hat{X}_{2 \cdot 1}$ are calculated to remove from X_2 the variance that it shares with X_1. These residuals indicate the number of publications relative to the number that would be expected on the basis of years of experience. Thus, the third person has the highest residual, indicating that he or she has published more than would be expected based on that person's years of experience. The formula for the partial slope of X_2 is

$$b_{Y2 \cdot 1} = \frac{\sum (X_2 - \hat{X}_{2 \cdot 1})(Y - \overline{Y})}{\sum (X_2 - \hat{X}_{2 \cdot 1})^2} \tag{3.5}$$

When the calculations are carried out (Table 3.2) the partial slope is found to equal 2. This is the true value for the effect of publications described in the original causal equation.

We now have the two partial slopes for the variables in the salary example. These slopes are equivalent to $b_{Y1 \cdot 2}$ and $b_{Y2 \cdot 1}$ in Equation 3.2, the least-squares multiple regression equation. In the multiple regression case, as in the bivariate case, the least-squares solution always predicts the mean of the dependent variable when the independent variables are at their means:

$$\overline{Y} = a_{Y \cdot 12} + b_{Y1 \cdot 2}\overline{X}_1 + b_{Y2 \cdot 1}\overline{X}_2$$

Therefore, the following formula can be used to determine the intercept:

$$a_{Y \cdot 12} = \overline{Y} - b_{Y1 \cdot 2}\overline{X}_1 - b_{Y2 \cdot 1}\overline{X}_2 \tag{3.6}$$

When the values of the partial slopes and the means are entered into this formula (Table 3.2), the value of the intercept is found to equal 20. This, again, is the same value that was given in the original causal equation. This completes the multiple regression equation, producing

$$\hat{Y} = a + b_1X_1 + b_2X_2 = 20 + 1X_1 + 2X_2 \tag{3.7}$$

Equation 3.7 is plotted in Figure 3.8. As can be seen, the multiple regression

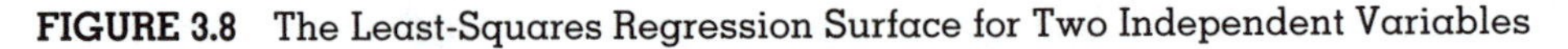

FIGURE 3.8 The Least-Squares Regression Surface for Two Independent Variables

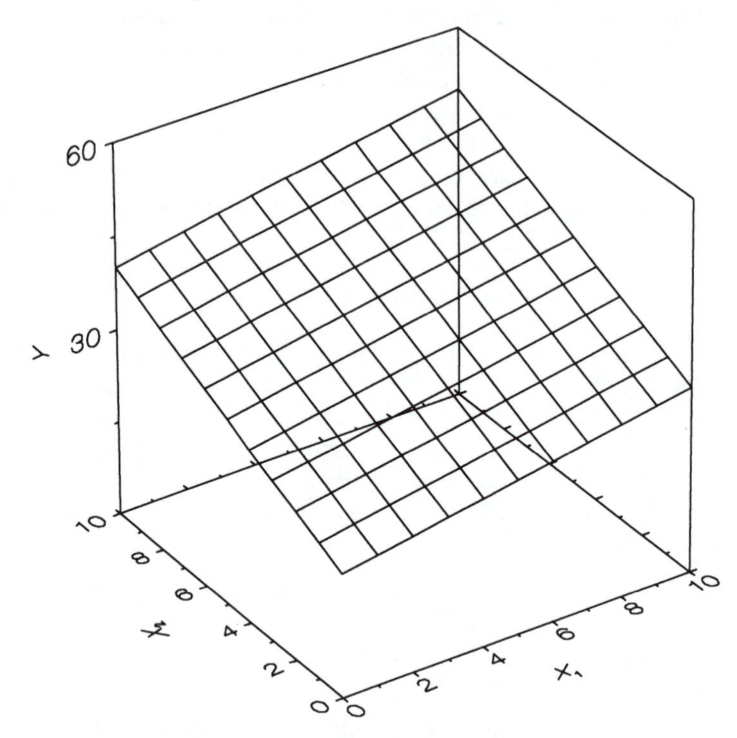

equation with two independent variables is a three-dimensional surface. The partial slope for X_1 can be seen most clearly as the slope of the line where the regression surface intersects the X_1Y plane at $X_2 = 0$ or at $X_2 = 10$. The partial slope for X_2 can be seen as the slope of the line formed by the intersection of the regression surface with the X_2Y plane, at either $X_1 = 0$ or $X_1 = 10$.

The part of the original causal equation that remains to be estimated is the error term ε. Equation 3.7 is used to compute predicted values of Y, i.e., $\hat{Y}_{12}$ (see Table 3.3). The subscript is used to indicate that these predictions are derived from an equation with two independent variables. Thus, the subscript 12 stands for X_1 and X_2. A dot notation, such as $Y \cdot 12$, is not used in this case because it would be redundant to use Y as a subscript of Y itself. The residuals $Y - \hat{Y}_{12}$ are then computed (Table 3.3). The values of the residuals are equal to the values of ε given in Table 3.1. Therefore, the least-squares solution has correctly estimated ε as well as the β's and α.

Table 3.3 also gives the sum of the squared residuals, i.e., $\Sigma (Y - \hat{Y}_{12})^2$, which equals 14. This is the quantity that is minimized by the least-squares solution. Thus, there is no other equation that would produce a value of $\Sigma (Y - \hat{Y}_{12})^2$

TABLE 3.3 Accuracy of Prediction in the Academic Salary Example

$\hat{Y}_{12}$	$Y - \hat{Y}_{12}$	$(Y - \hat{Y}_{12})^2$	$\hat{Y}_{12} - \bar{Y}$	$(Y - \bar{Y})(\hat{Y}_{12} - \bar{Y})$	$(\hat{Y}_{12} - \bar{Y})^2$
24	2	4	−11	99	121
24	−1	1	−11	132	121
41	−2	4	6	24	36
38	−1	1	3	6	9
48	2	4	13	195	169
175	0	14	0	456	456

$$s_{Y-\hat{Y}_{12}} = \sqrt{\frac{\sum (Y - \hat{Y}_{12})^2}{n}} = \sqrt{\frac{14}{5}} = 1.6733$$

$$\hat{s}_{Y-\hat{Y}_{12}} = \sqrt{\frac{\sum (Y - \hat{Y}_{12})^2}{n-3}} = \sqrt{\frac{14}{2}} = 2.6458$$

$$s_{\hat{Y}_{12}} = \sqrt{\frac{\sum (\hat{Y}_{12} - \bar{Y})^2}{n}} = \sqrt{\frac{456}{5}} = 9.5499$$

$$R_{Y \cdot 12} = r_{Y\hat{Y}_{12}} = \frac{s_{Y\hat{Y}_{12}}}{s_Y s_{\hat{Y}_{12}}} = \frac{\sum (Y - \bar{Y})(\hat{Y}_{12} - \bar{Y})/n}{s_Y s_{\hat{Y}_{12}}} = \frac{456/5}{(9.6954)(9.5499)} = .9850$$

$$R^2_{Y \cdot 12} = \frac{\sum (Y - \bar{Y})^2 - \sum (Y - \hat{Y}_{12})^2}{\sum (Y - \bar{Y})^2} = \frac{470 - 14}{470} = .9702$$

$$\hat{R}^2 = 1 - (1 - R^2)\frac{n-1}{n-k-1} = 1 - (1 - .9702)\frac{5-1}{5-2-1} = 1 - (.0298)\frac{4}{2} = .9404$$

smaller than 14. The formulas for the standard deviation of the residuals and the standard error of the residuals are

$$s_{Y-\hat{Y}_{12}} = \sqrt{\frac{\sum (Y - \hat{Y}_{12})^2}{n}} \tag{3.8}$$

$$\hat{s}_{Y-\hat{Y}_{12}} = \sqrt{\frac{\sum (Y - \hat{Y}_{12})^2}{n-3}} \tag{3.9}$$

These are the same formulas that were used in the bivariate case, except for $n - 3$ in the denominator of Equation 3.9 and the subscript for the predicted values of Y, which indicates that the predictions are generated by a multiple regression equation with two independent variables. The standard deviation of the residuals is 1.6733 (Table 3.3), which means that the standard deviation of the errors of prediction is \$1,673.30. If these five cases were a sample from a larger population, Equation 3.9 would be used to estimate the standard deviation of the residuals in the population, which again is called the *standard error of estimate*. If the standard deviation of the residuals is squared, of course, we get the variance of the residuals ($1.6733^2 = 2.7999$). This value is the portion of salary variance that is *unexplained* by experience and publications. The next questions, then, are how large is the explained variance relative to the total variance and how closely associated are the predicted and observed values.

Multiple Correlation

There is a very simple way to find out the strength of the association between the predicted and observed values of Y. We compute the Pearson product-moment correlation between these two variables according to the following formula:

$$R_{Y \cdot 12} = r_{Y\hat{Y}_{12}} = \frac{s_{Y\hat{Y}_{12}}}{s_Y s_{\hat{Y}_{12}}} = \frac{\sum (Y - \overline{Y})(\hat{Y}_{12} - \overline{Y})/n}{s_Y s_{\hat{Y}_{12}}} \quad \textbf{(3.10)}$$

This measure is called the **multiple correlation** and its symbol is R. The subscript of R indicates that Y is predicted by X_1 and X_2. The formula also indicates that the multiple correlation is equivalent to the bivariate correlation between the predicted and observed values of Y. Thus, R is equal to the covariance between the observed Y and the predicted Y, divided by the standard deviations of Y and the predicted Y. That is, R is a special type of standardized covariance. Notice that the covariance includes the difference between the predicted Y and the mean of Y (i.e., $\hat{Y}_{12} - \overline{Y}$). This is because the mean of the predicted scores is equal to the mean of Y itself, as is always the case with least-squares predictions. Therefore, the $(\hat{Y}_{12} - \overline{Y})$'s are the deviation scores of the predicted values. The formula for the standard deviation of the predicted scores is given in Table 3.3.

When the standard deviations and the covariance are entered into Equation 3.10, R is found to equal .9850. Although the maximum value of R is $+1$, the same as the maximum value for the bivariate r, the minimum value of R is 0. The multiple correlation can never be negative. This should make intuitive sense because a negative correlation between Y and the predicted Y would mean that as Y increases, the predicted values of Y decrease. These would be very bad predictions, indeed. The multiple R can also be interpreted as a special type of standardized slope: an increase of one standard deviation in the predicted scores is associated with an increase of $R = .985$ standard deviation in the observed scores. In our salary example, R is very large, indicating that the discretionary power of the salary official to make individual salary adjustments (i.e., ε) was quite minimal.

A second measure of multiple association is analogous to the coefficient of determination for the bivariate case (r^2). It is sometimes called the *multiple coefficient of determination*, but more frequently it is reported simply as R^2. Its formula is

$$R^2_{Y\cdot 12} = \frac{\sum (Y - \overline{Y})^2 - \sum (Y - \hat{Y}_{12})^2}{\sum (Y - \overline{Y})^2} \quad \textbf{(3.11)}$$

The formula is identical to that of the coefficient of determination except that the subscript of the predicted values indicates that the predicted values were generated by a multiple regression equation with two independent variables. Thus, the numerator is the total variation minus the unexplained variation and the denominator is the total variation of Y. When these quantities are entered into Equation 3.11, it is found that $R^2 = .9702$ (Table 3.3). This value, of course, is equal to the square of R found for Equation 3.10. Notice that as a correlation approaches 1 (and -1 in the case of a bivariate r) there is very little difference between the values of the correlation and the squared correlation. In the middle of the range of the correlation, however, the difference is dramatic, e.g., $R^2 = (.50)^2 = .25$.

Shrunken R^2

The sample R^2 is a biased estimate of the population R^2. The direction of the bias is positive; that is, the sample R^2 overestimates the value in the population. To understand why this is so, imagine a situation in which all of the independent variables have true r's with Y that equal zero in the population; thus, the population R would be zero. Because of random sampling errors, however, we would almost never get a sample r of exactly zero. Although sampling error might produce a positive sample r for some X's and a negative r for other X's, the sample r^2's for these r's will always be positive. Thus, there will be a tendency for some or all of the X's to have positive r^2's with Y because of sampling error. This in turn will cause R^2 to be greater than zero by chance. It is also true that when a population r^2 is small, there is more room for error on the high side than on the low side. Thus, the sample R^2 will tend to overestimate the population value even when the population value is not zero.

The following formula may be used to correct for this bias:

$$\hat{R}^2 = 1 - (1 - R^2)\frac{n - 1}{n - k - 1} \quad \textbf{(3.12)}$$

Equation 3.12 gives the shrunken R^2, which is sometimes simply called the adjusted R^2. In this formula, k stands for the number of independent variables in the regression equation. There are two important components in this equation, $(1 - R^2)$ and $(n - 1)/(n - k - 1)$. The first, $1 - R^2$, is the unexplained variance component. When it is subtracted from 1, we get the explained variance. Be-

TABLE 3.4 Illustrative Values of Shrunken R^2

$\frac{n - 1}{n - k - 1}$	k	R^2 .10	.20	.40
$n = 200$				
1.026	5	.0768	.1794	.3845
1.053	10	.0524	.1577	.3683
1.112	20	−.0006	.1106	.3330
$n = 100$				
1.053	5	.052	.157	.368
1.112	10	.001	.110	.333
1.253	20	−.128	−.003	.248

fore subtracting, however, $1 - R^2$ is adjusted by multiplying it by $(n - 1)/(n - k - 1)$. Since this latter term is always greater than 1, the multiplication always adjusts the unexplained variance upward before subtracting it from 1. Since the unexplained variance is adjusted upward, R^2 is adjusted downward or shrunken.

The greater the number of independent variables k for a given sample size, the greater will be $(n - 1)/(n - k - 1)$, and thus the greater will be the shrinkage. When there are many X's for a given sample size, there is more opportunity for R^2 to increase by chance. A second important element here is the value of R^2 itself; the greater the value of R^2, the less will be the shrinkage. When the sample explained variance is high, it is less likely to have occurred by chance. These points are illustrated in Table 3.4, where the shrunken $\hat{R}^2$ is given for three values of R^2, three values of k, and two values of n.

The values of the adjusting term $(n - 1)/(n - k - 1)$ are also given for each value of k. As k increases, this term increases. This table shows that as the number of independent variables increases, the shrinkage increases for each value of the sample R^2. However, there is less shrinkage for higher values of R^2, both proportionately and as differences. Thus, one of the lessons is that you can artificially increase your explained variance to a certain degree by adding more independent variables, but this artificial component of increase will be less when the variables are doing a good job of explaining variance, that is, when the explained variance is relatively high. Notice that the shrunken $\hat{R}^2$ can even become negative. Since it is impossible to explain a negative proportion of variance, a negative value may be reported as zero. Finally, the number of independent variables relative to n also affects the shrinkage. The greater the ratio n/k, the less will be the shrinkage. As the ratio of cases to independent variables decreases, the shrinkage will increase. Thus, we see that there is more shrinkage at each value of R^2 for $n = 100$ than for $n = 200$.

Table 3.3 gives the shrunken R^2 for the academic salary example. The shrinkage is from .9702 to .9404. Notice that although there is a very low ratio of cases to independent variables ($n/k = 5/2 = 2.5$), there is only a small amount of shrinkage because R^2 is so high.

We have now completed our coverage of the most fundamental statistics pertaining to the multiple regression equation: the partial regression slopes, which estimate the linear effect of each independent variable; the intercept, which is necessary to make predictions of Y; the residuals (errors of prediction) and their standard deviation; and the multiple correlation, which measures the total explanatory power of the set of independent variables. All of this material was presented as a straightforward extension of the principles of bivariate regression. The key step forward here was the notion of residualizing each independent variable on the other and then using the residualized X's to examine their unique contributions to Y. We have seen that residualizing is the method of gaining control over other causes of Y that may be sources of spurious associations. Thus, residualizing is the nonexperimental alternative to the use of randomized experimental designs. The next section will present several additional useful measures of partial association that are also based upon residualizing.

Standardized Measures of Partial Relationships

Standardized Partial Slopes

As with bivariate slopes, unstandardized partial regression slopes have no fixed upper and lower limits. The magnitude (absolute value) of the partial slope is directly related to the size of the standard deviation of the dependent variable relative to the standard deviation of the residualized independent variable. Thus, although the unstandardized slope indicates the effect of the independent variable, it is not easy to tell whether this effect is strong or weak. Furthermore, since the independent variables are usually measured on different scales (e.g., years of experiences and number of publications in our example), the effect of one variable cannot be directly compared with the effect of another. We can't compare the effect of a year of experience ($1,000) with the effect of one publication ($2,000) because years and publications are different units of measurement.

One solution to this problem is to convert the raw scores to standard scores (z scores) and conduct a multiple regression on these standard scores. The standard scores for each variable will each have a standard deviation equal to 1. Since the standard scores for each independent variable are based on the same unit of measurement (an increase of one standard score means an increase of one standard deviation), the regression coefficients can be compared between independent variables. The standard scores for each variable in our example are given at the top of Table 3.5. The cross-product of the z scores for each pair of variables is also shown. When the sum of cross-products is divided by n for each pair, we have the standardized covariances. As indicated by Equation 2.19, the standardized covariance is equal to the standardized bivariate regression slope. This slope is also equal to the correlation coefficient. The standardized coefficients (B's or r's) are given in the middle of Table 3.5. For example, the slope of the relationship between z_1 and z_2 (which is symmetric) is

$$B_{12} = B_{21} = r_{12} = \frac{\sum z_1 z_2}{n} = \frac{4.0302}{5} = .8060$$

The procedures for determining the standardized partial slopes are exactly the same as for the unstandardized slopes, except that they are applied to the standardized scores. Each independent variable is residualized with respect to the other, and then the dependent variable is regressed on the residualized independent variables. With respect to z_1, the above slope is first used to get predicted scores for z_1 (remember that the intercept is zero for standardized equations):

$$\hat{z}_1 = .8060 z_2$$

The difference between the observed z scores and the predicted z scores is then computed ($z_1 - \hat{z}_{1 \cdot 2}$) to get the residualized z scores. The slope for the regression of the z scores for Y on the residualized z scores for X_1 is equal to

$$B_{Y1 \cdot 2} = \frac{\sum z_Y (z_1 - \hat{z}_{1 \cdot 2})}{\sum (z_1 - \hat{z}_{1 \cdot 2})^2} \quad \textbf{(3.13)}$$

Notice in Table 3.5 that the sum of the residuals equals zero; thus, the mean of the residuals of standard scores is zero. The numerator in Equation 3.13, the sum of cross-products, is the covariation between the standardized dependent variable and the residuals of the standardized independent variable. The denominator in Equation 3.13 is the variation of the independent variable. Thus, this is the standard formula for a least-squares slope as applied to z scores, one of which has been residualized. B is used again to indicate a standardized slope, and the 2 following the dot in the subscript indicates that X_2 is held constant. The value of this slope (Table 3.5) indicates that a unit increase in z_1 is associated with a change of .3804 in z_Y when z_2 is held constant. Remembering that a unit change in a z score is equivalent to a raw score change of one standard deviation, we can say that when X_1 increases by one standard deviation and X_2 is held constant, Y increases by .3804 standard deviation.

TABLE 3.5 Standardized Slopes for the Academic Salary Example

$z_1 = \frac{X_1 - \overline{X}_1}{s_1} = \frac{0 - 5}{3.6875} = -1.3558$ for Case 1 $\quad z_2 = \frac{X_2 - \overline{X}_2}{s_2} \quad z_Y = \frac{Y - \overline{Y}}{s_Y}$

z_1	z_2	z_Y	z_1z_2	z_1z_Y	z_2z_Y
−1.3558	−.9487	−.9283	1.2862	1.2586	.8807
−.8135	−1.2649	−1.2377	1.0290	1.0069	1.5656
.0000	.9487	.4126	.0000	.0000	.3914
.8135	.0000	.2063	.0000	.1678	.0000
1.3558	1.2649	1.5471	1.7150	2.0976	1.9569
.0000	.0000	.0000	4.0302	4.5308	4.7946

$$B_{Y1} = r_{Y1} = \frac{\sum z_1 z_Y}{n} = \frac{4.5308}{5} = .9062 \qquad B_{Y2} = r_{Y2} = \frac{\sum z_2 z_Y}{n} = \frac{4.7946}{5} = .9589$$

$$B_{12} = B_{21} = r_{12} = \frac{\sum z_1 z_2}{n} = \frac{4.0302}{5} = .8060$$

$\hat{z}_1 = .8060z_2$

$\hat{z}_{1\cdot 2}$	$z_1 - \hat{z}_{1\cdot 2}$	$(z_1 - \hat{z}_{1\cdot 2})^2$	$z_Y(z_1 - \hat{z}_{1\cdot 2})$
−.7647	−.5911	.3494	.5487
−1.0196	.2061	.0425	−.2550
.7647	−.7647	.5848	−.3155
.0000	.8135	.6618	.1678
1.0196	.3362	.1130	.5202
.0000	.0000	1.7515	.6662

$$B_{Y1\cdot 2} = \frac{\sum z_Y(z_1 - \hat{z}_{1\cdot 2})}{\sum (z_1 - \hat{z}_{1\cdot 2})^2} = \frac{.6662}{1.7515} = .3804$$

$\hat{z}_2 = .8060z_1$

$\hat{z}_{2\cdot 1}$	$z_2 - \hat{z}_{2\cdot 1}$	$(z_2 - \hat{z}_{2\cdot 1})^2$	$z_Y(z_2 - \hat{z}_{2\cdot 1})$
−1.0928	.1414	.0208	−.1337
−.6557	−.6092	.3711	.7540
.0000	.9487	.9000	.3914
.6557	−.6557	.4299	−.1353
1.0928	.1721	.0296	.2663
.0000	.0000	1.7514	1.1427

$$B_{Y2\cdot 1} = \frac{\sum z_Y(z_2 - \hat{z}_{2\cdot 1})}{\sum (z_2 - \hat{z}_{2\cdot 1})^2} = \frac{1.1427}{1.7514} = .6524$$

$\hat{z}_Y = B_1z_1 + B_2z_2 = .3804z_1 + .6524z_2$

The same procedures are used to calculate the standardized partial slope for X_2, shown at the bottom of Table 3.5. The standardized regression equation is

$$\hat{z}_Y = B_1 z_1 + B_2 z_2 = .3804 z_1 + .6524 z_2$$

Comparing the sizes of the two *B*'s leads to the conclusion that the number of publications has a stronger effect on salary than the number of years of experience.

Notice in the table that the variation of the residuals (sum of squared residuals) equals 1.7514 for both independent variables. When this is divided by *n*, the variance of the residuals equals .3503, which is smaller than the variance of the z's. Although the variance of z scores is always 1, the variance of the residuals of the z scores will always be less than 1 (unless the two independent variables have a correlation of exactly zero). The variance is less than 1 because the shared variance between the two z's has been partialled out of each one, leaving only a fraction of the original variance, i.e., the unexplained variance. Since r^2 equals the proportion of the variance of each z that is explained by the other z, the unexplained variance of each z equals $1 - r^2 = 1 - .806^2 = .3503$. Since the variance of the residuals of the z scores is less than 1, these residuals are no longer standardized. A standardized variable always has a variance of 1. Thus, $z_1 - \hat{z}_{1\cdot2}$ *and* $z_2 - \hat{z}_{2\cdot1}$ *are residuals of standardized scores, but they are not standardized residuals.* Because of this, the variance of z_Y (which is 1) is greater than the variance of $z_1 - \hat{z}_{1\cdot2}$ (which is less than 1). When the variances of the variables that are being used in a regression are unequal, the regression slope does not have fixed upper and lower limits of +1 and −1, respectively. Thus, the standardized partial slope, unlike the standardized bivariate slope, may be greater than +1 or less than −1. It is called a standardized slope because it is the slope for an equation that uses standard scores not because it has fixed limits.

A situation in which a standardized slope is greater than 1 can be created by taking the original causal equation that was used to generate the values of *Y* and changing it to the following form:

$$Y = 20 - 1X_1 + 2X_2 + \varepsilon$$

The slope for experience is now −1 instead of 1. That is, as experience increases, holding constant publications, salary is caused to decline. If we use the same values for X_1, X_2, and the error term that are given in Table 3.1 but plug them into the new equation, a different set of *Y* values will be generated. The *X*'s will have the same means, standard deviations, and covariance, of course, but *Y* will have new values that will have a different mean, standard deviation, and covariance with each *X*. If we carry out the same residualizing procedure that we have been using, we will get unstandardized partial slopes of −1 and 2 for experience and publications, respectively. We will also get the following standardized partial slopes:

$$B_{Y1 \cdot 2} = -.8505$$

$$B_{Y2 \cdot 1} = 1.4587$$

The standardized slope for publications is greater than 1. You should verify this by carrying out the necessary calculations. First, calculate the new Y scores and convert them to z scores. Then compute the products of these z scores with the residualized z scores for X_2 given in Table 3.5 (these residualized z scores don't change because the original values of X_1 and X_2 have not been changed). Finally, enter the sum of products, along with the given variation of the residuals, into Equation 3.13 for the standardized partial slope.

This illustration of a standardized slope that is greater than 1 is somewhat unusual. Although such a value is legitimate, it does not appear to occur very frequently in empirical research. It is a rather extreme instance of a phenomenon called *suppression*. The above example will be discussed at some length in the final section of this chapter, which deals with redundancy and suppression.

The possibility of a value greater than 1, however, indicates that, strictly speaking, the standardized slope should not be used to assess the strength of the effect of a variable. That is, the B's that we have calculated above can't be used to tell whether the variables have strong or weak effects because there are no maximum or minimum values for this statistic. In practice, however, the standardized slopes are used more frequently than any other measure to assess the strength of the effects. This is usually valid because practically speaking (if not mathematically speaking) the range of B is -1 to $+1$. Furthermore, it is entirely valid to use B to make relative comparisons of the effects of the independent variables. This is because they are all rescaled to a common unit of measurement, z scores. Thus, we can conclude in both our original example and the new variation on the original data that the number of publications has a stronger effect than the number of years of experience because the absolute value of the B for publications is greater than the absolute value of the B for experience.

Semipartial Correlation

The next measure uses the residualized raw scores that were used to determine the unstandardized partial regression slopes. The semipartial correlation is defined as the correlation between Y and an independent variable from which the other independent variable has been partialled. The formula for the semipartial correlation (*sr*) for X_1 is

$$sr_1 = \frac{\sum (Y - \overline{Y})(X_1 - \hat{X}_{1\cdot 2})/n}{s_Y s_{X_1 - \hat{X}_{1\cdot 2}}} \qquad \textbf{(3.14)}$$

Since this statistic is a correlation, the numerator of Equation 3.14 contains the covariance between Y and the residuals of X_1 (i.e., $X_1 - \hat{X}_{1\cdot 2}$) and the denominator contains the standard deviations of these scores. Thus, the semipartial correlation is the standardized covariance between Y and the residuals of X_1. It is called a *semipartial* correlation (also sometimes called *part correlation*) because X_2 is partialled from X_1 but not from Y. Because *sr* is a correlation, it could also be interpreted as a special type of standardized slope (the slope for the standardized scores of Y regressed on the standardized residuals of X_1). In order to avoid confusion with the standardized partial slope discussed above, however, this interpretation will not be used here. Because the covariance between Y and the residuals of X_1 is divided by the standard deviations of these variables, *sr* is standardized to have a range of -1 to $+1$. Thus, it indicates the direction and strength of the relationship between Y and the unique variance of X_1.

The formula for sr_2 is given in Table 3.6. The values of the semipartial correlations, as calculated in the table, are $sr_1 = .2251$ and $sr_2 = .3861$. They indicate that publications have a stronger effect on salary than experience, a conclusion that is in agreement with the conclusion based on comparing their B's (Table 3.5). The semipartial correlation, however, will always be smaller than the standardized partial slope for any given variable, as is the case here.

TABLE 3.6 Semipartial Correlations for the Academic Salary Example

$$s_{X_1 - \hat{X}_{1\cdot 2}} = \sqrt{\frac{\sum (X_1 - \hat{X}_{1\cdot 2})^2}{n}} = \sqrt{\frac{23.82}{5}} = 2.1827$$

$$s_{X_2 - \hat{X}_{2\cdot 1}} = \sqrt{\frac{\sum (X_2 - \hat{X}_{2\cdot 1})^2}{n}} = \sqrt{\frac{17.5146}{5}} = 1.8716$$

$$sr_1 = \frac{\sum (Y - \overline{Y})(X_1 - \hat{X}_{1\cdot 2})/n}{s_Y s_{X_1 - \hat{X}_{1\cdot 2}}} = \frac{23.82/5}{(9.6954)(2.1827)} = .2251$$

$$sr_2 = \frac{\sum (Y - \overline{Y})(X_2 - \hat{X}_{2\cdot 1})/n}{s_Y s_{X_2 - \hat{X}_{2\cdot 1}}} = \frac{35.0294/5}{(9.6954)(1.8716)} = .3861$$

$$sr_1^2 = R^2_{Y\cdot 12} - r^2_{Y2} = .9702 - (.9589)^2 = .0507$$

$$sr_2^2 = R^2_{Y\cdot 12} - r^2_{Y1} = .9702 - (.9062)^2 = .1490$$

The reason for this will be clear when some alternate formulas are briefly noted later. With two independent variables, B and sr will give the same ranking of the independent variables in terms of their relative importance; that is, the variable with the greater B will also have the greater sr. With three or more independent variables, however, the ranking of the strengths of effects of the independent variables in terms of their B's may occasionally not agree with their ranking in terms of their sr's. This is an issue that has received little or no attention in the statistical literature and is only noted here to forewarn you of potential discrepancies in conclusions that may be drawn in using one measure rather than the other.

The squared semipartial correlation (sr^2) can be interpreted in an analogous fashion to the bivariate coefficient of determination (r^2). It is the proportion of variance in Y that is explained by a residualized independent variable or the proportion of variance *uniquely* explained by the independent variable. The sr^2 can be computed in the same fashion as the coefficient of determination: regress Y on the residualized X; use that bivariate equation to calculate predictions of Y; compute the residuals of Y; subtract the residual variation from the total variation of Y to get the explained variation; and divide the explained variation by the total variation of Y to get the proportion explained by the residualized X. Rather than go through all of these calculations, however, a quicker method will be used that leads to the same interpretation and adds another insightful perspective as well.

The squared semipartial r for X_1 can be defined and calculated as

$$sr_1^2 = R_{Y\cdot12}^2 - r_{Y2}^2 = .9702 - (.9589)^2 = .0507 \qquad \textbf{(3.15)}$$

The difference between the squared multiple correlation and the squared r for X_2 equals the squared semipartial r for X_1. As a check on this definition, note that $sr^2 = (.2251)^2 = .0507$. The definition provided by Equation 3.15 indicates that first we regress Y on X_2 to determine how much variance in Y can be explained by X_2 alone. Then we regress Y on both X_1 and X_2 to determine how much additional variance can be explained by adding X_1 to the equation along with X_2. The difference between the proportion explained by both variables (R^2) and that explained by X_2 alone is the additional proportion of variance explained by X_1. This additional amount of explained variance is the amount that X_1 explains uniquely, that is, independently of X_2. Thus, sr_1^2 equals the proportion of the total variance of Y that X_1 explains over and above the other independent variable(s).

The equation for sr_2^2 is

$$sr_2^2 = R_{Y\cdot12}^2 - r_{Y1}^2 = .9702 - (.9062)^2 = .1490$$

Thus, X_2 uniquely explains .149 of the variance of Y, that is, over and above the amount that can be explained by X_1 alone. The above formulas also imply that the squared semipartial correlation can also be interpreted as the decline in the proportion of explained variance that would occur if an independent variable were dropped from a multiple regression equation.

Notice, however, that the sum of the two *sr*'s does not equal R^2:

$$R^2_{Y \cdot 12} - (sr_1^2 + sr_2^2) = .9702 - (.0507 + .1490) = .9702 - .1997 = .7705$$

The ballantine shown in Figure 3.2 may help to explain this result. In the ballantine, area *a* is the proportion of the variance in *Y* that is uniquely shared with X_1. Therefore, $sr_1^2 = a$. Remember that the area of each circle is 1. Analogously, $sr_2^2 = b$. Area *c* is the area that is jointly shared with both independent variables. It cannot be uniquely assigned to either. Since the total explained variance (R^2) equals $a + b + c$, area *c* is equal to

$$c = R^2_{Y \cdot 12} - sr_1^2 - sr_2^2 = .7705$$

This unallocated area *c* is very large in this example because the two independent variables are highly correlated ($r = .806$), and each has a high zero-order *r* with *Y*. Thus, there is little unique variance in *Y* to assign to either independent variable after the variance associated with the other independent variable has been subtracted.

One way of allocating *c* is to specify a causal relationship between X_1 and X_2, if possible. In our example, it may be valid to specify that experience is a cause of publications. If so, then experience is antecedent to publications. Area *c* would be assigned to the antecedent variable, in this case X_1. The following equation would then be used to allocate the total explained variance in *Y* among the two independent variables:

$$R^2_{Y \cdot 12} = r^2_{Y1} + sr_2^2 = (.9062)^2 + (.2251)^2 = .8212 + .1490 = .9702$$

The proportion of the total explained variance allocated to experience would be equal to its squared zero-order correlation (.8212), and the proportion allocated to publications would be equal to its squared semipartial correlation. Experience now is attributed much more explanatory power than publications because it is believed to be causally antecedent to publications. It must be emphasized that this allocation cannot be "checked" for validity by examining the data. It is a specification that is external to the empirically observed covariances between the variables. If we do not believe that we can validly specify a causal order between the *X*'s, we are stuck. We can't allocate *c*. We must then be content to measure the effects of each *X* that are unique or independent of one another, the partial coefficients. This discussion of area *c* illustrates the crucial role that theory plays in specifying causal models in nonexperimental research.

It is important to note one other fact about area *c* in the ballantine that is clearly not implied by the picture. Area *c* can be negative. That is, the sum of the two squared *sr*'s can be greater than R^2. This is a counter-intuitive fact of statistical analysis; the sum of the parts is greater than the whole. This phenomenon is referred to as *suppression* and will be discussed later.

Partial Correlation

A partial correlation is a correlation between a residualized Y and a residualized X. To determine the partial r for X_1, we would partial X_2 from X_1, as we have already done. We would also partial X_2 out of Y, which has not been done yet. Y would be regressed on X_2 to get the equation shown at the top of Table 3.7. This equation is then used to get predicted Y scores. After calculating the residuals of Y, the partial correlation is calculated according to the following formula:

$$pr_1 = r_{Y1 \cdot 2} = \frac{\sum (Y - \hat{Y}_2)(X_1 - \hat{X}_{1 \cdot 2})/n}{s_{Y - \hat{Y}_2}\, s_{X_1 - \hat{X}_{1 \cdot 2}}} \qquad \textbf{(3.16)}$$

The numerator in Equation 3.16 is the covariance between the residualized variables, and the denominator contains the standard deviations of the residualized variables. Thus, the partial r is a standardized covariance between two residual variables. As such, like any correlation between two sets of scores, its limits are -1 and $+1$. A partial r indicates the correlation between the portions of Y and X_1, for example, that are each independent of X_2. Table 3.7 shows that $pr_1 = .7936$. The table also gives the formula and calculations for pr_2, which equals .9129. These values are much larger than the values of the semipartial r's. In fact, the partial r will always be higher than the semipartial r. An examination of the squared partial r should make it clear why this is so.

A formula for the squared partial r is

$$pr_1^2 = \frac{R^2_{Y \cdot 12} - r^2_{Y2}}{1 - r^2_{Y2}} \qquad \textbf{(3.17)}$$

The numerator is identical to the formula for the squared semipartial correlation. It indicates how much variance X_1 explains over and above that explained by X_2. The denominator is the proportion of Y that is unexplained by X_2. It is this amount of Y that is left to be explained by X_1; $1 - r^2_{Y2}$ is the maximum amount of Y that can be uniquely explained by X_1. Thus, the squared partial r indicates the proportion of the variance in Y that cannot be explained by one independent variable but can be explained by the other independent variable. This is why *pr* is larger than *sr*; it does not measure how much of the total variance of Y is uniquely explained by an independent variable, as does the semipartial r, but instead measures how well the independent variable explains that fraction of the variance of Y that is not explained by the other in-

TABLE 3.7 Partial Correlation for the Academic Salary Example

$\hat{Y} = 20.3 + 2.94X_2$

$\hat{Y}_2$	$Y - \hat{Y}_2$	$(Y - \hat{Y}_2)^2$	$(Y - \hat{Y}_2)(X_1 - \hat{X}_{1\cdot 2})$
26.18	−.18	.0324	.3924
23.24	−.24	.0576	−.1824
43.82	−4.82	23.2324	13.5924
35.00	2.00	4.0000	6.0000
46.76	3.24	10.4976	4.0176
75.00	.00	37.8200	23.8200

$$s_{Y-\hat{Y}_2} = \sqrt{\frac{\sum (Y - \hat{Y}_2)^2}{n}} = \sqrt{\frac{37.82}{5}} = 2.7503$$

$$pr_1 = r_{Y1\cdot 2} = \frac{\sum (Y - \hat{Y}_2)(X_1 - \hat{X}_{1\cdot 2})/n}{s_{Y-\hat{Y}_2} s_{X_1 - \hat{X}_{1\cdot 2}}} = \frac{23.82/5}{(2.7503)(2.1827)} = .7936$$

$\hat{Y} = 23.09 + 2.382X_1$

$\hat{Y}_1$	$Y - \hat{Y}_1$	$(Y - \hat{Y}_1)^2$	$(Y - \hat{Y}_1)(X_2 - \hat{X}_{2\cdot 1})$
23.090	2.91	8.4681	1.3267
27.854	−4.854	23.5613	9.3512
35.000	4.000	16.0000	12.0000
42.146	−5.145	26.4813	10.6702
46.910	3.090	9.5481	1.6813
75.000	.000	84.0588	35.0294

$$s_{Y-\hat{Y}_1} = \sqrt{\frac{\sum (Y - \hat{Y}_1)^2}{n}} = \sqrt{\frac{84.0588}{5}} = 4.1002$$

$$pr_2 = r_{Y2\cdot 1} = \frac{\sum (Y - \hat{Y}_1)(X_2 - \hat{X}_{2\cdot 1})/n}{s_{Y-\hat{Y}_1} s_{X_2 - \hat{X}_{2\cdot 1}}} = \frac{35.0294/5}{(4.1002)(1.8716)} = .9129$$

$$pr_1^2 = \frac{R^2_{Y\cdot 12} - r^2_{Y2}}{1 - r^2_{Y2}} = \frac{.9702 - (.9589)^2}{1 - (.9589)^2} = .6299$$

$$pr_2^2 = \frac{R^2_{Y\cdot 12} - r^2_{Y1}}{1 - r^2_{Y1}} = \frac{.9702 - (.9062)^2}{1 - (.9062)^2} = .8333$$

dependent variable. The value of pr_1^2 is .6299 (Table 3.7). This means that experience explains about 63% of the variance in salary that is not explained by publications. The partial r^2 for publications explains about 83% of the salary variance that cannot be explained by experience.

Partial r's do not appear to be as much in vogue today as they once were. This correlation, however, has been presented as one last logical step in the residualizing procedures that we have been carrying out. As with the B's and sr's, the pr's may on occasion rank-order the independent variables in terms of relative explanatory importance differently from the B's and sr's, when there are three or more independent variables.

Lastly, formulas for the *unstandardized* slopes that consist of regressing the residuals of Y on the residualized X's are

$$b_{Y1\cdot 2} = b_{(Y-\hat{Y}_2)(X_1-\hat{X}_{1\cdot 2})} = \frac{\sum (X_1 - \hat{X}_{1\cdot 2})(Y - \hat{Y}_2)}{\sum (X_1 - \hat{X}_{1\cdot 2})^2} = \frac{23.82}{23.82} = 1.00$$

$$b_{Y2\cdot 1} = b_{(Y-\hat{Y}_1)(X_2-\hat{X}_{2\cdot 1})} = \frac{\sum (X_2 - \hat{X}_{2\cdot 1})(Y - \hat{Y}_1)}{\sum (X_2 - \hat{X}_{2\cdot 1})^2} = \frac{35.0294}{17.5146} = 2.00$$

Notice that the values of the slopes are the same as those given by Equations 3.4 and 3.5. Since both approaches give the same values, we will stick to Equations 3.4 and 3.5, which are simpler because they do not involve residualizing Y.

Additional Formulas

Table 3.8 presents some additional formulas that may be used to calculate all of the statistics covered in this chapter more rapidly. All that is needed are the three zero-order r's and the standard deviations. You should examine these formulas to see how they are used. Although they are useful, they do not as clearly indicate the meaning of the statistics as do the formulas that have been covered above. The formulas for the measures of partial relationships (b, B, sr, pr) that have been stressed all contain residualized variables to emphasize that this is the method for holding constant other causes in nonexperimental research.

Notice also that the numerators of the formulas for B, sr, and pr are all identical. Thus, they differ only in their denominators, and even here there is one term that is identical in all three formulas: $(1 - r_{12}^2)$. By comparing sr and B, we can see that sr will always be smaller than B because its denominator $\sqrt{1 - r_{12}^2}$ will always be larger than the denominator of B $(1 - r_{12}^2)$. Also, pr will always be larger than sr because its denominator contains an additional term $\sqrt{1 - r_{Y2}^2}$, for example, that is a fraction and thus makes the denominator of pr smaller than that of sr. There is no rule, however, to determine whether pr will be larger or smaller than B.

TABLE 3.8 Additional Formulas for the Regression and Correlation Statistics for the Academic Salary Example

$$B_{Y1\cdot 2} = \frac{r_{Y1} - r_{Y2}r_{12}}{1 - r_{12}^2} = \frac{.9062 - (.9589)(.8060)}{(1 - .8060^2)} = .3805$$

$$B_{Y2\cdot 1} = \frac{r_{Y2} - r_{Y1}r_{12}}{1 - r_{12}^2} = \frac{.95892 - (.9062)(.8060)}{(1 - .8060^2)} = .6522$$

$$b_{Y1\cdot 2} = \frac{s_Y}{s_1} B_{Y1\cdot 2} = \frac{9.6954}{3.6875}(.3860) = 1$$

$$b_{Y2\cdot 1} = \frac{s_Y}{s_2} B_{Y2\cdot 1} = \frac{9.6954}{3.1623}(.6522) = 2$$

$$sr_1 = \frac{r_{Y1} - r_{Y2}r_{12}}{\sqrt{1 - r_{12}^2}} = \sqrt{1 - r_{12}^2}\, B_{Y1\cdot 2} = \sqrt{1 - .8060^2}\,(.3805) = .2252$$

$$sr_2 = \frac{r_{Y2} - r_{Y1}r_{12}}{\sqrt{1 - r_{12}^2}} = \sqrt{1 - r_{12}^2}\, B_{Y2\cdot 1} = \sqrt{1 - .8060^2}\,(.6522) = .3860$$

$$pr_1 = \frac{r_{Y1} - r_{Y2}r_{12}}{\sqrt{1 - r_{Y2}^2}\sqrt{1 - r_{12}^2}} = \frac{sr_1}{\sqrt{1 - r_{Y2}^2}} = \frac{.2252}{\sqrt{1 - .9589^2}} = .7937$$

$$pr_2 = \frac{r_{Y2} - r_{Y1}r_{12}}{\sqrt{1 - r_{Y1}^2}\sqrt{1 - r_{12}^2}} = \frac{sr_2}{\sqrt{1 - r_{Y1}^2}} = \frac{.3860}{\sqrt{1 - .9062^2}} = .9129$$

$$R_{Y\cdot 12}^2 = B_1 r_{Y1} + B_2 r_{Y2} = (.3805)(.9062) + (.6522)(.9589) = .9702$$

Finally, these formulas for B, sr, and pr are applicable only to the case of two independent variables. The formulas for b and R^2, however, can be extended to include cases with three or more independent variables.

A Multiple Regression Example: Earnings, Education, and Experience

The various multiple regression and correlation statistics that have been discussed and illustrated with the hypothetical academic salary example will now be illustrated with data from the 1977 wave of the 1973–1977 Quality of Employment Survey (Quinn and Staines 1979). The analysis is restricted to white males who were between 25 and 64 years of age at the time of the survey and worked 35 or more hours per week. Additionally, only wage and salaried workers were included (self-employed males were excluded). The dependent variable is the annual salary of the workers, in hundreds of dollars (Y = EARN). The independent variables are the number of years of schooling completed (X_1 = EDUC: *0 years* = 0; *1–7 years* = 4; *8 years* = 8; *9–11 years* = 10; *12 years* = 12; *13–15 years* = 14; *16 years* = 16; *17–19 years* = 18) and the total number of years worked since age 15 (X_2 = EXPER).[2] This example is analogous to the academic salary example because both use occupational earnings as the dependent variable and experience as one of the independent variables. Furthermore, from the perspective of human capital theory in economics, years of education is treated as a measure of productivity and thus is somewhat analogous to publications (an indicator of productivity in the academic salary example).

The descriptive statistics at the top of Table 3.9 indicate that the 547 men in this analysis had an average income from their occupations of $11,641, an average of 22 years of work experience, and a mean education of about 12.5 years of schooling. Education is positively correlated with earnings (see Figure 3.9a). Interpreting this correlation as a standardized bivariate regression slope, we can say that as education increases by one standard deviation (3.7 years), earnings increase by about .375 of a standard deviation ($5,196). Experience, on the other hand, has only a negligible positive correlation with earnings (see Figure 3.9b). Education and experience, however, are negatively correlated (see Figure 3.9c); if two men's educations differ by one standard deviation, the more educated man will have .323 of a standard deviation less experience, on the average. Since the two independent variables are correlated, we should not use the bivariate (zero-order) correlations to assess the strength of each variable's effect on income. We should use the partial regression/correlation coefficients for this purpose.

Figure 3.10 shows two three-dimensional plots of the data points. In plot (a), it is somewhat difficult to know where in the three-dimensional space various points fall. Dropping the spikes from the points to the horizontal plane helps, but with over 500 plotted points it is still difficult to tell where some points are located. For example, looking at Cases 400 and 552, which do you think has more education? In Figures 3.10 and 3.9 several outliers have been highlighted. In Figure 3.9a, 552 and 557 are at the same point because they have equal earnings and education, but in Figure 3.9b it can be seen that 552 has more experience than 557. In Figure 3.9c, cases 483 and 202 are at the highest level of education and near the top in experience. Various measures of outliers will be presented after we have examined the regression results. Before turning to the figures for the least-squares regression coefficients, can you tell from Figure

2. Women and blacks have been excluded from the analysis because research has repeatedly shown that they receive significantly smaller monetary returns on their educational attainment and experience.

TABLE 3.9 Regression of Earnings of Males on Years of Schooling and Experience

		Mean	*Std. Dev.*	*Variance*	
Y	EARN	116.411	51.955	2699.322	
X_1	EDUC	12.468	3.361	11.296	
X_2	EXPER	21.934	10.924	119.334	
n	547				
			Correlations		
		EARN	*EDUC*	*EXPER*	
Y	EARN	1.000	.375	.059	
X_1	EDUC	.375	1.000	−.323	
X_2	EXPER	.059	−.323	1.000	
	R	.4203			Sum of Squares
	R^2	.1767		Regression	260398.512
	Shrunken R^2	.1737		Residual	1213433.937
	Std. Error	47.229		Total	1473832.449
		b	*B*	*sr*	*pr*
X_1	EDUC	6.798	.440	.416	.417
X_2	EXPER	.954	.201	.190	.205
	Constant	10.720			

$$sr_1^2 = R_{Y\cdot 12}^2 - r_{Y2}^2 = .1767 - (.059)^2 = .173$$

$$sr_2^2 = R_{Y\cdot 12}^2 - r_{Y1}^2 = .1767 - (.375)^2 = .036$$

$$pr_1^2 = \frac{R_{Y\cdot 12}^2 - r_{Y2}^2}{1 - r_{Y2}^2} = \frac{.1767 - (.059)^2}{1 - (.059)^2} = .174$$

$$pr_2^2 = \frac{R_{Y\cdot 12}^2 - r_{Y1}^2}{1 - r_{Y1}^2} = \frac{.1767 - (.375)^2}{1 - (.375)^2} = .042$$

3.10 whether experience or education has an effect on earnings? There are too many data points to see the relationships very well in these plots. Therefore, we will examine several subplots containing only subsets of the cases.

Figures 3.11a through d show the relationships between earnings and experience for four, eight, ten, and twelve years of education, respectively. There appears to be little or no evidence for an effect of experience on earnings in these plots. But in Figures 3.11e through g there is clearly a positive relationship between experience and earnings. These last three plots are for workers who have completed some college. There is also some evidence in all of the plots of a decrease in earnings for those with the greatest amount of experience. Methods for investigating nonlinear effects will be taken up in Chapter 6. We

FIGURE 3.9 (a & b) Bivariate Scatterplots for Earnings, Education, and Experience, Jittered

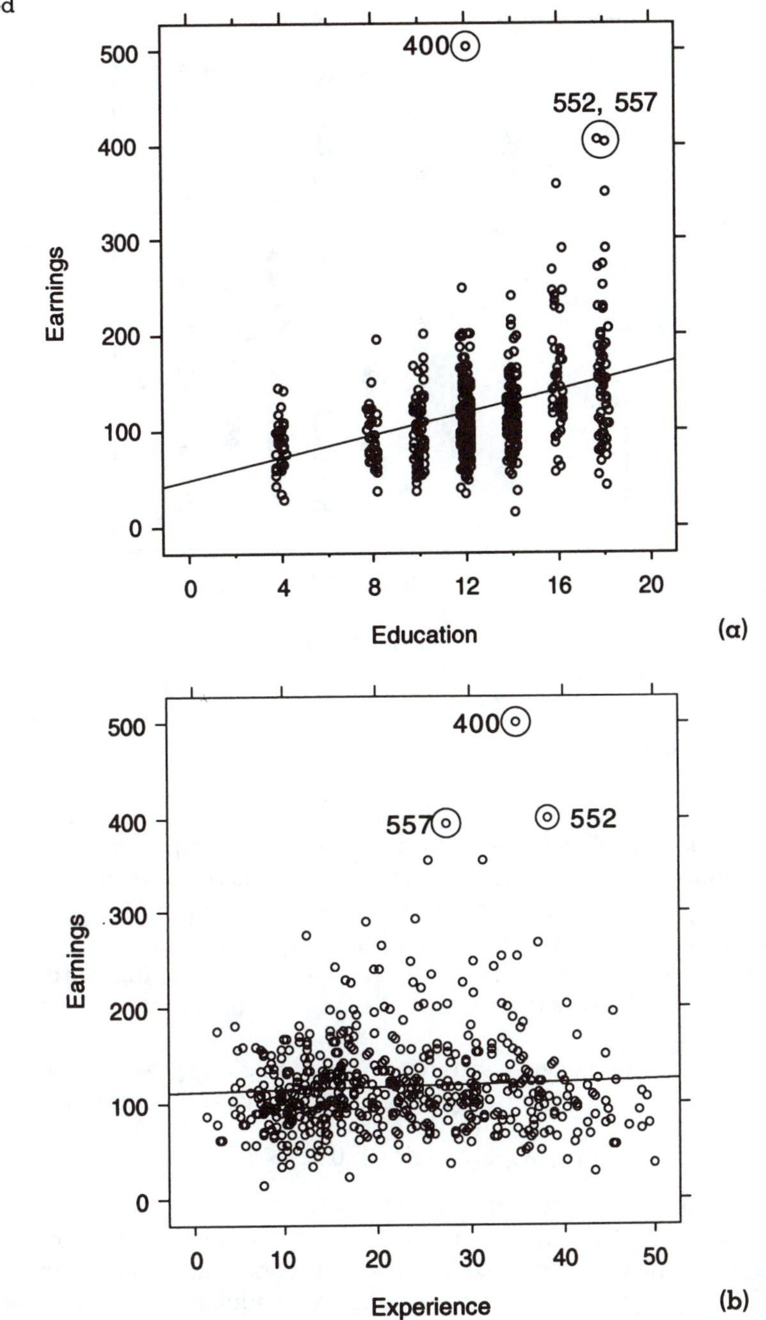

FIGURE 3.9 (c)

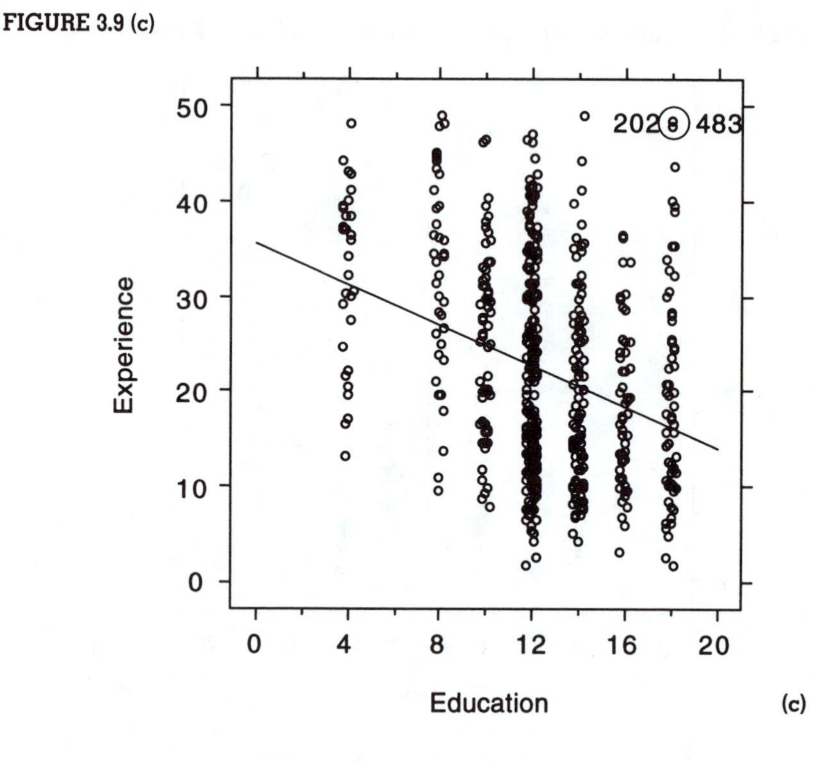

can also see in the plots that the height of the earnings profiles increases as education increases. That is, there is a positive relationship between education and earnings. Finally, we can see that the earnings-experience profiles tend to be more concentrated in the sector of the horizontal plane corresponding to high experience when education is low, but as education increases, the profiles move closer and closer to the low-experience sector. That is, there is a negative relationship between education and experience, as the correlation coefficients in Table 3.9 show.

Now we examine the least-squares linear multiple regression solution. Table 3.9 gives the following least-squares multiple regression equation:

$$\hat{Y} = a_{Y \cdot 12} + b_{Y1 \cdot 2}X_1 + b_{Y2 \cdot 1}X_2 = 10.72 + 6.798X_1 + .954X_2$$

The partial slopes indicate that an increase in experience of one year, with education held constant, causes an increase in earnings of $95, and an additional year of schooling, with experience held constant, increases income by $680. Although everyone knows (or assumes) that education pays off, the slope

FIGURE 3.10 (a & b) Three-Dimensional Scatterplots for Earnings, Education, and Experience

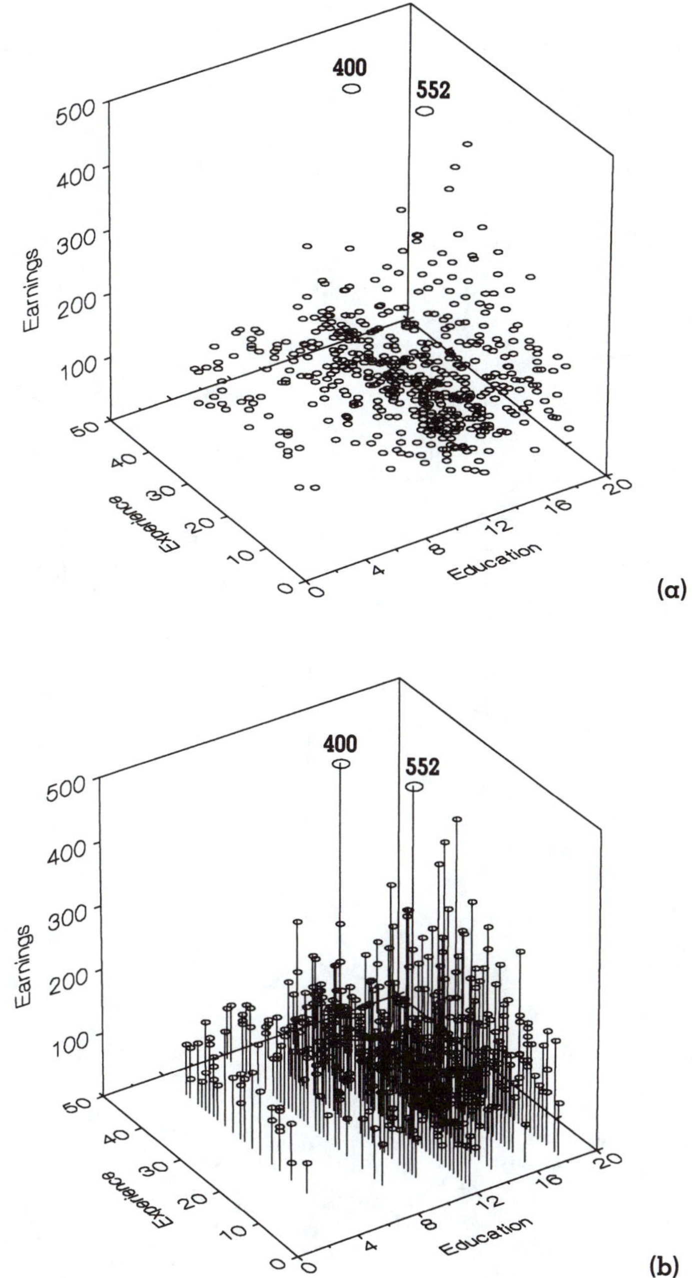

FIGURE 3.11 (a & b) Plots for Earnings and Experience, Holding Education Constant

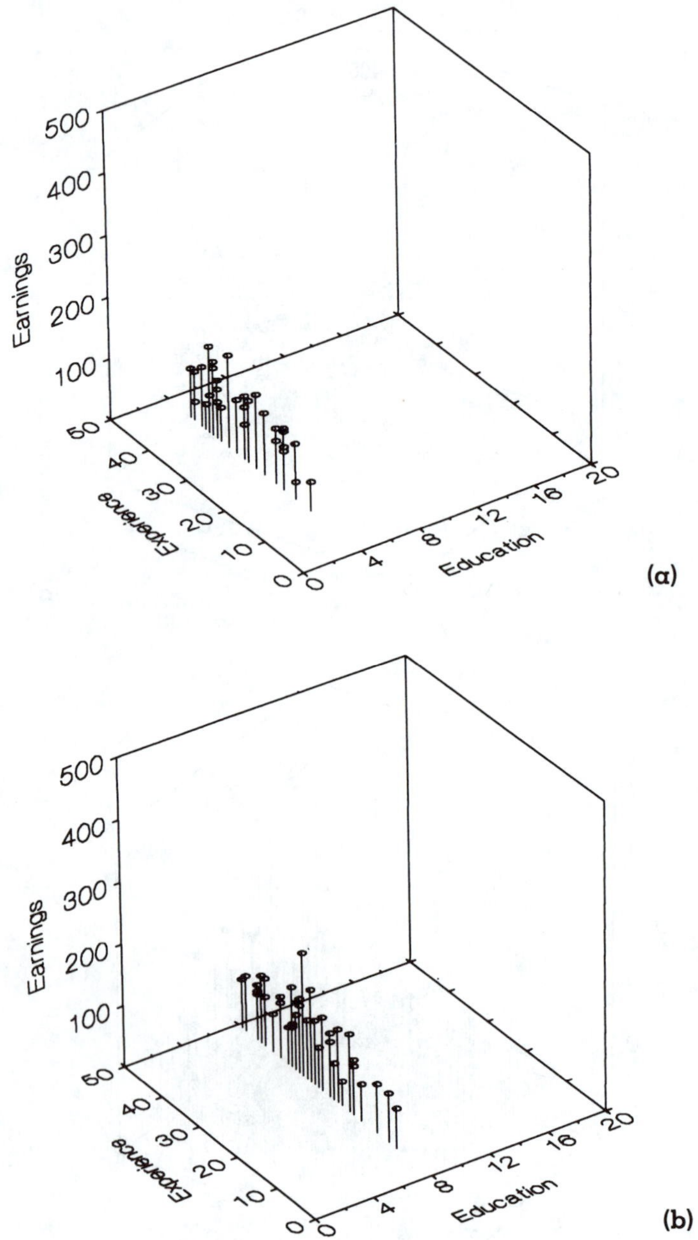

FIGURE 3.11 (c & d)

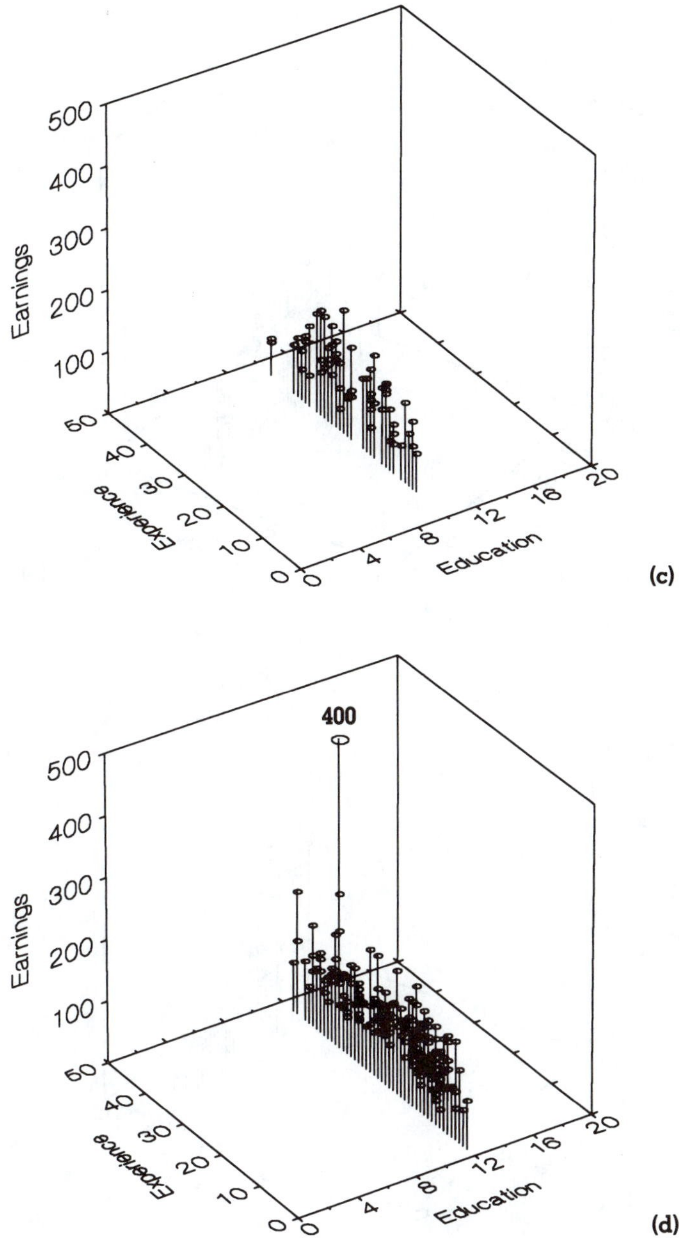

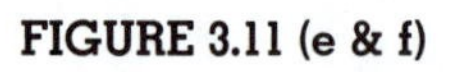

FIGURE 3.11 (e & f)

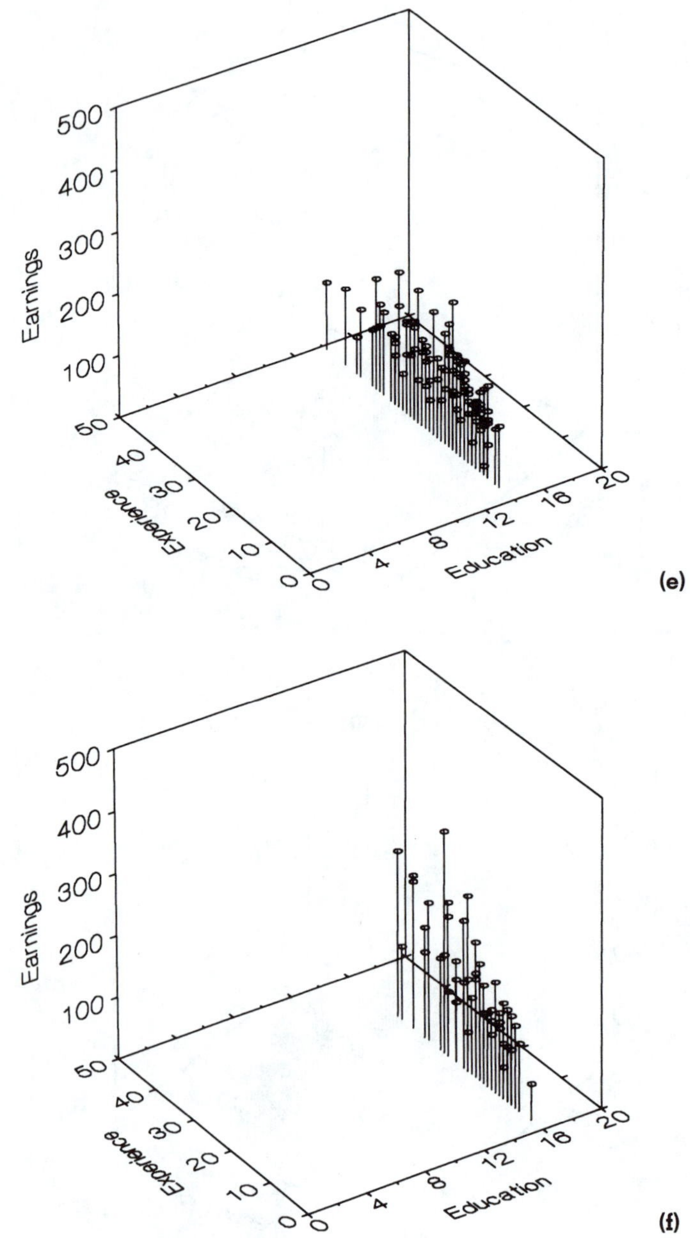

FIGURE 3.11 (g)

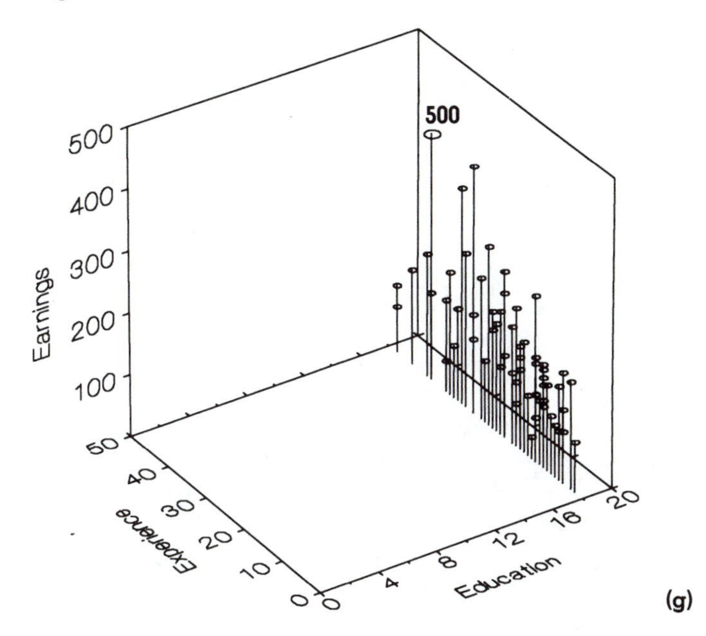

for schooling provides a quantitative estimate of the earnings value of schooling, independent of labor force experience (for white males). And the slope for experience indicates that labor force experience also pays off; if two men have the same number of years of schooling, the one with more experience will earn more (no age discrimination indicated by this model). The Y-intercept equals \$1,072, which is the predicted income when both experience and education equal zero.

Figure 3.12 shows the plane defined by the least-squares equation for earnings regressed on education and experience. The plane is a linear surface that is fitted to the data. This means that education is estimated to have a positive linear effect on earnings at every level of experience. This is indicated by the six parallel lines in the plane at values of experience of 0, 10, 20, 30, 40, 50. The slope of the earnings-education line is identical at each of the levels of experience, and that slope equals 6.798, the partial regression slope. There are also six parallel lines running in the opposite direction at values of education of 0, 4, 8, 12, 16, and 20. These six lines also have equal slopes, but these slopes represent the change in earnings per year of increase in experience.

Table 3.9 also gives several statistics for evaluating the accuracy of the predictions. The standard error of estimate indicates that the standard deviation

FIGURE 3.12 The Multiple-Regression Surface for Earnings, Education, and Experience

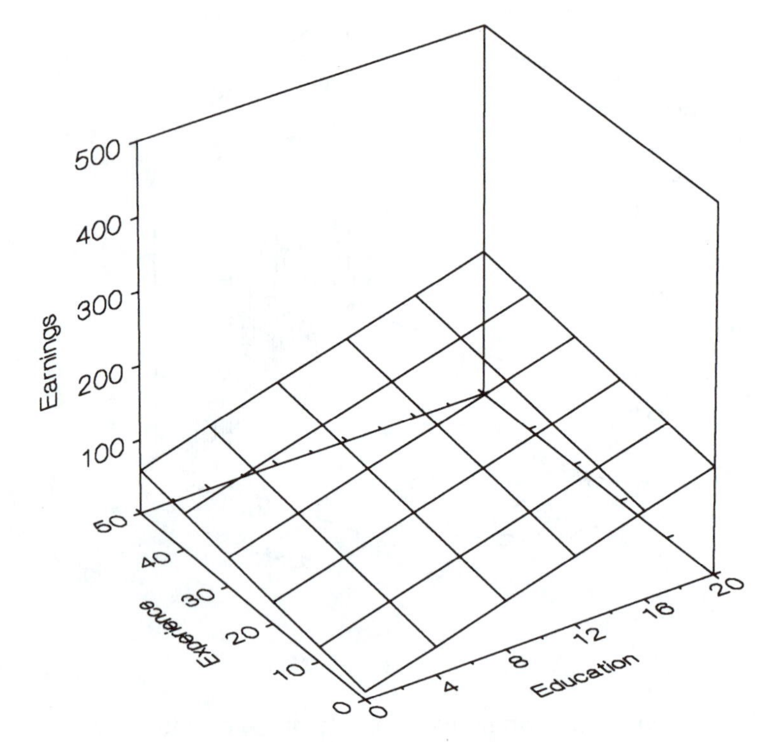

of the residuals is about \$4,723. If the standard error is squared ($47.229^2 = 2230.58$), we see that the unexplained income variance is about \$223,058. This is the variance of incomes caused by other variables that are not included in the equation. This unexplained variance appears to be a big chunk of the total variance, which equals \$269,932. We may compute the proportion of the total variance that is explained by education and experience as follows:

$$R^2 = \frac{\text{Total Sum of Squares} - \text{Residual Sum of Squares}}{\text{Total Sum of Squares}}$$

$$= \frac{1473832.449 - 1213433.937}{1473832.449} = .1767$$

This indicates that .1767 (17.67%) of the variance (or variation) of income is explained by experience and education together. Thus, more than three-fourths of the income variance is explained by other factors. In a world of multiple causation, it should probably not surprise us that only a little more than one-sixth of the variance is explained by experience and education; the fact that only two variables can explain this much variance might be viewed by many observers as quite remarkable. The shrunken R^2 is not much smaller than R^2 because there is such a high ratio of the number of cases to the number of independent variables (547/2 = 273.5).

The square root of R^2 equals the multiple R, which indicates that when the predicted scores $\hat{Y}$ increase by one standard deviation, the observed Y's increase by .42 standard deviations. The value of the multiple R—approximately two-fifths of its maximum value—would seem to indicate that experience and education are better predictors of earnings than the value of R^2 indicates, which is only one-sixth of its maximum value. Although R^2 is more frequently used to measure the explanatory power of the independent variables (probably because of its intuitively appealing interpretation), both R and R^2 are equally valid for this purpose. The fact that they may sometimes suggest different conclusions (as in this case) indicates that there is a certain amount of ambiguity to assessing explanatory power with which we must learn to live.

Finally, it is often desirable to make some comparison of the explanatory power of the independent variables. We cannot compare the unstandardized partial slopes of experience and education because, although they are both measured in terms of years, years of schooling and years of experience are not exactly comparable quantities. The standardized partial regression slopes (the B's), however, may be compared because they are based on standardized (z) scores. The standardized slopes (Table 3.9) indicate that when experience increases by one standard deviation and education is held constant, income will increase by .201 standard deviation; an increase of one standard deviation in education, on the other hand, leads to an increase in income of .440 standard deviation, with experience held constant. Thus, education appears to be the stronger predictor of income. The semipartial correlations (sr's) and the partial correlations (pr's) also lead to the same conclusion. (Remember, B, sr, and pr will always rank the independent variables in the same order in the case of only two X's but will not always agree when there are three or more X's.)

The squared semipartial correlation for education indicates that education explains .173 of the variance in income over and above the variance that can be explained by experience alone (Table 3.9). The proportion of the variance in income that is uniquely explained by experience is only $sr^2 = .036$. The sum of the two sr^2's equals .209, which is greater than R^2. It was briefly noted in the section on the semipartial correlation that this more-or-less counter-intuitive result may sometimes occur and that it is one aspect of a phenomenon referred to as *suppression*. The conditions under which suppression occurs will be examined in much greater detail in a subsequent section. However, this example illustrates, as was pointed out in the section on semipartial correlation, that the sr^2's alone cannot be used to allocate the explained variance into components caused by each of the independent variables. Sometimes the sum of the sr^2's

TABLE 3.10 Outliers and Influential Cases

Case	Cook's D	DFBETA			Resid.	Mahal.
		a	b_1	b_2		
552	.128	−5.39	.092	.305	229.69	7.6
400	.097	−1.80	.081	.057	374.30	1.5
557	.081	−3.99	.049	.269	240.19	4.0
553	.057	−3.46	.048	.220	187.33	4.8
483	.041	3.03	−.058	−.157	−103.90	12.2

is greater than the total explained variance (as in this case), and sometimes it is less than the total explained variance (the more common case). Only when the X's are completely independent of one another (uncorrelated) will the sr^2's sum to R^2.

In order to allocate the explained variance between the independent variables, we must be able to specify validly a causal order between the X's. In this case we can validly specify that education may be a cause of experience but experience cannot cause education. That is, the longer a person stays in school, the less labor force experience the person will have.[3] This causal order allows us to use the following equation to allocate the explained variance:

$$R^2_{Y\cdot 12} = r^2_{Y1} + sr^2_2 = (.375)^2 + .036 = .177$$

In this specification the independent variable that is causally antecedent to the other independent variable, i.e., education, is given credit for all of the variance it can explain alone, i.e., $r^2_{Y1} = .141$. The other independent variable (experience) is then given credit for all of the variance that it can explain over and above that explained by the first variable, i.e., $sr^2_2 = .036$. The sum of the two components equals the total explained variance. Under the specified causal order, education is again found to have more explanatory power than experience.

Table 3.10 presents measures of outliers for the five most influential cases in the sample. They are ranked in terms of a new measure of influence, **Cook's D**. Our previous measure of influence was DFBETA, the change in the regres-

3. Not all of the relationship between education and experience is attributable to the effect of education on experience. Some of the relationship can be traced to a common cause, i.e., age. Different age cohorts may have different amounts of education because of period effects that determined the demand for education at the time the individuals were being socialized and attending school. There was less demand for education at the time older respondents were growing up. Age also has an effect on experience; the older a man is, holding constant education, the more work experience he will have. Thus, age has a negative effect on education and a positive effect on experience, causing experience and education to be negatively correlated. The allocation process used here actually attributes to education too much of the explained variance that education and experience share.

sion slope when a case is deleted from the regression equation. However, there are now two slopes and so neither one can be the sole indicator of influence. Cook's D, however, is a measure of the total influence on the multiple regression equation of deleting a particular case. D takes into account the changes in both slopes plus the change in the intercept. The formula is too complex to present at this point. Suffice it to say that the greater the value of Cook's D, the greater is the influence of a case. Table 3.10 also shows the actual changes that occur in the intercepts and slopes when each case is deleted, i.e., the DFBETAs. The measure that was presented for regression outliers, the residual of Y, is also included in Table 3.10. Finally, Mahalanobis's distance is also given, a measure of X outliers that was not used previously. Instead of being a measure of the distance of a case from the mean of a single X (such as z_X^2, the measure that was used for bivariate relationships), Mahalanobis's distance indicates the distance of a case from the means of all the X's (but not Y).

The most influential case is number 552, a man with 18 years of education, 39 years of experience, and annual earnings of $40,000. This case is highlighted in Figures 3.9, 3.10, and 3.11. This case does not have the largest residual nor the largest value of Mahalanobis, but it does have the largest Cook's D because it has sizable values for both the residual and Mahalanobis. DFBETA shows that if number 552 were deleted, the slope for experience would decrease by .092 and the slope for education would decrease by .305. These are the largest changes in the slopes.

The case with by far the largest residual is number 400 (see the plots). This man earns $37,430 more than predicted by his experience and education. This is mainly due to the fact that although he has a lot of experience (35 years), he has only an average amount of education (12 years). The relatively high earnings compared to the average education leads to a huge residual. Despite the fact that Mahalanobis's distance is not very great for this case (it is not a large X outlier), Cook's D indicates it is the second most influential case.

Finally, Case 483 is the largest X outlier, according to Mahalanobis's distance. This man is extreme on both education (18 years) and experience (48 years). Although number 483 has a lot of potential leverage on the regression equation, the residual for this case is not among the very largest (residual equals −$10,390). Still, due to the very large Mahalanobis's distance, Case 483 is the fifth most influential case in the sample.

Multiple Regression/Correlation with Three or More Independent Variables

The formulas for two independent variables can be extended in a straightforward fashion to cases with three or more independent variables. The multiple regression equation with full subscript notation for three independent variables is

$$\hat{Y} = a_{Y\cdot 123} + b_{Y1\cdot 23}X_1 + b_{Y2\cdot 13}X_2 + b_{Y3\cdot 12}X_3 \tag{3.18}$$

The subscript notation for these unstandardized slopes has two variables to the right of the dot. Thus, we would interpret the slope for X_1 as the change in Y per unit increase in X_1, with X_2 and X_3 held constant. This equation suggests that the only difference between the formulas with three or more X's and those with two X's will lie in the number of the symbols to the right of the dots in the subscripts. The formulas for the unstandardized partial slopes are

$$b_{Y1\cdot 23} = \frac{\sum (X_1 - \hat{X}_{1\cdot 23})(Y - \overline{Y})}{\sum (X_1 - \hat{X}_{1\cdot 23})^2}$$

$$b_{Y2\cdot 13} = \frac{\sum (X_2 - \hat{X}_{2\cdot 13})(Y - \overline{Y})}{\sum (X_2 - \hat{X}_{2\cdot 13})^2}$$

$$b_{Y3\cdot 12} = \frac{\sum (X_3 - \hat{X}_{3\cdot 12})(Y - \overline{Y})}{\sum (X_3 - \hat{X}_{3\cdot 12})^2}$$

The term in each formula for the predicted X scores indicates that these predictions come from a multiple regression equation with one X predicted by the other two. Thus, the slope for X_1 implies that X_1 would be regressed on X_2 and X_3 in order to get the predicted scores $\hat{X}_{1\cdot 23}$. These predictions would then be used to get the residuals $X_1 - \hat{X}_{1\cdot 23}$. The partial slope would thus be defined as the slope of Y on X_1 from which both X_2 and X_3 have been partialled.

The formulas for the other statistics, presented for X_1 only, are

$$R_{Y\cdot 123} = r_{Y\hat{Y}_{123}} = \frac{\sum (Y - \overline{Y})(\hat{Y}_{123} - \overline{Y})/n}{s_Y s_{\hat{Y}_{123}}}$$

$$R^2_{Y\cdot 123} = \frac{\sum (Y - \overline{Y})^2 - \sum (Y - \hat{Y}_{123})^2}{\sum (Y - \overline{Y})^2}$$

$$B_{Y1\cdot 23} = \frac{\sum z_Y(z_1 - \hat{z}_{1\cdot 23})}{\sum (z_1 - \hat{z}_{1\cdot 23})^2}$$

$$sr_1 = \frac{\sum (Y - \overline{Y})(X_1 - \hat{X}_{1\cdot 23})/n}{s_Y s_{X_1 - \hat{X}_{1\cdot 23}}}$$

$$sr_1^2 = R^2_{Y\cdot 123} - R^2_{Y\cdot 23}$$

$$pr_1 = r_{Y1\cdot 23} = \frac{\sum (Y - \hat{Y}_{23})(X_1 - \hat{X}_{1\cdot 23})/n}{s_{Y-\hat{Y}_{23}}\, s_{X_1-\hat{X}_{1\cdot 23}}}$$

$$pr_1^2 = \frac{R^2_{Y\cdot 123} - R^2_{Y\cdot 23}}{1 - R^2_{Y\cdot 23}}$$

Again, these are defining formulas that are used to enhance interpretation of the statistics as opposed to calculations. Multiple regression statistics are almost never calculated with desk calculators. A desk-calculator solution, however, can be found in Cohen and Cohen (1983, Appendix 2). Canned computer solutions, such as those provided by SPSS, are typically used. Matrix algebra methods that are used by computer solutions are briefly introduced in Appendix 3A.

Redundancy and Suppression

When two or more independent variables are used in a multiple regression equation, the partial coefficients for each variable are likely to have values different from the bivariate coefficients. In this section, we will discuss the different ways in which the standardized partial slopes (*B*'s) and semipartial correlations (*sr*'s) may differ from the bivariate standardized slope or correlation. *B* and *sr* may have the same sign as *r* but a smaller absolute value, they may have the same sign as *r* but a larger absolute value, or they may have the opposite sign from *r* and be either smaller or larger in absolute value. The case where *B*/*sr* has the same sign but a smaller value than *r* is referred to as **redundancy**. The case where *B*/*sr* has the same sign and is larger than *r* and the case where *B*/*sr* has a different sign from *r* are both referred to as **suppression**.

Two Independent Variables

We will examine the patterns of correlations between *Y* and two *X*'s that are associated with redundancy and suppression. When discussing the patterns of *r*'s for the three-variable case that are associated with redundancy and suppression, it will be helpful to refer to the computational formulas for the partial *B*'s:

$$B_{Y1\cdot 2} = \frac{r_{Y1} - r_{Y2}r_{12}}{1 - r_{12}^2}$$

$$B_{Y2\cdot 1} = \frac{r_{Y2} - r_{Y1}r_{12}}{1 - r_{12}^2}$$

The numerator of each formula is very important for understanding when redundancy and suppression occur. The numerator shows that in order to get the partial *B* for one independent variable it is necessary to subtract from its correlation with *Y* the product of the correlation of the second independent variable with Y times the correlation between the two independent variables. The size and sign of the product term relative to the correlation of the first variable with Y determines whether redundancy or suppression occurs. To see how this works, Figure 3.13 shows three possible patterns of correlations between the variables.

Pattern (a) in Figure 3.13 consists of three positive correlations. Pattern (b) consists of a positive correlation between Y and one independent variable, a negative correlation between Y and the second independent variable, and a negative correlation between the two independent variables. Pattern (c) consists of a negative correlation between Y and each independent variable and a positive correlation between the two independent variables. The thing that the three patterns have in common is that for each one $r_{Y1}r_{12}$ has the same sign as r_{Y2} and $r_{Y2}r_{12}$ has the same sign as r_{Y1}; that is, the correlation of each X with Y has the same sign as the product of the other two correlations. Thus, the second term in the numerator of the formula for each B has the same sign as the first term in the numerator. This fact leads to the following rule: redundancy exists for Type 1 correlation patterns when

$$|r_{Y1}| > |r_{Y2}r_{12}| \quad \text{and} \quad |r_{Y2}| > |r_{Y1}r_{12}|$$

Since both terms in the numerator have the same sign for Type 1 patterns, if the first correlation is larger in absolute value than the product of the second two correlations, the difference will have the same sign as the first r and will be smaller in absolute value than the first r (closer to zero). Thus, B will be smaller than r and will have the same sign, i.e., redundancy. It must be emphasized that this rule must hold true for both independent variables.

The following correlations are examples of Type 1 correlation patterns for which this rule holds.

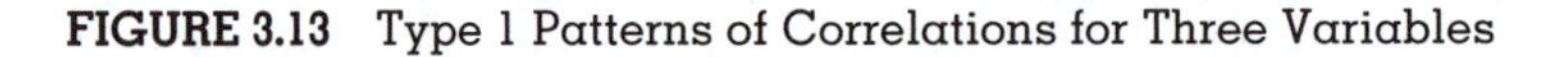

FIGURE 3.13 Type 1 Patterns of Correlations for Three Variables

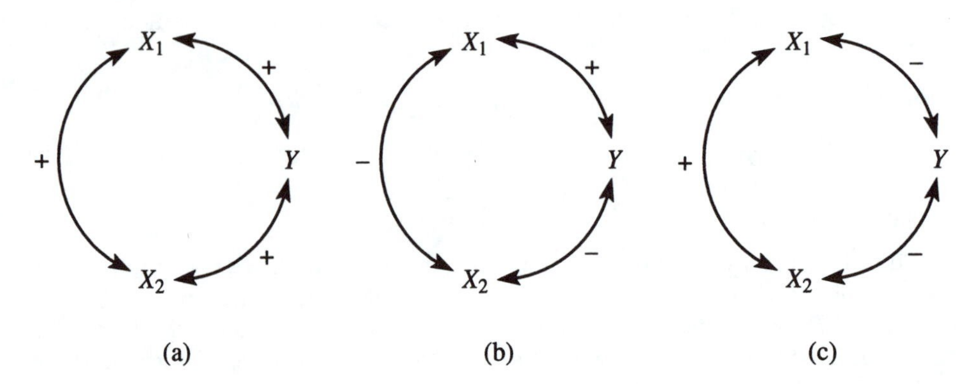

r_{Y1}	r_{Y2}	r_{12}	$B_{Y1\cdot2} = r_{Y1} - r_{Y2}r_{12}/(1 - r_{12}^2)$	$B_{Y2\cdot1} = r_{Y2} - r_{Y1}r_{12}/(1 - r_{12}^2)$
.6	.7	.5	$\frac{.6 - (.7)(.5)}{1 - .25} = \frac{.25}{.75} = .33$	$\frac{.7 - (.6)(.5)}{1 - .25} = \frac{.40}{.75} = .53$
.6	−.7	−.5	$\frac{.6 - (-.7)(-.5)}{1 - .25} = \frac{.25}{.75} = .33$	$\frac{.7 - (.6)(-.5)}{1 - .25} = \frac{-.40}{.75} = -.53$
−.6	−.7	.5	$\frac{.6 - (-.7)(.5)}{1 - .25} = \frac{-.25}{.75} = -.33$	$\frac{-.7 - (-.6)(.5)}{1 - .25} = \frac{-.40}{.75} = -.53$

The original example data for experience, publications, and salary was also an example of redundancy. The correlations and partial B's were as follows:

$$r_{Y1} = .9062, \quad r_{Y2} = .9589, \quad r_{12} = .8060, \quad B_{Y1\cdot2} = .3805, \quad B_{Y2\cdot1} = .6522$$

The correlations are all positive, and the product of the two r's in the numerator of each formula for B is smaller than the first r.

It should be mentioned now that the three Type 1 patterns in Figure 3.13 are essentially equivalent to one another because when we reverse-score an appropriately chosen variable in patterns (b) and (c), the correlations will all become positive. A variable may be reverse-scored by applying the formula $X' = (X_{max} + X_{min}) - X$ to each value of X. X' will have the same range as X, but cases with the maximum score on X will have the minimum score on X', cases with the minimum score on X will have the maximum score on X', and intermediate values will be appropriately reoriented as well. Reverse-scoring will change the sign of a variable's correlation with all other variables without changing the absolute value of the correlations. In pattern (b), for example, if X_2 were reverse-scored, it would then have a positive r with both Y and X_1. And if Y were reversed in pattern (c), it would have a positive r with both X_1 and X_2. Thus, with the Type 1 patterns, it is possible to score the variables so that all three correlations are positive. This is not possible with the other patterns that we will examine. You do not need to reverse-score any variables, however. The point is that the three patterns are all manifestations of a single, underlying Type 1 pattern.

These patterns of correlations are called *redundancy* because part of the relationship between each independent variable and the dependent variable is redundant with the other independent variable. They each share some variance with Y that the other variable shares with Y. Thus, when each variable is statistically held constant, the partial relationships become smaller than the zero-order correlations. Redundancy means that the zero-order relationships overestimate the effect of each independent variable on Y. This is also true for the semipartial r; it will also be smaller in absolute value than the zero-order r when the first rule for the Type 1 patterns of correlations holds.

Finally, when we have redundancy, area c in the ballantine will be positive and will not be allocated to either independent variable; the variance that the two X's share with Y will not be assigned to either. Thus, when redundancy exists, the following relationship holds:

$$sr_1^2 + sr_2^2 \leq R^2$$

The sum of the squared semipartial r's, the variance that each X uniquely shares with Y, will be less than or equal to the total variance explained by the two variables together. Since the correlation between the X's will never be exactly equal, practically speaking, the above sum will be less than R^2. This pattern, redundancy, appears to be the most frequent outcome in empirical research.

When a Type 1 pattern of correlations exists, suppression may also occur. The second rule for Type 1 correlations is that suppression exists when

$$|r_{Y1}| < |r_{Y2}r_{12}| \quad \text{or} \quad |r_{Y2}| < |r_{Y1}r_{12}|$$

This rule states that when the absolute value of the correlation of either independent variable with Y is less than the absolute value of the product of the other two correlations, suppression exists. Since the sign of the first r in the numerator of the formula for the partial B is the same as the sign of the product term, if the magnitude of the correlation is less than the magnitude of the product, the sign of B will be opposite that of r. This is one type of suppression.

It is not possible, however, for the second rule to hold for both independent variables. Therefore,

$$\text{If } |r_{Y1}| < |r_{Y2}r_{12}| \text{ then } |r_{Y2}| > |r_{Y1}r_{12}|$$

and

$$\text{if } |r_{Y2}| < |r_{Y1}r_{12}| \text{ then } |r_{Y1}| > |r_{Y2}r_{12}|$$

This rule indicates that if the absolute value of r between Y and one of the independent variables is less than the absolute value of the product of the other two r's, then the absolute value of r between Y and the second X must be greater than the product. This is true because r_{12} is included in the product term of the numerator for both partial B's. This situation leads to the other type of suppression for the second X; its partial B will have the same sign and will be larger than its zero-order r with Y. Let us look at the following situation

$$|r_{Y1}| < |r_{Y2}r_{12}| \quad \text{and} \quad |r_{Y2}| > |r_{Y1}r_{12}|$$

In this case, $B_{Y1\cdot2}$ will be opposite in sign from r_{Y1} because r_{Y1} will be less than the product of the other two r's. It might appear, however, that redundancy might exist for X_2 because the numerator of the formula for its partial B (i.e., $r_{Y2} - r_{Y1}r_{12}$) will be smaller than r_{Y2}, in absolute value. It must be remembered, however, that the denominator of the formula contains $1 - r_{12}^2$, which in this case will be a sufficiently small fraction to cause $B_{Y2\cdot1}$ to be larger than r_{Y2}. In this situation, suppression occurs for both variables, but for one variable it is *opposite-sign* suppression (B has a different sign from r) and for the other variable it is *same-sign* suppression (B has the same sign and is larger than r).

We may examine an illustration of this type of suppression by returning to the hypothetical data on experience, publications, and salary. We will keep the original values of X_1, X_2, and ε, but we will change the causal formula so that X_1 has an effect of -1 instead of 1. This formula will generate new values of Y as well as new correlations of Y with the X's. The data and correlations are shown below, along with the B's, which are the same values that were reported in the earlier section on standardized partial slopes. Note that in the numerator for $B_{Y1 \cdot 2}$ the product term is greater than r_{Y1}; thus, opposite-sign suppression exists. The product in the numerator of $B_{Y2 \cdot 1}$, on the other hand, is smaller than r_{Y2}, and thus the sign doesn't change. The denominator, however, is sufficiently small to cause the value of $B_{Y2 \cdot 1}$ to be larger than r_{Y2} (same-sign suppression).

X_1	X_2	ε	Y
0	2	+2	26
2	1	−1	19
5	8	−2	29
8	5	−1	21
10	9	+2	30

$$Y = 20 - 1X_1 + 2X_2 + \varepsilon$$

$$r_{Y1} = .3252,\ r_{Y2} = .7731,\ r_{12} = .8060$$

$$B_{Y1 \cdot 2} = \frac{.3252 - (.7731)(.8060)}{1 - (.8060)^2} = \frac{.3252 - .6231}{1 - .6496} = \frac{-.2979}{.3504} = -.8503$$

$$B_{Y2 \cdot 1} = \frac{.7731 - (.3252)(.8060)}{1 - (.8060)^2} = \frac{.7731 - .26211}{1 - .6496} = \frac{.5110}{.3504} = 1.4583$$

The discovery of suppression is a very interesting outcome of multiple regression because it truly reveals conclusions that are very different from those that would be reached on the basis of bivariate relationships. The bivariates, in the above example, would indicate that both experience and publications have positive effects on salary. With respect to experience, such a finding would imply that as their experience increases professors learn skills that are independent of their publishing skills yet that are valuable enough in the academic setting to warrant higher pay. Yet, when the number of publications is held constant, it is found that experience reduces salary, suggesting that nonpublishing skills may become increasingly obsolete as experience increases. Whatever the interpretation, we would say that the relationship between publications and experience, along with the effect of publications on salary, is suppressing the bivariate relationship between experience and salary.

Why does this occur? First, the negative effect of experience on salary tends to create a negative correlation (or covariance) between the two variables, the true sign of the effect. But experience is positively correlated with publications, which has a positive effect on salary. This positive effect creates a positive correlation between publications and salary, and since experience is positively correlated with publications, it also tends to create a positive correlation between experience and salary. In this case, the positive correlation between experience and salary created by its correlation with publications is greater than the negative correlation created by its own negative effect on salary. Thus, the bivariate correlation actually turns out to be of opposite sign from the sign of the true independent effect of experience.

Whenever suppression occurs for one independent variable (when there are two) it also occurs for the other. Thus, experience is also suppressing the bivariate relationship between publications and salary. In this case, it is not suppressing the true sign of the effect of publications; instead, it is suppressing the magnitude or strength of the effect. Thus, the bivariate relationship would underestimate the true monetary value of publications.

Whenever suppression occurs, the following relationship between the semipartial *r*'s and the multiple *R* exists:

$$sr_1^2 + sr_2^2 \geq R^2_{Y \cdot 12}$$

The sum of the squared semipartial *r*'s is actually greater than the total proportion of the variance explained by the independent variables. Using the formulas in Table 3.6 to calculate these statistics gives

$$sr_1^2 + sr_2^2 = .2534 + .7455 = .9989$$

$$R^2_{Y \cdot 12} = .8513$$

While the sum of the squared *sr*'s suggests that virtually 100% of the variance is explained, the true value is about 85.13%. This implies that area c in the ballantine is negative. Since it is impossible to have a negative area, suppression shows the limitations of pictures as compared to mathematics for understanding relationships between variables. The strength of the ballantine is that it gives a good intuitive insight into the case of redundancy, but it is inadequate in the case of suppression.

The meaning of the squared semipartial *r*'s still holds; they indicate the amount that each variable can explain over and above the variance that can be explained by the other variable alone. Thus,

$$R^2_{Y \cdot 12} = sr_1^2 + r^2_{Y2} = .2534 + .5977 = .8511$$

$$R^2_{Y \cdot 12} = sr_2^2 + r^2_{Y1} = .7455 + .1058 = .8513$$

The reason that the sum of the squared *sr*'s is greater than the squared multiple *R* can be understood by remembering that these statistics indicate how much

FIGURE 3.14 Type 2 Patterns of Correlations for Three Variables

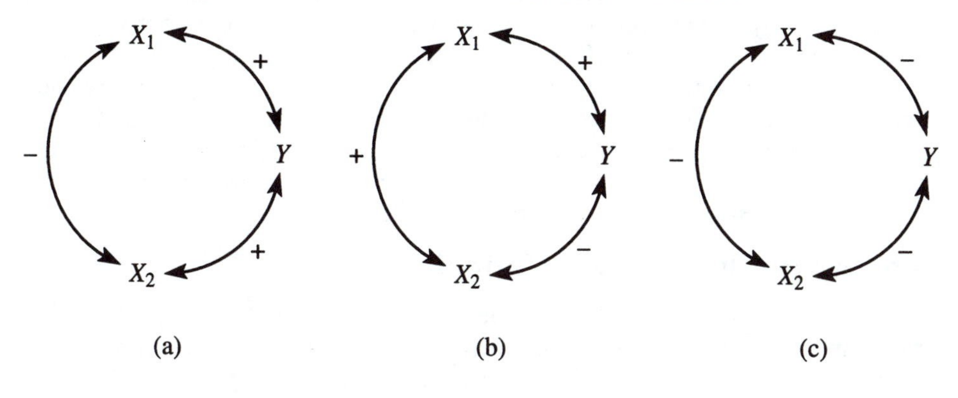

each variable explains independently of the other, that is, with the other variable held constant. In reality, the other variable is not held constant. Each variable varies relative to the other in such a way that each offsets or suppresses some of the variance that the other would share with Y. Thus, the sum of the "independent" contributions of each is actually greater than what they explain when varying together (covarying) in their suppressive relationship. It is still valid, however, to use the *sr*'s to compare the strength of the effects of each independent variable. And one of the formulas listed immediately above can be used to allocate the explained variance between the two variables—as was discussed in the section on semipartial correlation—*if* one of the independent variables can be specified as causally antecedent to the other.

A second set of patterns of correlations is shown in Figure 3.14. Referring again to the numerator of the formulas for the standardized *B*'s, in each of the patterns in Figure 3.14, the product term is opposite in sign from that of the first correlation: the sign of $r_{Y2}r_{12}$ is opposite of the sign of r_{Y1}, and the sign $r_{Y1}r_{12}$ is opposite of that of r_{Y2}. Since the product is subtracted from the zero-order *r*, the numerator of *B* for each independent variable will always be greater than its *r* with Y. Thus, the partial *B* will have the same sign and be greater in absolute value than the zero-order *r*, i.e., same-sign suppression. Same-sign suppression occurs for both independent variables when a Type 2 pattern of correlations exists. Thus, each independent variable is suppressing the size (but not the sign) of the bivariate relationship between the other independent variable and Y.

As with the Type 1 patterns, if one of the variables is reverse-scored in pattern (b) and pattern (c), patterns (b) and (c) will each have the same pattern of *r*'s as (a). You should determine which variable has to be reverse-scored in each pattern to give pattern (a). Also, as with Type 1 suppression, Type 2 suppression causes the sum of the squared semipartial correlations to be greater than the squared multiple *R*.

The example involving income, experience, and education (see Table 3.9) illustrates this type of suppression. Both experience and education were positively correlated with income (the dependent variable), but there was a negative correlation between experience and education. Therefore, the analysis of income, experience, and education was an example of pattern (a) in Figure 3.14. In this case, the partial *B*, *sr*, and *pr* for experience and for education each had the same sign and was larger than the zero-order *r*, and the sum of the sr^2's for experience and education was larger than R^2.

Three or More Independent Variables

With three or more independent variables, there may be Type 1 patterns for some pairs of independent variables and Type 2 patterns for other pairs. The potential combinations are too complex to diagram. There may also be suppressive relationships for some pairs and redundant relationships for other pairs. As implied by this, a single independent variable may be in a suppressive relationship with some variables and in a redundant relationship with others. Whether each variable's bivariate relationship with *Y* is suppressed or not depends upon the number and strength of suppressive relationships as compared to the number and strength of redundant relationships. If the redundant relationships outweigh the suppressive relationships, redundancy will occur for that variable; its partial *B* will have the same sign as and be smaller in absolute value than its zero-order *r*. If the suppressive relationships outweigh the redundant ones, suppression will occur; the partial *B* will be of opposite sign from the zero-order *r*, or it will have the same sign and be greater in absolute value.[4] Thus, to determine whether an independent variable is redundant with the total set of other independent variables or is suppressed by the set of other independent variables, simply compare its partial *B* with its bivariate *r*.

Summary

In this chapter we have focused primarily on the interpretation of five multiple regression and correlation statistics: the unstandardized partial regression slope, the standardized partial regression slope, the semipartial correlation, the partial correlation, and the multiple correlation, including the squared semipartial, partial, and multiple correlations. These five basic statistics can each be interpreted as a bivariate regression or correlation between one or two transformed variables.

The key step in this conceptualization in the case of an equation with two independent variables is the computation of the residual of each independent

4. With three or more *X*'s, however, it is possible for sr_i to be less than r_{yi} even though B_i is greater than r_{yi}. The decision about whether redundancy or suppression exists would then depend on whether you were using *B* or *sr* to measure the strength of the effect.

variable after it has been regressed on the other independent variable. This is important because the residualized independent variable is uncorrelated with the other independent variable. Therefore, the regression of the dependent variable on the residualized independent variable gives the change in the dependent variable per unit increase in the independent variable *with the other independent variable held constant*. This is the unstandardized partial regression slope. If the same procedure is applied to standardized scores (z scores), the standardized partial regression slope is obtained.

Unlike the bivariate standardized slope, however, the standardized partial slope does not have fixed upper and lower limits. A statistic with such limits may be obtained by computing the Pearson correlation between the dependent variable and the residualized independent variable, which is named the semipartial correlation. The squared semipartial correlation gives the proportion of the variance in the dependent variable that is uniquely associated with the independent variable. If we compute the residuals of the dependent variable after regressing it on the second independent variable and then correlate these residuals with the residuals of the first independent variable, we get the partial correlation. And the squared partial correlation gives us the proportion of the variance in the dependent variable that is not explained by the second independent variable but is explained by the first independent variable. Finally, the Pearson correlation between the dependent variable and the values of the dependent variable predicted by the multiple regression equation gives us the multiple correlation coefficient, the square of which tells us the proportion of the variance of the dependent variable that is explained by all of the independent variables as a set. The adjusted squared multiple correlation will sometimes be needed as a correction for an upward bias that results when the ratio of the number of cases to the number of independent variables is small.

Multiple regression is required when the independent variables are correlated with one another in order to get unbiased (nonspurious) estimates of their effects. The most common outcome of controlling for other causes is for the standardized partial slope and semipartial correlation to have the same sign but be smaller in absolute value in comparison to the bivariate correlation. This outcome is referred to as redundancy. Less common outcomes occur when these partial statistics are larger in absolute value or of opposite sign from the bivariate correlation; both of these cases are referred to as suppression.

The meaning and interpretation of the multiple regression and correlation statistics when there are more than two independent variables are straightforward extensions of the case of two independent variables. The statistics presented in this chapter are sometimes called descriptive statistics because they summarize relationships between measures or observations in a sample of cases. These least-squares statistics have been developed, however, because they allow us to make efficient estimates or inferences about the population from which the sample was drawn. We turn to the assumptions and rules of inferential statistics in the next chapter.

References

Berry, William D., and Stanley Feldman. 1985. *Multiple Regression in Practice.* Beverly Hills: Sage.

Cohen, Jacob, and Patricia Cohen. 1983. *Applied Multiple Regression/Correlation Analysis for the Behavioral Sciences* (2nd edition). Hillsdale, New Jersey: Lawrence Erlbaum Associates.

Quinn, Robert P., and Grahm Staines. 1979. *Quality of Employment Survey, 1973–1977: Panel.* Ann Arbor, Michigan: Inter-University Consortium for Political and Social Research.

APPENDIX 3A

Matrix Algebra for Multiple Regression

Matrices and Vectors

A matrix is defined as a rectangular array of numbers. An example array with three rows and two columns may be written as

$$\mathbf{A} = \begin{bmatrix} 3 & 8 \\ 1 & 5 \\ 6 & 7 \end{bmatrix}$$

Names of matrices are written in bold letters and the numbers are enclosed in brackets to show that the array is treated as a single entity. Arrays are also sometimes enclosed in parentheses, as in

$$\mathbf{B} = \begin{pmatrix} 2 & 8 & 4 \\ 9 & 1 & 5 \end{pmatrix}$$

The number of rows and the number of columns of a matrix is called the *order* of the matrix. **A** is a matrix of order 3 × 2 (pronounced 3-by-2) and **B** is a 2 × 3 matrix. The number of rows is always written first. It is sometimes convenient to write the order of a matrix as a subscript of the matrix name, as in $\mathbf{A}_{3\times 2}$ and $\mathbf{B}_{2\times 3}$.

The number or element in the *i*th row and *j*th column is indicated by a_{ij} and b_{ij} for matrices **A** and **B**, respectively. Thus, matrices **A** and **B** may be represented as

$$\mathbf{A}_{3\times 2} = \begin{bmatrix} a_{11} & a_{12} \\ a_{21} & a_{22} \\ a_{31} & a_{32} \end{bmatrix}$$

$$\mathbf{B}_{2\times 3} = \begin{pmatrix} b_{11} & b_{12} & b_{13} \\ b_{21} & b_{22} & b_{23} \end{pmatrix}$$

If k equals the number of independent variables and n equals the number of sample cases for which the variables are observed, then $\mathbf{X}_{n\times k}$ is the raw score matrix for the independent variables. The raw score matrix for the academic salary example (Table 3.1) is

$$\mathbf{X}_{5\times 2} = \begin{bmatrix} 0 & 2 \\ 2 & 1 \\ 5 & 8 \\ 8 & 5 \\ 10 & 9 \end{bmatrix}$$

The first column of **X** contains the observed values of X_1 (experience), and the second column gives the values of X_2 (publications).

The *transpose* of a matrix **A** is symbolized by **A**′. If the order of **A** is $r \times c$, then the order of **A**′ will be $c \times r$; that is, the number of rows of **A**′ will equal the number of columns of **A** and the number of columns of **A**′ will equal the number of rows of **A**. **A**′ is created by setting $x'_{ij} = x_{ji}$. This has the effect of making the first row of **A**′ equal to the first column of **A**, the second row of **A**′ equal to the second column of **A**, and in general, the *ith* row of **A**′ equal to the *ith* column of **A**. For example, **X** (as previously given) and the transpose of **X** are

$$\mathbf{X}_{5\times 2} = \begin{bmatrix} 0 & 2 \\ 2 & 1 \\ 5 & 8 \\ 8 & 5 \\ 10 & 9 \end{bmatrix} \qquad \mathbf{X}'_{2\times 5} = \begin{bmatrix} 0 & 2 & 5 & 8 & 10 \\ 2 & 1 & 8 & 5 & 9 \end{bmatrix}$$

If the number of rows r equals the number of columns c, then the matrix is a *square matrix*. The *main diagonal* of a square matrix is the set of elements on the diagonal from the upper left corner to the lower right corner. Thus, $i = j$ for each element on the principal diagonal. For example,

$$\mathbf{A}_{3\times 3} = \begin{bmatrix} a_{11} & a_{12} & a_{13} \\ a_{21} & a_{22} & a_{23} \\ a_{31} & a_{32} & a_{33} \end{bmatrix}$$

A square matrix **A** is *symmetric* if $a_{ij} = a_{ji}$ for all i and j. Thus, if $\mathbf{A}_{3\times 3}$ is symmetric, $a_{12} = a_{21}$, $a_{13} = a_{31}$, and $a_{23} = a_{32}$. This characteristic indicates that $\mathbf{A}' = \mathbf{A}$ when **A** is symmetric. Frequently encountered symmetric matrices in regression analysis are the variance-covariance matrix and the correlation matrix. The correlation matrix in the earnings, education, and experience example (Table 3.9) is

$$\mathbf{C}_{3\times 3} = \begin{array}{c} \begin{matrix} \quad Y & \quad X_1 & \quad X_2 \end{matrix} \\ \begin{bmatrix} 1.000 & .375 & .059 \\ .375 & 1.000 & -.323 \\ .059 & -.323 & 1.000 \end{bmatrix} \end{array}$$

Since the correlation between two variables is equal to the covariance between the standardized scores, **C** can also be thought of as the variance-covariance matrix for the z scores. Since the variance of z equals 1.000, it can be seen that the main diagonal of **C** contains the variances of the standardized variables: e.g., $\text{VAR}(z_Y) = c_{11} = 1.000$.

If all of the elements that are not on the main diagonal (the off-diagonal elements) are equal to zero , the square matrix is called a *diagonal* matrix. A frequently used diagonal matrix is the *identity* matrix **I**, which has 1's on the main diagonal:

$$\mathbf{I} = \begin{bmatrix} 1 & 0 & 0 \\ 0 & 1 & 0 \\ 0 & 0 & 1 \end{bmatrix}$$

A *vector* is an array of numbers consisting of either one row or one column. The earnings variable in Table 3.1 is a column vector:

$$\mathbf{Y}_{5\times 1} = \begin{bmatrix} 26 \\ 23 \\ 39 \\ 37 \\ 50 \end{bmatrix}$$

The transpose of $\mathbf{Y}_{5\times 1}$ is a row vector:

$$\mathbf{Y}'_{1\times 5} = [26 \quad 23 \quad 39 \quad 37 \quad 50]$$

If an array consists of only a single number, it is called a *scalar*. Alternatively, a scalar is a 1×1 matrix. For example, $n = 5$ is a scalar. Scalars are not written in boldface.

Matrix Operations

Addition and Subtraction

Two or more matrices (or vectors) may be added or subtracted if they have the same dimensionality, that is, if they have the same number of rows and columns. The addition and subtraction is carried out for each of the corresponding elements in the matrices:

$$\mathbf{A} + \mathbf{B} = \begin{bmatrix} a_{11} & a_{12} \\ a_{21} & a_{22} \\ a_{31} & a_{32} \end{bmatrix} + \begin{bmatrix} b_{11} & b_{12} \\ b_{21} & b_{22} \\ b_{31} & b_{32} \end{bmatrix} = \begin{bmatrix} a_{11} + b_{11} & a_{12} + b_{12} \\ a_{21} + b_{21} & a_{22} + b_{22} \\ a_{31} + b_{31} & a_{32} + b_{32} \end{bmatrix}$$

To perform subtraction, simply change the above (+) signs to (−) signs:

$$\mathbf{C} = \mathbf{A} - \mathbf{B} = \begin{bmatrix} 3 & 8 \\ 1 & 5 \\ 6 & 7 \end{bmatrix} - \begin{bmatrix} 2 & 9 \\ 8 & 1 \\ 4 & 5 \end{bmatrix} = \begin{bmatrix} 1 & -1 \\ -7 & 4 \\ 2 & 2 \end{bmatrix}$$

Multiplication

To multiply a matrix or a vector by a scalar, simply multiply each element by the scalar:

$$n\mathbf{A} = n\begin{bmatrix} a_{11} & a_{12} \\ a_{21} & a_{22} \\ a_{31} & a_{32} \end{bmatrix} = \begin{bmatrix} n \cdot a_{11} & n \cdot a_{12} \\ n \cdot a_{21} & n \cdot a_{22} \\ n \cdot a_{31} & n \cdot a_{32} \end{bmatrix}$$

The rules for multiplying two matrices, two vectors, or one matrix and one vector are more complicated than the operations described thus far. In order to carry out the multiplication **AB** (or **ab**), *the number of columns of the first matrix* (or *vector*) *must be equal to the number of rows of the second matrix* (or *vector*). For example, $\mathbf{A}_{3\times2}\mathbf{B}_{2\times3}$ can be multiplied but $\mathbf{A}_{3\times2}\mathbf{B}_{3\times2}$ cannot.

The method for multiplying two vectors with the correct dimensionality is

$$\mathbf{a}_{1\times3}\mathbf{b}_{3\times1} = [a_1 \quad a_2 \quad a_3]\begin{bmatrix} b_1 \\ b_2 \\ b_3 \end{bmatrix} = a_1b_1 + a_2b_2 + a_3b_3$$

As indicated above, the product **ab** consists of the *sum of products* of elements in the two vectors, a column vector *premultiplied* by a row vector. The result of multiplying the two vectors together is a scalar. For an example, consider the deviation scores $y = Y - \overline{Y}$ from Table 3.1. In order to get the sum of squares of Y (i.e., the variation of Y), the following vector multiplication is carried out:

$$\mathbf{y'y} = [-9 \quad -12 \quad 4 \quad 2 \quad 15]\begin{bmatrix} -9 \\ -12 \\ 4 \\ 2 \\ 15 \end{bmatrix}$$

$$= (-9)(-9) + (-12)(-12) + (4)(4) + (2)(2) + (15)(15) = 470$$

$$= \sum (Y - \overline{Y})^2$$

Matrix multiplication also involves computing the sum of products between a row of the first array and a column of the second array. Now, however, there may be several rows in the first matrix and several columns in the second matrix. The product of two matrices **AB** consists of taking the sum of products between each row of **A** and each column of **B**, as follows:

$$\mathbf{A}_{3\times2}\mathbf{B}_{2\times3} = \begin{bmatrix} a_{11} & a_{12} \\ a_{21} & a_{22} \\ a_{31} & a_{32} \end{bmatrix}\begin{bmatrix} b_{11} & b_{12} & b_{13} \\ b_{21} & b_{22} & b_{23} \end{bmatrix}$$

$$= \begin{bmatrix} a_{11}b_{11} + a_{12}b_{21} & a_{11}b_{12} + a_{12}b_{22} & a_{11}b_{13} + a_{12}b_{23} \\ a_{21}b_{11} + a_{22}b_{21} & a_{21}b_{12} + a_{22}b_{22} & a_{21}b_{13} + a_{22}b_{23} \\ a_{31}b_{11} + a_{32}b_{21} & a_{31}b_{12} + a_{32}b_{22} & a_{31}b_{13} + a_{32}b_{23} \end{bmatrix}$$

$$= \mathbf{P}_{3\times 3}$$

The order of the resulting product matrix equals the number of rows of the first matrix **A** and the number of columns of the second matrix **B**, which in this case will be 3×3. In this case, the product **AB** is not of the same order as the product **BA**, which is

$$\mathbf{B}_{2\times 3}\mathbf{A}_{3\times 2} = \begin{bmatrix} b_{11} & b_{12} & b_{13} \\ b_{21} & b_{22} & b_{23} \end{bmatrix} \begin{bmatrix} a_{11} & a_{12} \\ a_{21} & a_{22} \\ a_{31} & a_{32} \end{bmatrix} = \mathbf{P}_{2\times 2}$$

In general, an $n \times k$ matrix times a $k \times m$ matrix gives an $n \times m$ matrix:

$$\mathbf{A}_{n\times k}\mathbf{B}_{k\times m} = \mathbf{P}_{n\times m}$$

For an example of matrix multiplication, let $\mathbf{x}_{5\times 2}$ be a matrix of the deviation scores $x_1 = X_1 - \overline{X}_1$ and $x_2 = X_2 - \overline{X}_2$ given in Table 3.1. $\mathbf{x}'_{2\times 5}\mathbf{x}_{5\times 2}$ is the *sums of squares and cross products* matrix given below.

$$\mathbf{x'x} = \begin{bmatrix} -5 & -3 & 0 & 3 & 5 \\ -3 & -4 & 3 & 0 & 4 \end{bmatrix} \begin{bmatrix} -5 & -3 \\ -3 & -4 \\ 0 & 3 \\ 3 & 0 \\ 5 & 4 \end{bmatrix}$$

$$= \begin{bmatrix} (-5)(-5)+(-3)(-3)+0\cdot 0+3\cdot 3+5\cdot 5 & (-5)(-3)+(-3)(-4)+0\cdot 3+3\cdot 0+5\cdot 4 \\ (-3)(-5)+(-4)(-3)+3\cdot 0+0\cdot 3+4\cdot 5 & (-3)(-3)+(-4)(-4)+3\cdot 3+0\cdot 0+4\cdot 4 \end{bmatrix}$$

$$= \begin{bmatrix} 25 + 9 + 0 + 9 + 25 & 15 + 12 + 0 + 0 + 20 \\ 15 + 12 + 0 + 0 + 20 & 9 + 16 + 9 + 0 + 16 \end{bmatrix} = \begin{bmatrix} 68 & 47 \\ 47 & 50 \end{bmatrix}$$

$$= \begin{bmatrix} \sum (X_1 - \overline{X}_1)^2 & \sum (X_1 - \overline{X}_1)(X_2 - \overline{X}_2) \\ \sum (X_2 - \overline{X}_2)(X_1 - \overline{X}_1) & \sum (X_2 - \overline{X}_2)^2 \end{bmatrix}$$

If we multiply the scalar $1/n$ times $\mathbf{x'x}$, we will get the *variance–covariance* matrix

$$\frac{1}{n}\mathbf{x'x} = \begin{bmatrix} \frac{68}{5} & \frac{47}{5} \\ \frac{47}{5} & \frac{50}{5} \end{bmatrix} = \begin{bmatrix} 13.6 & 9.4 \\ 9.4 & 10.0 \end{bmatrix} = \begin{bmatrix} s_1^2 & s_{12} \\ s_{12} & s_2^2 \end{bmatrix}$$

The Multiple Regression Model

The causal or multiple regression model $Y = \alpha + \beta_1 X_1 + \beta_2 X_2 + \cdots + \beta_k X_k + \varepsilon$ can be represented as the matrix equation

$$\mathbf{Y}_{n\times 1} = \mathbf{X}_{n\times(k+1)}\boldsymbol{\beta}_{(k+1)\times 1} + \boldsymbol{\varepsilon}_{n\times 1} \tag{3A.1}$$

In Equation 3A.1, k equals the number of independent variables and n equals the number of sample cases. **Y** is a column vector of the observed values of Y, **X** is a matrix containing a column of 1's followed by k columns of observations for the k independent variables, ε is a column vector of the disturbances, and $\boldsymbol{\beta}$ is a column vector containing the intercept and k slopes,

$$\boldsymbol{\beta} = \begin{bmatrix} \alpha \\ \beta_1 \\ \beta_2 \\ \vdots \\ \beta_k \end{bmatrix}$$

Without subscripts, Equation 3A.1 is

$$\mathbf{Y} = \mathbf{X}\boldsymbol{\beta} + \boldsymbol{\varepsilon} \tag{3A.2}$$

Thus, Equation 3A.2 is a simplified way to represent a multiple regression equation. You will notice that Equation 3A.2 also differs from the usual notation in that $\boldsymbol{\beta}$ follows rather than precedes **X**. This is required for the proper matrix multiplication, as will be seen below.

The academic salary example will be used to illustrate Equation 3A.2. The equation for this example is $Y = 20 + 1X_1 + 2X_2 + \varepsilon$. The necessary matrices are

$$\mathbf{Y} = \begin{bmatrix} 26 \\ 23 \\ 39 \\ 37 \\ 50 \end{bmatrix} \quad \mathbf{X} = \begin{bmatrix} 1 & 0 & 2 \\ 1 & 2 & 1 \\ 1 & 5 & 8 \\ 1 & 8 & 5 \\ 1 & 10 & 9 \end{bmatrix} \quad \boldsymbol{\varepsilon} = \begin{bmatrix} +2 \\ -1 \\ -2 \\ -1 \\ +2 \end{bmatrix} \quad \boldsymbol{\beta} = \begin{bmatrix} 20 \\ 1 \\ 2 \end{bmatrix}$$

The reason for the column of 1's in **X** will become apparent as the matrix multiplication is carried out below. The matrix equation, shown in full, is

$$\begin{bmatrix} Y_1 \\ Y_2 \\ Y_3 \\ Y_4 \\ Y_5 \end{bmatrix} = \begin{bmatrix} 1 & X_{11} & X_{21} \\ 1 & X_{12} & X_{22} \\ 1 & X_{13} & X_{23} \\ 1 & X_{14} & X_{24} \\ 1 & X_{15} & X_{25} \end{bmatrix} \begin{bmatrix} \alpha \\ \beta_1 \\ \beta_2 \end{bmatrix} + \begin{bmatrix} \varepsilon_1 \\ \varepsilon_2 \\ \varepsilon_3 \\ \varepsilon_4 \\ \varepsilon_5 \end{bmatrix}$$

$$= \begin{bmatrix} 1 & 0 & 2 \\ 1 & 2 & 1 \\ 1 & 5 & 8 \\ 1 & 8 & 5 \\ 1 & 10 & 9 \end{bmatrix} \begin{bmatrix} 20 \\ 1 \\ 2 \end{bmatrix} + \begin{bmatrix} +2 \\ -1 \\ -2 \\ -1 \\ +2 \end{bmatrix} = \begin{bmatrix} 20 \cdot 1 + 1 \cdot 0 + 2 \cdot 2 \\ 20 \cdot 1 + 1 \cdot 2 + 2 \cdot 1 \\ 20 \cdot 1 + 1 \cdot 5 + 2 \cdot 8 \\ 20 \cdot 1 + 1 \cdot 8 + 2 \cdot 5 \\ 20 \cdot 1 + 1 \cdot 10 + 2 \cdot 9 \end{bmatrix} + \begin{bmatrix} +2 \\ -1 \\ -2 \\ -1 \\ +2 \end{bmatrix}$$

$$= \begin{bmatrix} 20 + 1 \cdot 0 + 2 \cdot 2 + 2 \\ 20 + 1 \cdot 2 + 2 \cdot 1 - 1 \\ 20 + 1 \cdot 5 + 2 \cdot 8 - 2 \\ 20 + 1 \cdot 8 + 2 \cdot 5 - 1 \\ 20 + 1 \cdot 10 + 2 \cdot 9 + 2 \end{bmatrix} = \begin{bmatrix} 26 \\ 23 \\ 39 \\ 37 \\ 50 \end{bmatrix}$$

$$= \begin{bmatrix} \alpha + \beta_1 X_{11} + \beta_2 X_{21} + \varepsilon_1 \\ \alpha + \beta_1 X_{12} + \beta_2 X_{22} + \varepsilon_2 \\ \alpha + \beta_1 X_{13} + \beta_2 X_{23} + \varepsilon_3 \\ \alpha + \beta_1 X_{14} + \beta_2 X_{24} + \varepsilon_4 \\ \alpha + \beta_1 X_{15} + \beta_2 X_{25} + \varepsilon_5 \end{bmatrix} = \begin{bmatrix} Y_1 \\ Y_2 \\ Y_3 \\ Y_4 \\ Y_5 \end{bmatrix}$$

The column of 1's in **X** is necessary in order for the intercept α to be included in the expression for each case. Because α is included in the $\boldsymbol{\beta}$ vector, β_0 is sometimes used in place of α, which gives $Y = \beta_0 + \beta_1 X_1 + \beta_2 X_2 + \cdots + \beta_k X_k + \varepsilon$ as the regression equation and allows $\boldsymbol{\beta}$ to contain only β's:

$$\boldsymbol{\beta} = \begin{bmatrix} \beta_0 \\ \beta_1 \\ \beta_2 \\ \vdots \\ \beta_k \end{bmatrix}$$

Matrix Inversion

Rules for adding, subtracting, and multiplying matrices and vectors have been presented. Matrix division, however, is an undefined operation. Although it is not possible to divide matrices, there is a matrix operation that is analogous to division. It involves multiplication with an *inverse* matrix. In order to understand the use of inverse matrices it is helpful to remember that dividing a by b is equivalent to multiplying a by the *reciprocal* of b:

$$\frac{a}{b} = \frac{1}{b} a$$

For example, $\frac{10}{5} = \frac{1}{5} 10 = .2(10) = 2$. We call $\frac{1}{b}$ the *reciprocal* of b. Reciprocals may be written

$$\frac{1}{b} = b^{-1}$$

and

$$b^{-1}b = \frac{1}{b}b = \frac{b}{b} = 1$$

In matrix notation, $\mathbf{B}^{-1}$ is analogous to b^{-1}, but $\mathbf{B}^{-1}$ is called the *inverse* of **B** instead of the reciprocal of **B**. The definition of an inverse matrix states that if

$$\mathbf{CB} = \mathbf{I}$$

then **C** is the inverse of **B**. For example, if

$$\mathbf{CB} = \begin{bmatrix} 4 & -1 \\ -3 & 1 \end{bmatrix} \begin{bmatrix} 1 & 1 \\ 3 & 4 \end{bmatrix} = \begin{bmatrix} 1 & 0 \\ 0 & 1 \end{bmatrix} = \mathbf{I}$$

then **C** is the inverse of **B**, i.e., $\mathbf{C} = \mathbf{B}^{-1}$. Thus, we can write

$$\mathbf{B}^{-1}\mathbf{B} = \mathbf{I}$$

It is also true that $\mathbf{BB}^{-1} = \mathbf{I}$. Thus, a matrix that is premultiplied by its inverse or postmultiplied by its inverse gives the identity matrix **I**. The matrix algebra operation that is analogous to a/b is $\mathbf{B}^{-1}\mathbf{A}$.

Not all matrices have an inverse. First, the inverse of **B** exists only if **B** is a square matrix. Second, not all square matrices have an inverse. If a square matrix does not have an inverse, it is said to be *singular*. We will note below the meaning of singular matrices in the context of regression analysis. Matrices that do have inverses are said to be *nonsingular*.

In order to find the inverse of a matrix (if it exists) a new concept called the *determinant* of a matrix will be introduced. The determinant of **B** may be denoted by either det **B** or $|\mathbf{B}|$. The determinant of a 2×2 matrix equals

$$\det \mathbf{B} = |\mathbf{B}| = \begin{vmatrix} b_{11} & b_{12} \\ b_{21} & b_{22} \end{vmatrix} = b_{11}b_{22} - b_{21}b_{12}$$

For example,

$$|\mathbf{B}| = \begin{vmatrix} 1 & 1 \\ 3 & 4 \end{vmatrix} = 1 \cdot 4 - 3 \cdot 1 = 1$$

This indicates that a determinant equals a single number, that is, it is a scalar.

The determinant of the correlation matrix **R** for two independent variables is

$$|\mathbf{R}| = \begin{vmatrix} 1 & r_{12} \\ r_{21} & 1 \end{vmatrix} = 1 \cdot 1 - r_{21}r_{12} = 1 - r_{12}^2$$

In the case of the academic salary example,

$$|\mathbf{R}| = \begin{vmatrix} 1 & .806 \\ .806 & 1 \end{vmatrix} = 1 - (.806)^2 = .35$$

This indicates that as the correlation between the two X's approaches unity (that is, as it approaches perfect collinearity), the determinant will approach zero. A perfect correlation between the two variables indicates linear dependency and produces a determinant of zero. When the determinant equals 0, the matrix is said to be singular because, as will be seen below, the inverse does not exist. This point can be generalized (without proof) to a correlation matrix for any number of independent variables. As the determinant approaches zero, it indicates increasing linear dependency between one variable and the others or between subsets of the variables. If one variable or a subset of variables is a perfect linear function of the others, there is perfect multicollinearity and the determinant will be zero. Thus, the determinant of the correlation matrix $\mathbf{R}_{k\times k}$ may serve as an inverse indicator of multicollinearity among the k independent variables.

For square matrices with more than two rows and two columns it becomes much more complex to find the determinant. The procedures for finding the determinants of larger matrices will not be presented here but may be found in Namboodiri (1984).

The final concept needed to find the inverse matrix is called the *adjoint* matrix or adj **B**. In the case of a 2 × 2 matrix the adjoint matrix is defined as

$$\text{adj } \mathbf{B} = \begin{bmatrix} b_{22} & -b_{12} \\ -b_{21} & b_{11} \end{bmatrix}$$

This means that the two elements on the main diagonal of **B** are interchanged and the off-diagonal elements are multiplied by -1. For the example **B** used previously, the adjoint is

$$\text{adj } \mathbf{B} = \begin{bmatrix} 4 & -1 \\ -3 & 1 \end{bmatrix}$$

And for the correlation matrix **R**, the adjoint would be

$$\text{adj } \mathbf{R} = \begin{bmatrix} 1 & -r_{12} \\ -r_{21} & 1 \end{bmatrix} = \begin{bmatrix} 1.000 & -.806 \\ -.806 & 1.000 \end{bmatrix}$$

Rules for finding adjoints of larger matrices will not be given here either (see Namboodiri 1984).

Now the inverse of **B** may be calculated as

$$\mathbf{B}^{-1} = \frac{1}{|\mathbf{B}|} \text{adj } \mathbf{B}$$

$$= \frac{1}{1}\begin{bmatrix} 4 & -1 \\ -3 & 1 \end{bmatrix} = \begin{bmatrix} 4 & -1 \\ -3 & 1 \end{bmatrix}$$

In this case, the inverse is equal to the adjoint because the determinant is 1; this, however, is not generally the case. We can see that the calculated inverse is verified by the definition of an inverse:

$$\mathbf{B}^{-1}\mathbf{B} = \begin{bmatrix} 4 & -1 \\ -3 & 1 \end{bmatrix}\begin{bmatrix} 1 & 1 \\ 3 & 4 \end{bmatrix} = \begin{bmatrix} 4\cdot 1 + (-1)3 & 4\cdot 1 + (-1)4 \\ (-3)1 + 1\cdot 3 & (-3)1 + 1\cdot 4 \end{bmatrix} = \begin{bmatrix} 1 & 0 \\ 0 & 1 \end{bmatrix} = \mathbf{I}$$

And the inverse of the correlation matrix is

$$\mathbf{R}^{-1} = \frac{1}{|\mathbf{R}|}\text{adj }\mathbf{R} = \frac{1}{.35}\begin{bmatrix} 1.000 & -.806 \\ -.806 & 1.000 \end{bmatrix} = \begin{bmatrix} 2.857 & -2.303 \\ -2.303 & 2.857 \end{bmatrix}$$

Matrix Solution for *b*

Now we can examine the matrix solution for the least-squares multiple regression equation

$$\mathbf{b} = (\mathbf{X}'\mathbf{X})^{-1}\mathbf{X}'\mathbf{Y} \tag{3A.3}$$

X is the $n \times (k + 1)$ raw score matrix for the k independent variables, and **Y** is the $n \times 1$ raw score matrix for the dependent variable. The formula says to premultiply **X** by its transpose, which gives a matrix of sums of squares and cross-products (SSCP) for the raw scores (not deviation scores). The transpose of **X** is also multiplied by the **Y** vector, which gives a matrix of sums of cross-products of the X's with Y. Then the inverse of the SSCP matrix is multiplied by **X′Y**, which is analogous to dividing the covariation between X and Y by the variation of X. You should be able to see a similarity between this matrix equation for multiple regression and the simple formula for the bivariate regression slope.

Equation 3A.3 will be illustrated with the academic salary example data. The data matrices are

$$\mathbf{Y} = \begin{bmatrix} 26 \\ 23 \\ 39 \\ 37 \\ 50 \end{bmatrix} \quad \mathbf{X} = \begin{bmatrix} 1 & 0 & 2 \\ 1 & 2 & 1 \\ 1 & 5 & 8 \\ 1 & 8 & 5 \\ 1 & 10 & 9 \end{bmatrix} \quad \mathbf{X}' = \begin{bmatrix} 1 & 1 & 1 & 1 & 1 \\ 0 & 2 & 5 & 8 & 10 \\ 2 & 1 & 8 & 5 & 9 \end{bmatrix}$$

and in terms of these matrices, Equation 3A.3 becomes

$$\mathbf{b} = \left(\begin{bmatrix} 1 & 1 & 1 & 1 & 1 \\ 0 & 2 & 5 & 8 & 10 \\ 2 & 1 & 8 & 5 & 9 \end{bmatrix}\begin{bmatrix} 1 & 0 & 2 \\ 1 & 2 & 1 \\ 1 & 5 & 8 \\ 1 & 8 & 5 \\ 1 & 10 & 9 \end{bmatrix}\right)^{-1}\begin{bmatrix} 1 & 1 & 1 & 1 & 1 \\ 0 & 2 & 5 & 8 & 10 \\ 2 & 1 & 8 & 5 & 9 \end{bmatrix}\begin{bmatrix} 26 \\ 23 \\ 39 \\ 37 \\ 50 \end{bmatrix}$$

Notice, however, that **X′X** will be a 3 × 3 matrix. Because the rules for taking the inverse of a 3 × 3 matrix have not been presented in this appendix, we will not be able to demonstrate the use of this equation to calculate the **b** matrix.

If we use a deviation score formula, however, instead of a raw score formula, we will be able to solve for **b**. Let **x** be a matrix of deviation scores for X_1 and X_2 (i.e., $x = X - \overline{X}$), and let **y** be the vector of deviation scores for Y (i.e., $y = Y - \overline{Y}$) (see Table 3.1):

$$\mathbf{x} = \begin{bmatrix} -5 & -3 \\ -3 & -4 \\ 0 & 3 \\ 3 & 0 \\ 5 & 4 \end{bmatrix} \mathbf{y} = \begin{bmatrix} -9 \\ -12 \\ 4 \\ 2 \\ 15 \end{bmatrix}$$

A column of 1's is not included in **x** because the intercept equals zero for the deviation score regression equation. The deviation score matrix formula for $\mathbf{b}_{2\times 1}$ is

$$\mathbf{b} = (\mathbf{x}'\mathbf{x})^{-1}\,\mathbf{x}'\mathbf{y}$$

$$= \left(\begin{bmatrix} -5 & -3 & 0 & 3 & 5 \\ -3 & -4 & 3 & 0 & 4 \end{bmatrix} \begin{bmatrix} -5 & -3 \\ -3 & -4 \\ 0 & 3 \\ 3 & 0 \\ 5 & 4 \end{bmatrix} \right)^{-1} \begin{bmatrix} -5 & -3 & 0 & 3 & 5 \\ -3 & -4 & 3 & 0 & 4 \end{bmatrix} \begin{bmatrix} -9 \\ -12 \\ 4 \\ 2 \\ 15 \end{bmatrix}$$

It was shown previously that **x′x** gives the following SSCP matrix:

$$\mathbf{x}'\mathbf{x} = \begin{bmatrix} -5 & -3 & 0 & 3 & 5 \\ -3 & -4 & 3 & 0 & 4 \end{bmatrix} \begin{bmatrix} -5 & -3 \\ -3 & -4 \\ 0 & 3 \\ 3 & 0 \\ 5 & 4 \end{bmatrix} = \begin{bmatrix} 68 & 47 \\ 47 & 50 \end{bmatrix}$$

$$= \begin{bmatrix} \sum x_1^2 & \sum x_1 x_2 \\ \sum x_2 x_1 & \sum x_2^2 \end{bmatrix}$$

The determinant of the SSCP matrix is

$$|\mathbf{x}'\mathbf{x}| = \sum x_1^2 \sum x_2^2 - \sum x_1 x_2 \sum x_1 x_2$$
$$= 68 \cdot 50 - 47 \cdot 47 = 3400 - 2209 = 1191$$

The adjoint of the SSCP matrix is

$$\text{adj}\,(\mathbf{x'x}) = \begin{bmatrix} \sum x_2^2 & -\sum x_1x_2 \\ -\sum x_2x_1 & \sum x_1^2 \end{bmatrix} = \begin{bmatrix} 50 & -47 \\ -47 & 68 \end{bmatrix}$$

Now, with the determinant and the adjoint, the inverse of the SSCP matrix can be calculated:

$$(\mathbf{x'x})^{-1} = \frac{1}{|\mathbf{x'x}|}\,\text{adj}\,(\mathbf{x'x})$$

$$= \frac{1}{\sum x_1^2 \sum x_2^2 - \left(\sum x_1x_2\right)^2} \begin{bmatrix} \sum x_2^2 & -\sum x_1x_2 \\ -\sum x_2x_1 & \sum x_1^2 \end{bmatrix}$$

$$= \frac{1}{1191} \begin{bmatrix} 50 & -47 \\ -47 & 68 \end{bmatrix}$$

$$= \begin{bmatrix} \dfrac{50}{1191} & \dfrac{-47}{1191} \\ \dfrac{-47}{1191} & \dfrac{68}{1191} \end{bmatrix}$$

The sum of cross products between Y and the X's is

$$\mathbf{x'y} = \begin{bmatrix} -5 & -3 & 0 & 3 & 5 \\ -3 & -4 & 3 & 0 & 4 \end{bmatrix} \begin{bmatrix} -9 \\ -12 \\ 4 \\ 2 \\ 15 \end{bmatrix}$$

$$= \begin{bmatrix} (-5)(-9) + (-3)(-12) + 0\cdot 4 + 3\cdot 2 + 5\cdot 15 \\ (-3)(-9) + (-4)(-12) + 3\cdot 4 + 0\cdot 2 + 4\cdot 15 \end{bmatrix}$$

$$= \begin{bmatrix} 45 + 36 + 0 + 6 + 75 \\ 27 + 48 + 12 + 0 + 60 \end{bmatrix} = \begin{bmatrix} 162 \\ 147 \end{bmatrix}$$

$$= \begin{bmatrix} \sum x_1 y \\ \sum x_2 y \end{bmatrix}$$

Finally, the partial regression slopes can be solved as follows:

$$\mathbf{b} = (\mathbf{x}'\mathbf{x})^{-1}\,\mathbf{x}'\mathbf{y}$$

$$= \begin{bmatrix} \dfrac{\sum x_2^2}{\sum x_1^2 \sum x_2^2 - \left(\sum x_1x_2\right)^2} & \dfrac{-\sum x_1x_2}{\sum x_1^2 \sum x_2^2 - \left(\sum x_1x_2\right)^2} \\ \dfrac{-\sum x_2x_1}{\sum x_1^2 \sum x_2^2 - \left(\sum x_1x_2\right)^2} & \dfrac{\sum x_1^2}{\sum x_1^2 \sum x_2^2 - \left(\sum x_1x_2\right)^2} \end{bmatrix} \begin{bmatrix} \sum x_1 y \\ \sum x_2 y \end{bmatrix}$$

$$= \begin{bmatrix} \dfrac{50}{1191} & \dfrac{-47}{1191} \\ \dfrac{-47}{1191} & \dfrac{68}{1191} \end{bmatrix} \begin{bmatrix} 162 \\ 147 \end{bmatrix}$$

$$= \begin{bmatrix} \dfrac{50 \cdot 162 - 47 \cdot 147}{1191} \\ \dfrac{-47 \cdot 162 + 68 \cdot 147}{1191} \end{bmatrix} = \begin{bmatrix} \dfrac{1191}{1191} \\ \dfrac{2382}{1191} \end{bmatrix} = \begin{bmatrix} 1 \\ 2 \end{bmatrix}$$

$$= \begin{bmatrix} b_1 \\ b_2 \end{bmatrix}$$

The slopes from the deviation score matrix are identical to those computed by the method of residualization in Table 3.2. The intercept can be computed as usual: $a = \overline{Y} - b_1 \overline{X}_1 - b_2 \overline{X}_2$.

Standardized Slopes. A matrix formula for the standardized slopes is

$$\mathbf{B}_{k \times 1} = \mathbf{R}^{-1}_{k \times k}\,\mathbf{r}_{k \times 1}$$

where $\mathbf{R}$ is the correlation matrix for the independent variables and $\mathbf{r}$ is a vector of the correlations of Y with each X. The following formulas for the determinant and adjoint of $\mathbf{R}$ were given previously:

$$|\mathbf{R}| = \begin{vmatrix} 1 & r_{12} \\ r_{21} & 1 \end{vmatrix} = 1 - r_{12}^2$$

$$\text{adj } \mathbf{R} = \begin{bmatrix} 1 & -r_{12} \\ -r_{21} & 1 \end{bmatrix}$$

Inserting the determinant and adjoint in the formula for the inverse of $\mathbf{R}$ gives

$$\mathbf{R}^{-1} = \frac{1}{|\mathbf{R}|} \text{adj } \mathbf{R} = \frac{1}{1 - r_{12}^2} \begin{bmatrix} 1 & -r_{12} \\ -r_{21} & 1 \end{bmatrix} = \begin{bmatrix} \dfrac{1}{1 - r_{12}^2} & \dfrac{-r_{12}}{1 - r_{12}^2} \\ \dfrac{-r_{21}}{1 - r_{12}^2} & \dfrac{1}{1 - r_{12}^2} \end{bmatrix}$$

If we now enter the inverse in the formula for **B** and multiply, we get

$$\mathbf{B} = \mathbf{R}^{-1}\mathbf{r} = \begin{bmatrix} \dfrac{1}{1 - r_{12}^2} & \dfrac{-r_{12}}{1 - r_{12}^2} \\ \dfrac{-r_{21}}{1 - r_{12}^2} & \dfrac{1}{1 - r_{12}^2} \end{bmatrix} \begin{bmatrix} r_{Y1} \\ r_{Y2} \end{bmatrix} = \begin{bmatrix} \dfrac{r_{Y1}}{1 - r_{12}^2} - \dfrac{r_{12}r_{Y2}}{1 - r_{12}^2} \\ \dfrac{-r_{21}r_{Y1}}{1 - r_{12}^2} + \dfrac{r_{Y2}}{1 - r_{12}^2} \end{bmatrix}$$

$$= \begin{bmatrix} \dfrac{r_{Y1} - r_{12}r_{Y2}}{1 - r_{12}^2} \\ \dfrac{r_{Y2} - r_{21}r_{Y1}}{1 - r_{12}^2} \end{bmatrix} = \begin{bmatrix} B_1 \\ B_2 \end{bmatrix}$$

The formulas for the B's in **B** are equivalent to those given in Table 3.8 and those used in the discussion of redundancy and suppression.

We can now solve for the standardized slopes in the academic salary example by substituting the inverse of **R**, which was previously determined, into the matrix formula for the B's:

$$\mathbf{B} = \mathbf{R}^{-1}\mathbf{r} = \begin{bmatrix} 2.857 & -2.303 \\ -2.303 & 2.857 \end{bmatrix} \begin{bmatrix} .906 \\ .959 \end{bmatrix}$$

$$= \begin{bmatrix} 2.857(.906) - 2.303(.959) \\ -2.303(.906) + 2.857(.959) \end{bmatrix} = \begin{bmatrix} .38 \\ .65 \end{bmatrix}$$

Reference

Namboodiri, Krishnan. 1984. *Matrix Algebra: An Introduction*. Beverly Hills: Sage Publications.

4 Tests of Statistical Significance

In Chapters 2 and 3 we discussed least-squares regression and correlation coefficients that describe various aspects of the linear relationships between dependent and independent variables. These statistics are often computed from a probability sample of observations in order to make inferences or decisions about parameters of the population from which the sample was drawn. The most common hypothesis to be tested in regression analyses is whether a linear regression slope equals zero in the population.

In order to understand how such decisions are made, in the first two sections of this chapter we emphasize two basic sampling concepts: the sampling distribution and the standard error of the least-squares slope. Some of the characteristics of the sampling distribution are first illustrated with a small, concrete example. Then the properties of the theoretical sampling distribution of the slope are described in general terms (i.e., the mean, the standard deviation, and the shape of the distribution). The characteristics of the sampling distribution are based on several assumptions about the error term in the population equation, assumptions that are discussed in some detail in order to provide a firm foundation for hypothesis testing. If these assumptions are true, the Gauss-Markov theorem provides the justification for using the least-squares criterion to compute the regression slope.

Knowledge about the sampling distribution is used to define a *t* statistic that can be used, for example, to test the hypothesis that the slope equals zero in the population (a two-tailed test). Formulas for the standard errors of both the bivariate and partial slopes are discussed and used to compute the sample *t* statistic. If the probability of obtaining the sample t value is very small when the population slope equals zero, we reject the hypothesis that the population slope equals zero. The rejection of this hypothesis becomes problematic, however, when there is high multicollinearity—that is, when the independent variables are highly intercorrelated. Although multicollinearity is a problem that must be understood in order to make intelligent use of multiple regression, it does not lead to biased decision making. With one exception, however, violations of the regression assumptions lead to biased hypothesis tests. The nature of each bias is described.

Next, we present a general formula for a very flexible F statistic that can be used in hypothesis testing. The F statistic can be used as an alternative to the t statistic for testing the hypothesis that the population slope equals zero. The F statistic also allows us to test hypotheses about sets of variables, such as whether the squared multiple correlation is zero and whether a subset of variables has any relationship to the dependent variable.

In the final section of the chapter we consider, in some detail, how to detect and correct for the violation of an important regression assumption, the assumption that the error variance is homoskedastic. Both graphic and statistical methods of detecting heteroskedasticity are discussed, and then two methods of correcting for significant heteroskedasticity are described. The first involves correcting the standard error of the least-squares slope so that an unbiased t test can be conducted. The second method involves using weighted least-squares instead of ordinary least-squares to estimate the population slope. Weighted least-squares can provide a more efficient estimate of the population slope when heteroskedasticity is present.

Illustrative Sampling Distribution of *b*

The *population* equation for a linear causal model with one independent variable is

$$Y = \alpha + \beta X + \varepsilon \qquad \textbf{(4.1)}$$

Since the population is usually too large to measure X and Y for all elements (cases) in the population, β is usually estimated from a sample of n cases drawn from the population. The sample is selected by some probability method, such as **simple random sampling**. In simple random sampling every element in the population has an equal chance (probability) of being selected. After the sample is drawn and X and Y are measured, the least-squares bivariate regression slope b is calculated and used as the best estimate of β.

When a random sample is selected without replacement from a population of N elements, there will be a number of distinct combinations of n elements that might be selected into the sample. One sample is considered to be the same as another sample if and only if they contain exactly the same n cases.

The number of distinct combinations of sample elements when, for example, $n = 5$ and $N = 10$ equals

$$\frac{N!}{n!(N-n)!} = \frac{10 \cdot 9 \cdot 8 \cdot 7 \cdot 6 \cdot 5 \cdot 4 \cdot 3 \cdot 2 \cdot 1}{(5 \cdot 4 \cdot 3 \cdot 2 \cdot 1)(5 \cdot 4 \cdot 3 \cdot 2 \cdot 1)} = 252$$

But if the population is larger—for example, $N = 50$—and the sample size remains $n = 5$, the number of distinct samples increases to 2,118,760! Thus, it can be seen that for large populations, the number of distinct combinations of elements that might be selected in a sample of size n will be incredibly large.[1]

Many of the combinations of n cases that might be selected will have sample values of b that are equal or nearly equal to the population parameter β. Because random sampling is used, however, a few of the samples that might result will have values of b that are much larger than β, and a few will have values of b that are much smaller than β. In fact, there will be a wide range of values of the sample b that might result by chance, although some values will be much more likely than others. The fact that some values of the sample statistic b will be different from the population parameter β is called **sampling error**.

The distribution of the values of b for all distinct combinations of elements that might occur in a random sample of size n is called the **sampling distribution of *b***. The concept of a sampling distribution will be illustrated with the hypothetical example from Chapter 3 dealing with the academic marketplace. We will use the following bivariate equation in which X = Experience and Y = Earnings:

$$Y = 20 + 1X + \varepsilon$$

Using the values of X and ε from Table 3.1, the above equation is used to generate the following values of Y for each value of X:

Case	**X**	ε	**Y**
1	0	+2	22
2	2	−1	21
3	5	−2	23
4	8	−1	27
5	10	+2	32

Now imagine that these five cases represent the entire population of values of X and Y. Thus, $N = 5$, and the population parameter that we wish to estimate

1. It should be noted, however, that not every possible sample can be used to estimate the population slope. If every element in the sample happened to have the same value of X, then the variance of X would be zero and the slope could not be computed. If X had a reasonable amount of variance in the population, the probability of drawing a sample where the variance of X is zero would be very small.

is $\beta = 1$. This example is unrealistic in two senses. First, we would not normally take a sample from a population of only five elements. Second, we would never know the values of X and Y for all elements of the population from which we wish to take a sample; otherwise, there would be no need to take a sample. These extreme simplifications, however, are helpful in illustrating the nature of a sampling distribution.

Let us first assume that we wish to take a sample of $n = 2$ from this population. There are $5!/(2!\,3!) = 10$ distinct pairs of elements that could be selected by chance in a random sample of $n = 2$ from this population. These are listed below, along with the value of the least-squares regression slope b that has been computed for each pair of cases.[2]

$n = 2$

Cases	b
1, 2	−.500
1, 3	.200
1, 4	.625
1, 5	1.000
2, 3	.667
2, 4	1.000
2, 5	1.375
3, 4	1.333
3, 5	1.800
4, 5	2.500
Sum	10.000

You should verify that you know how to calculate these slopes.

In a random sample of $n = 2$, any one of these ten different pairs of elements might have been selected and the corresponding value of b used as an estimate of the population parameter $\beta = 1$. Each of the distinct pairs has an equal probability of being selected. You can see that two of the samples produced $b = 1$, which would have correctly estimated the population parameter if either one of these samples had been selected. The other eight samples would have produced b's that differed from β, some by not too much (e.g., less than $\pm .4$) and a couple by a large amount (± 1.5). This difference is the sampling error. In a realistic research instance, you would not be looking at the results for ten samples and you would not know the value of β. You would have only one of the above ten sample b's, and you would not know whether it was right on target, close, or way off the mark.

The above values of b represent the sampling distribution of b for a sample size of $n = 2$ from a population of size N. The sampling distribution of b is shown in Figure 4.1, where the height of the bar above each sample value of b indi-

2. The sample slope b can be computed with Equation 2.6. More simply, it equals the difference in Y for the pair of coordinates divided by the corresponding difference in X.

FIGURE 4.1 Sampling Distribution for $n = 2$

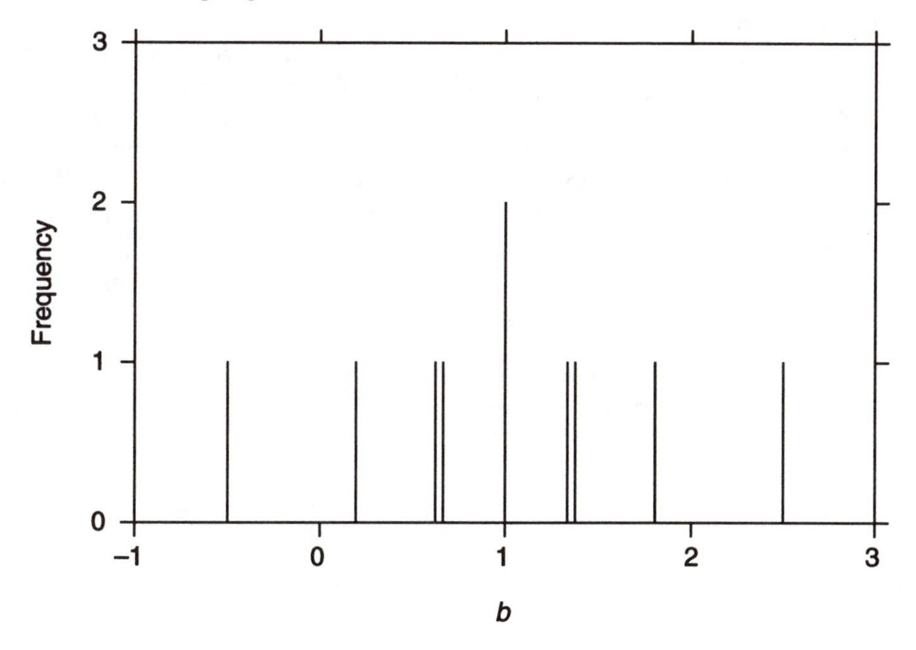

cates its frequency. This distribution is symmetric around the value $b = 1$. The most frequently occurring value is $b = 1$. The farther away you move from $b = 1$ in either direction, the fewer are the sample values that are encountered. Looking back at the ten sample values of b, we see that the sum is equal to 10. Therefore, the mean of the b's is 1. Thus, $\bar{b} = \beta$. This indicates that the mean of the sampling distribution of b equals the population parameter β. The distribution of the b's is symmetric around the population β, and the b's are more likely to be clustered near β than far from it. Thus, the sampling distribution is roughly "bell"-shaped.

In addition to central tendency (i.e., the mean), a second important characteristic of any distribution is a measure of dispersion, such as the standard deviation. Using the usual definitional formula for the standard deviation but substituting b for X, we get

$$s_b = \sqrt{\frac{\sum (b - \bar{b})^2}{10}} = .79$$

Thus, the standard amount by which the sample slopes from all of the distinct samples of size $n = 2$ deviate from their mean is about .8. The standard deviation of a sampling distribution is called the **standard error**; in this case, it is called the *standard error of b*. It indicates the amount that b is likely to deviate from β due to sampling error (i.e., chance).

It will be instructive to look at one more sampling distribution for our hypothetical salary data. If we take samples of size $n = 3$ from the same population of $N = 5$ cases, there will also be ten distinct combinations of cases that each have an equal probability of being selected as our sample, i.e., $5!/(3!\,2!) = 10$. These ten samples, along with the values of the least-squares b for each sample, are given below. The sum of these ten sample slopes is 10, and thus the mean of b is also 1, just as it was for samples of size $n = 2$. For both $n = 2$ and $n = 3$, the mean of the sampling distribution of b is equal to β.

$n = 3$

Cases	*b*
1, 2, 3	.237
1, 2, 4	.712
1, 2, 5	1.107
1, 3, 4	.582
1, 3, 5	1.000
1, 4, 5	.893
2, 3, 4	1.000
2, 3, 5	1.418
2, 4, 5	1.288
3, 4, 5	1.763
Sum	10.000

The bar graph of the sampling distribution for $n = 3$ is shown in Figure 4.2. It is symmetric and "bell"-shaped, just as the sampling distribution for $n = 2$ was. The distribution of b's for $n = 3$, however, is not as dispersed as that for $n = 2$. It is clustered more closely around the mean of the distribution, indicating that sampling errors are not as great. For example, there are no sample values of b above 2 or below 0, as there were for the distribution when $n = 2$. The standard deviation of the sample b's for $n = 3$ is only $s_b = .41$, as compared to a value of .79 when $n = 2$. Thus, the standard error was cut in half by increasing the sample size from 2 to 3.

Theoretical Sampling Distribution of *b*

Characteristics of the Distribution of *b*

In the above hypothetical examples we were able to see the actual values of the least-squares b's for all distinct samples, as well as the shape of the distri-

FIGURE 4.2 Sampling Distribution for $n = 3$

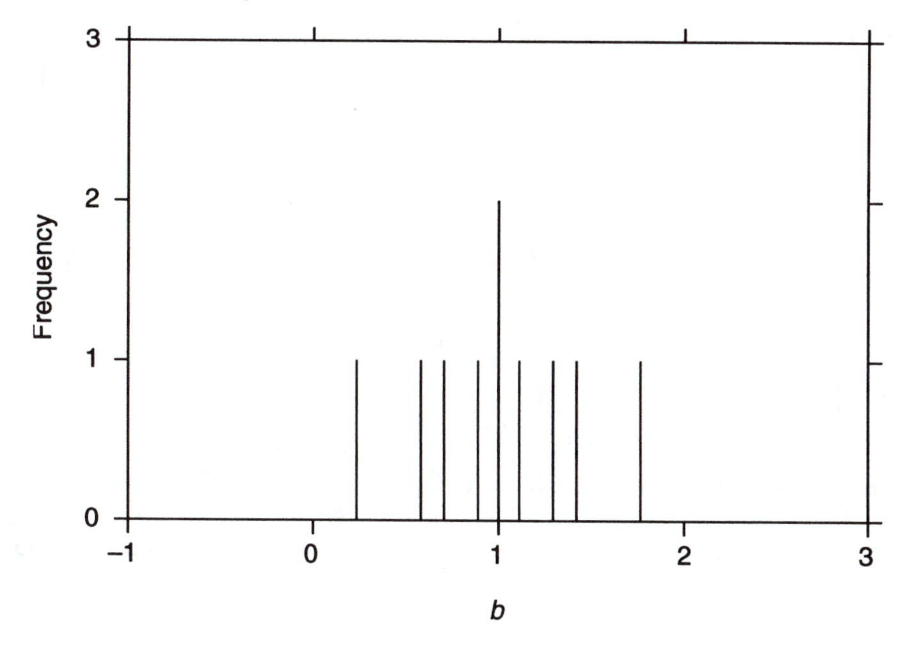

bution (i.e., the frequency or percentage for each value of b). In applied regression analysis, however, the sampling distribution of b for any pair of real X's and Y's is not an empirical distribution that can be directly observed. The sampling distribution of b is a theoretical distribution derived by use of mathematical theory. It does not list or describe all of the numerical values that the sample slopes may have. This is impossible because the population values of X and Y are unknown. If we knew the values of X and Y for all elements in the population, we would not select a sample but would use the population values to compute b directly. The theoretical sampling distribution of b, however, does provide useful information about the mean, standard deviation, and shape of the distribution of sample slopes. This theoretical knowledge about the aggregate characteristics of all the possible sample values is immensely useful in helping us to determine the probability that any single sample value is in error.

A diagram of the theoretical sampling distribution of b is shown in Figure 4.3. When certain standard assumptions (to be described below) are made about ε in Equation 4.1, the sampling distribution of b for samples of size n (Figure. 4.3) has the following characteristics:

FIGURE 4.3 The Sampling Distribution of the Least-Squares Slope *b*

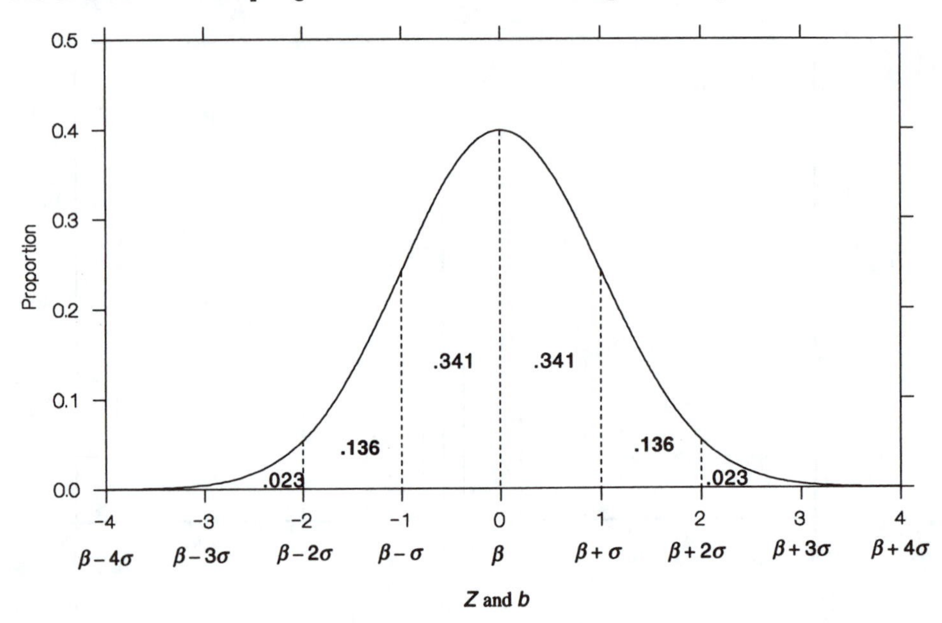

$$\mu_b = \beta \qquad \textbf{(4.2)}$$

$$\sigma_b = \frac{\sigma_\varepsilon}{s_X\sqrt{n}} \qquad \textbf{(4.3)}$$

$$b \sim N \qquad \textbf{(4.4)}$$

The Greek symbols in Equations 4.2 and 4.3 indicate that these formulas contain true population parameters; β and σ_ε are parameters of the population of all *elements* from which we wish to draw a sample, whereas μ_b and σ_b are parameters of the true sampling distribution, the distribution of *b*'s from the population of all distinct *samples* of size *n*.

Equation 4.2 indicates that the mean of the sampling distribution of *b* is equal to β, the population slope. (The Greek letter μ, named Mu, stands for a population mean.) That is why β is at the middle of the distribution in Figure 4.3. Although we don't know what the value of β is, it is comforting to know that the mean of the slopes for all possible samples that might be selected is equal to the parameter that we desire to estimate. This characteristic means that *b is an unbiased estimate of* β. "Unbiased" clearly does not mean that our computed sample b will be equal to β. It means that if we think of β as a target at which we are shooting with our sample b's, the b's will be centered around β rather than tending to be on one side or the other. Thus, the sample b's are just as likely to be greater than β as they are to be less than β; on the average, they will be right on target.

Equation 4.3, the second characteristic of the sampling distribution, gives a formula for the standard deviation of the b's, which you will remember is called the *standard error* of b (because the variance of b represents sampling error).[3] (The Greek letter σ, named Sigma, indicates a population standard deviation.) Equation 4.3 shows several factors that influence the amount of sampling error. It shows that the greater the standard deviation of ε (the numerator), the greater will be the standard error of b. This means that the smaller the variance created by the other causes of Y, the more reliably we will be able to estimate the effect of X. Thus, the effects of other causes of Y (i.e., summarized by ε) increase sampling error with respect to b, making us less certain about how well we are estimating β.

Equation 4.3 also shows that the standard error of b is inversely related to the square root of n. Thus, we can reduce sampling error in our estimate of β by increasing the sample size. We saw this in our hypothetical examples of sample sizes of $n = 2$ and $n = 3$. However, note that it is the square root of n that reduces the standard error. Thus, there are diminishing returns to increasing sample size. The reduction in sampling error resulting from a given increase in sample size will be greater when the additional cases are added to a small sample than when they are added to a large sample.

Finally, Equation 4.3 shows that the standard error of b is inversely related to the sample standard deviation of X. The greater the variance of X, the smaller will be the sampling error. Thus, when we have a wide range of values of X with which to estimate β, the estimates of β will bounce around less from sample to sample than when we must use a narrow band of X values to make an estimate.

To summarize, Equation 4.3 shows that the standard error of the sampling distribution is positively related to the standard deviation of ε and inversely related to both the standard deviation of X and sample size.[4] Although the sample size and the standard deviation of X are known, the standard deviation of ε is not known because σ_ε is a population parameter. Thus, although the identity indicated by Equation 4.3 is true (given certain assumptions to be described below), we can't use the formula to compute the standard error of b directly. Equation 4.3 can, however, be modified slightly to allow us to estimate

3. Although our emphasis is on the slope, the parameters of the sampling distribution of a are $\mu_a = \alpha$ and $\sigma_a = \sigma_\varepsilon \sqrt{1/n + \overline{X}^2/\Sigma (X - \overline{X})^2}$.

4. Equation 4.3 readily simplifies to $\sigma_b^2 = \sigma_\varepsilon^2/\Sigma (X - \overline{X})^2$, which is often presented by other texts. We prefer Equation 4.3 because it explicitly shows the contributions of s_X and n.

the standard error of b empirically with a single sample of elements from the population.[5]

The third characteristic of the sampling distribution (Equation 4.4) is that b has a normal distribution (N means *normal* rather than population size). This characteristic of b will be used when conducting hypothesis tests. Figure 4.3 shows some characteristics of the normal distribution of b. For instance, about two-thirds of the potential sample b's will be within one standard deviation from the mean of these b's (i.e., $\beta \pm \sigma$). Also, about 5% of the b's will be *more* than ± 2 standard deviations from the mean (the exact value is $\pm 1.96\sigma$). The Z scores shown under the horizontal axis are equal to

$$Z_b = \frac{b - \beta}{\sigma_b} \qquad \textbf{(4.5)}$$

This is a standardized variable because the mean of the slopes β is subtracted from each possible sample b and the resulting deviation score $(b - \beta)$ is divided by the standard deviation of the slopes. Since b is normally distributed, Z also has a normal distribution. Therefore, a capital Z is used to indicate that this is a *standard normal* score.

Gauss-Markov Theorem. Not only is b unbiased (Equation 4.2) and normally distributed (Equation 4.4) with a standard error given by Equation 4.3, the *Gauss-Markov* theorem states that the least-squares regression equation provides the *B*est *L*inear *U*nbiased *E*stimate (BLUE) of the population slope. *Best* means that the least-squares slope will be a more efficient estimator of β than any other unbiased linear estimator of β. The *efficiency* of a sample statistic (i.e., an estimator) refers to the size of its standard error; the smaller the standard error, the greater the efficiency. More efficient estimators are those whose values vary less from sample to sample. The fact that the least-squares estimator is more efficient than any other unbiased method of estimating β is the major justification for choosing the regression line that minimizes the sum of squared errors of prediction (the squared residuals). Remember that in Chapter 2 we compared the fit of the least-squares line with that of an "educated guess" line. Although both lines fit the example data equally well in terms of the sum of the absolute values of the residuals, it was said that the least-squares line was superior because it was less susceptible to sampling errors. We now know that this claim is based on the Gauss-Markov theorem, which states that it minimizes

5. Equation 4.3 is correct when X is a *fixed* variable, that is, when the values of X are under the control of the investigator so that the same values would be observed in each sample (i.e., s_X would be constant from sample to sample). In the social sciences it is more often the case that X is not controlled by the investigator. In this case, X is a random or *stochastic* variable; its values are a random sample of those occurring naturally in the population from which the sample is drawn. When X is stochastic, the sampling variance of b is $\sigma_b^2 = \sigma_\varepsilon^2/\sigma_X^2(n - 2)$ instead of the square of Equation 4.3 (Greene 1990, p. 324). Whether X is fixed or stochastic, however, the formula that will be used for *estimating* σ_b is the same.

the standard deviation of the sampling distribution, i.e., the standard error. Thus, the least-squares b will have a smaller standard error than an estimator of β that minimizes the absolute values of the residuals.

Assumptions About ε

The characteristics of the sampling distribution of b that have been described depend upon a series of assumptions about the error term ε in the population equation $Y = \alpha + \beta X + \varepsilon$. First, in order to ensure that b is an unbiased estimator, we must assume that

(A.1) ε is uncorrelated with X

This assumption is necessary in nonexperimental research.[6] If X and ε were correlated, that would mean our measured independent variable X was correlated with one or more other causes of Y that are not included in the equation. In that case, we would not be able to control or hold them constant in order to estimate the effect of X that is independent of the other causes (ε). Thus, if the omitted causes of Y are correlated with X because they are causes of X (i.e., ε is a common cause of X and Y), b is a biased estimate of β. This problem is also sometimes referred to as *spuriousness*. Furthermore, if the omitted causes were merely correlated with X but were neither causes of X nor caused by X, b would also be a biased estimate of β. However, if the other cause(s) of Y were intervening variables between X and Y, b would not be a biased estimate of β. In this case, however, b would represent the total effect of X, that is, the direct effect that is unmediated by the other causes plus the indirect effect that passes through the intervening variables. This latter point is not mentioned by most texts, which usually make a blanket statement that it must be assumed that X is not correlated with ε. Thus, the more precise assumption is that *the variance in ε that is not caused by X is uncorrelated with X*. This issue can be only briefly introduced at this point and will be discussed in more detail in Chapter 8.

In order to justify Equation 4.3 for the standard error of b, there are two additional assumptions that must be made about ε. First, it is assumed that

(A.2) ε has the same variance for all values of X

The conditional variance of ε is constant across all values of X. If ε has a constant variance for all values of X, it is said to be *homoskedastic*. Homoskedasticity means that the predicted values of Y are equally good (or poor) at all levels of X; that is, the average squared error of prediction does not vary as X changes (Figure 4.4a). If the variance of ε is greater for some values of X than for others, *heteroskedasticity* exists. For example, heteroskedasticity would exist if the

6. In this case, X is *stochastic*, i.e., the observed values of X are not under the control of the investigator, as they are in an experimental design. When X is *fixed* by the investigator, the assumption that X and ε are uncorrelated is not necessary.

FIGURE 4.4 (a & b) Illustrative Patterns of Homoskedasticity and Heteroskedasticity

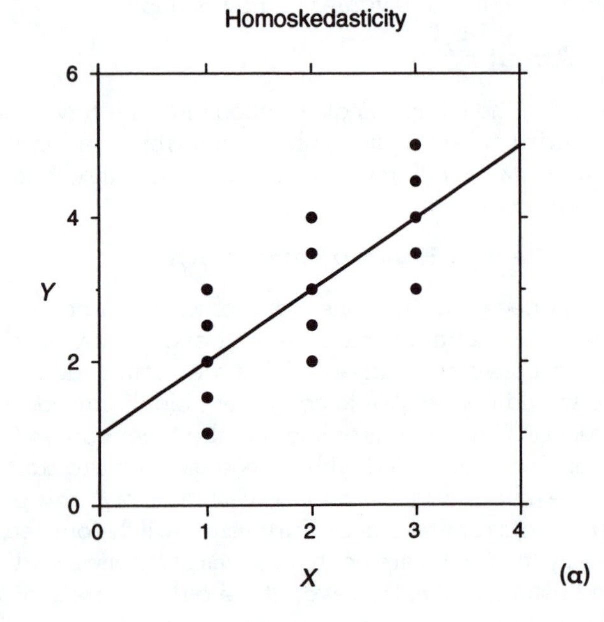

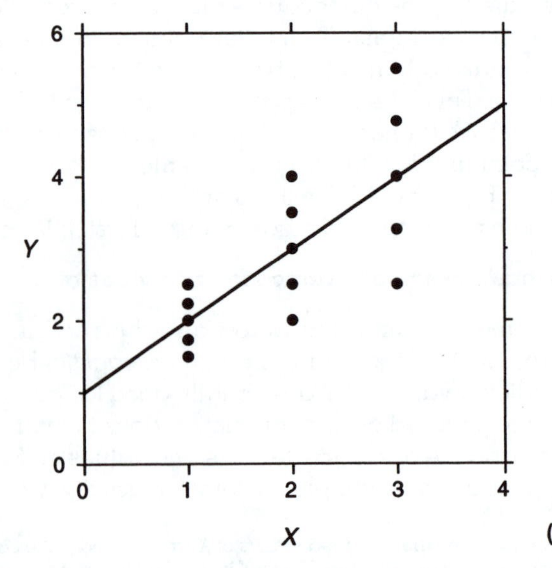

equation makes poorer predictions of Y (large variance of ε) for high values of X than for low values of X (Figure 4.4b); in this case there is a positive correlation between X and the variance of ε. Or the variance of ε might be greater for low levels of X than for high levels, i.e., a negative correlation between X and the variance of ε. Other patterns of heteroskedasticity also are possible.

Notice in Figures 4.4a and b that the slope of the line is the same for the heteroskedastic pattern of ε's as it is for the homoskedastic pattern. This is intended to illustrate the more general point that even if heteroskedasticity exists, the least-squares b will still be an *unbiased* estimate of β. That is, the mean of the sampling distribution of b will still equal β even if the homoskedasticity assumption is violated.

One problem created by heteroskedasticity is that Equation 4.3 no longer correctly describes the standard error of b. The true standard error of b may either be larger or smaller than the magnitude indicated by Equation 4.3; or, looked at from the opposite direction, Equation 4.3 will either overstate or understate the true standard deviation of the sampling distribution. If, for example, there is a positive correlation between X and the variance of ε, the value of the standard error as given by Equation 4.3 will be too small; that is, the sample estimator of β will not be as efficient as Equation 4.3 would indicate. On the other hand, if there is a negative correlation between X and the variance of ε, the standard error given by Equation 4.3 will be too large. In sum, Equation 4.3 is an incorrect formula for the standard error of the least-squares b when heteroskedasticity is present.

There is also another problem that results from violation of the homoskedasticity assumption. When heteroskedasticity is present, the Gauss-Markov theorem is no longer valid. That is, the least-squares estimate of β is no longer BLUE. There may be other linear unbiased methods of estimating β that provide more efficient estimates, i.e., that have smaller standard errors. Even if we correct Equation 4.3 for the standard error, there are other estimation techniques, such as *weighted least-squares* (to be described later), that will provide more efficient estimates than *ordinary least-squares.* To summarize the difficulties created by heteroskedasticity, the usual formula for the least-squares standard error is wrong, and the least-squares slope is no longer the most efficient (i.e., best) unbiased estimate of β.

A second assumption that must be made in order to justify Equation 4.3 is that

(A.3) the values of ε are independent of one another

When the independence assumption is violated, it is called *autocorrelation* or *serial correlation.* In order to understand autocorrelation, it is necessary to imagine some systematic basis for forming pairs of elements such that observations in each pair would tend to have similar values of ε (called positive autocorrelation) or such that observations in each pair would have dissimilar values of ε (negative autocorrelation). Temporal, physical, or social proximity are probably the most common factors that cause autocorrelation in a sample of observations or elements. Time-series designs, in which a single object is observed repeatedly over an extended period of time, is a research design in which autocorrelation is typically found. Observations taken adjacent to one another in time are likely to have similar ε's because the various unmeasured causes of Y that are summarized by ε tend to change slowly from one point in time to the next. Cluster sampling designs, in which geographic clusters of elements are selected in the first stage and then several elements are selected from each cluster, are also characterized by autocorrelation. People living in the same cluster, such as a neighborhood or block, are likely to have similar values of ε because of the existence of residential segregation based on socioeconomic status, race, and ethnicity. Another situation in which autocorrelation would exist is in samples composed of pairs of individuals involved in some type of social relationship, such as married couples. No matter how many independent variables are included in the equation, there would probably be additional unmeasured sources of similarity between husbands and wives that would cause ε to be correlated for couples. The key factor producing autocorrelation is nonindependence between some of the observations as a result of some type of proximity.

The consequences of violation of the nonautocorrelation assumption are the same as for violation of the homoskedasticity assumption. When autocorrelation is present, the least-squares b is still an unbiased estimator of β. Equation 4.3, however, will either overstate or understate the standard error of b, depending upon the nature of the autocorrelation. Also, the least-squares estimate of β will not be the most efficient linear unbiased estimator, i.e., b will not be the BLUE estimator.

The third characteristic of the sampling distribution (Equation 4.4) is that b is normally distributed. It is based on the assumption that

(A.4) ε **is normally distributed**

Thus, if ε is not normally distributed, b will not, in general, be normally distributed. However, b will still be the most efficient linear unbiased estimator of β (i.e., BLUE) if the homoskedasticity and independence assumptions hold. Equation 4.3 for the standard error of b will also still be valid. Moreover, if ε is not normally distributed, the sampling distribution of b will become increasingly similar to a normal distribution as *sample size increases*, as long as the other assumptions are true. That is, the sampling distribution of b will be approximately normal for large samples when ε is not normally distributed. Thus, when we are working with the relatively large samples that are often characteristic of social science research (where there may be several hundred cases or more), the sampling distribution of b will be approximately normal no matter what kind of distribution ε has.

t Tests

Bivariate Slopes

Since the true value of the standard error of *b* is unknown, it must be estimated from the single sample of observations that is available. The estimated standard error of *b* is

$$s_b = \frac{\hat{s}_{Y-\hat{Y}}}{s_X\sqrt{n}} \tag{4.6}$$

An alternate formula is

$$s_b = \frac{s_Y\sqrt{1 - r^2}}{s_X\sqrt{n - 2}} \tag{4.7}$$

These formulas use Roman letters to indicate that they are sample estimates of the standard error. Equation 4.6 is produced by substituting the sample statistic $\hat{s}_{Y-\hat{Y}}$ (standard error of estimate) for the population parameter in Equation 4.3. Although Equation 4.6 is actually the estimated standard error, unless stated otherwise, it will be referred to subsequently as simply the standard error.

The standard error of *b* can be used in a *t* test to evaluate the probability that sampling error may be causing the sample *b* to deviate from the population β. A *t* test can be defined for *any* sample statistic according to the following general formula:

$$t = \frac{\text{(Sample Statistic)} - \text{(Population Parameter)}}{\text{(Estimated Standard Error of the Statistic)}}$$

This is just a formula for a standardized score; in this case mean value of the statistic for all possible samples (the population parameter) is subtracted from the value of the statistic in a single sample, and the resulting deviation score is divided by the estimated standard deviation for all samples.

This formula can be used to get the *t* statistic for the mean as follows:

$$t = \frac{\overline{X} - \mu_X}{s_{\overline{X}}}$$

The formula can also be used to get *t* for the difference between two means:

$$t = \frac{(\overline{X}_1 - \overline{X}_2) - (\mu_1 - \mu_2)}{s_{(\overline{X}_1 - \overline{X}_2)}}$$

Most importantly for our purposes, it can be used to get *t* for the sample slope *b*, which is

$$t = \frac{b - \beta}{s_b} \tag{4.8}$$

As a special type of standard score, this t indicates how many standard deviations the sample b is above or below the mean of the sampling distribution of b.

Although b is distributed normally and Z in Equation 4.5 also has a normal distribution, t does not have a normal distribution. This is because instead of dividing by the true standard error of b (σ_b), the denominator of Equation 4.8 is the estimated standard error of b, which adds a degree of imprecision to the t statistic. Thus, the t distribution is flatter than the normal distribution. For example, there is a greater proportion of values greater than ± 2 standard deviations from the mean in the t distribution than in a normal distribution. Furthermore, there is a whole family of t distributions, one for each sample size n. As n gets smaller, the t distribution becomes increasingly flatter. But as n gets larger, it becomes increasingly similar to the normal distribution. When n is greater than 200, there is little difference between a t distribution and a normal distribution.

The t statistic is used most frequently in regression to test the null hypothesis that the population slope is zero. The null hypothesis and a frequent alternative hypothesis are

$$H_0: \beta = 0$$

$$H_1: \beta \neq 0$$

The alternative hypothesis H_1 states that the effect of X on Y (β) may be either positive or negative; it does not predict the sign, or direction, of the effect. The test for this hypothesis is called a **two-tailed** test.

The t statistic for testing this hypothesis is computed under the assumption that H_0 is true, such that

$$t = \frac{b - \beta}{s_b} = \frac{b - 0}{s_b} = \frac{b}{s_b} \qquad \textbf{(4.9)}$$

This t indicates the number of standard deviations that b is away from the mean of the sampling distribution, if $\beta = 0$. The t statistic is used to gain some protection against a *Type 1 error*, which is rejecting the null hypothesis when it is true. We can never be totally certain that if we reject the null hypothesis we will not be making a Type 1 error. Therefore, there is always some risk in drawing such a conclusion or in making such an inference. Social scientists, however, want to keep the probability of drawing an erroneous conclusion quite small. The most common probability of a Type 1 error with which social scientists are willing to live is .05; they want to have no more than a 5% chance of being wrong if they reject the null hypothesis. The maximum risk that the investigator is willing to take is called the *α level*, which is not to be confused with the intercept α in Equation 4.1. In formal hypothesis testing, the α level is stated prior to computing the test statistic t.

FIGURE 4.5 Cumulative Probability for a Standard Normal Score of 1.65

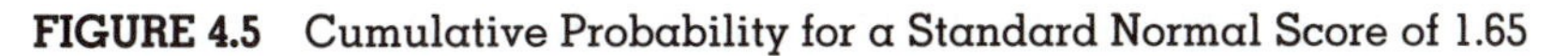

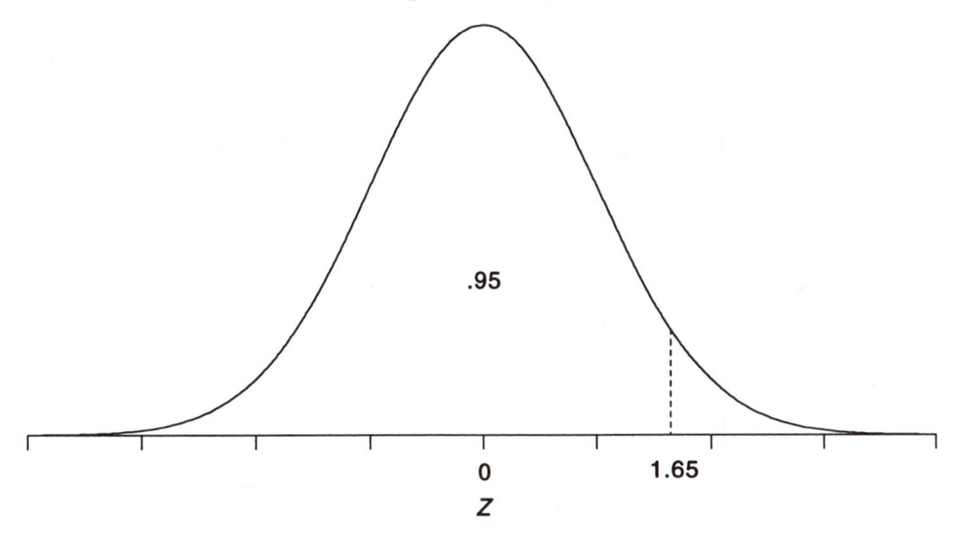

The t score that has been computed for a sample b can be used to find the probability of a Type 1 error. This probability (for a two-tailed test) is equal to the proportion of all sample b's that would be more than $\pm t$ standard deviations from the mean of the sampling distribution of b (which is hypothesized to equal 0). You can also think of this probability as the proportion of all b's that would have an absolute value greater than or equal to the absolute value of the sample b when the true slope equals 0. Since there is a different t distribution for each sample size n, however, most t tables give only a few critical values of t and their associated probabilities (such as .05 and .10). Thus, it is usually not possible to determine the exact probability of a Type 1 error from the t table but only to learn whether $p \leq .05$ or $p \leq .10$, for example. Most computer programs (such as SPSS) will print the exact probability, which is called the **significance level**.

Since the t distribution is approximately normal when n is large, the standard normal distribution may be used to get a very close approximation to the exact probability of a Type 1 error. Some standard normal tables report the cumulative probabilities for values of Z ranging from about -4 to $+4$. Figure 4.5 illustrates what these probabilities mean.

If you looked up $Z = 1.65$ in a standard normal table, it would show a probability of .95 for that value. This indicates that 95% of the scores in a standard normal distribution would have values less than or equal to 1.65, as shown in Figure 4.5. If n is large, a value of $t = 1.65$ for the sample slope would have approximately this same probability. Thus, you could conclude that $1 - p$

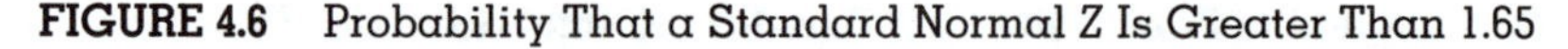

FIGURE 4.6 Probability That a Standard Normal Z Is Greater Than 1.65

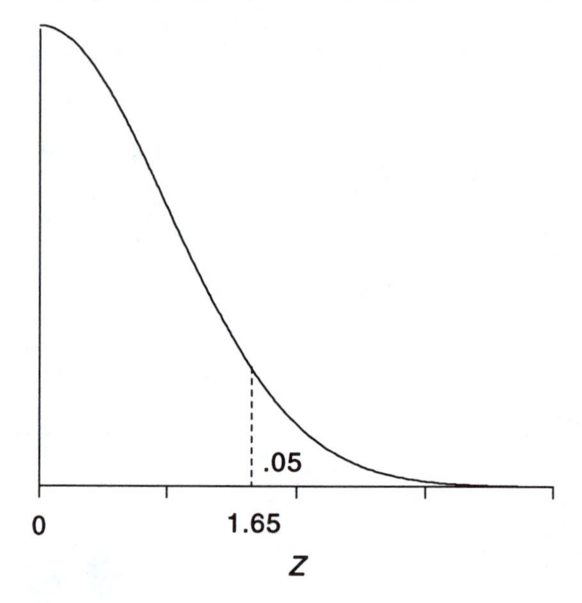

= .05 is the proportion of all b's that would be more than t = 1.65 standard deviations *above* the mean of the sampling distribution, if the mean is equal to zero ($\beta = 0$). Since this is a relatively small proportion, you could be relatively sure that $\beta \neq 0$.

Instead of reporting the cumulative normal probability for both negative and positive values of Z, some tables report the proportion of all standard normal scores that are greater than Z, for positive values of Z only (see Table A.1 in the Appendix). Figure 4.6 illustrates this for Z = 1.65. Since the normal distribution is symmetrical, the probability of a standard normal Z being less than some negative value of Z is the same as that for a positive Z of the same absolute value. If t = 1.65 for your sample b, then you can use such a table to find that the probability of getting a positive sample b this large by chance when β = 0 is .05. If you want the two-tailed probability—that is, the probability of getting either a positive sample b or a negative sample b as large as the absolute value of the observed sample b—then the probability would be doubled. Thus, if t = 1.65 for the sample b, the two-tailed probability would be .10.

Figure 4.7 illustrates the standard normal percentiles for a two-tailed test with α = .05. If H_1 is $\beta \neq 0$, that is, if you would like to be able to reject the null hypothesis in favor of either a negative slope or a positive slope and your sample t = 1.96 or t = −1.96, then the exact probability of making a Type 1 error is .025 + .025 = .05, if n is large. The probability of incorrectly rejecting

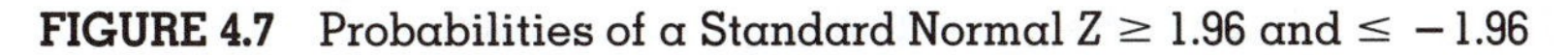

FIGURE 4.7 Probabilities of a Standard Normal $Z \geq 1.96$ and ≤ -1.96

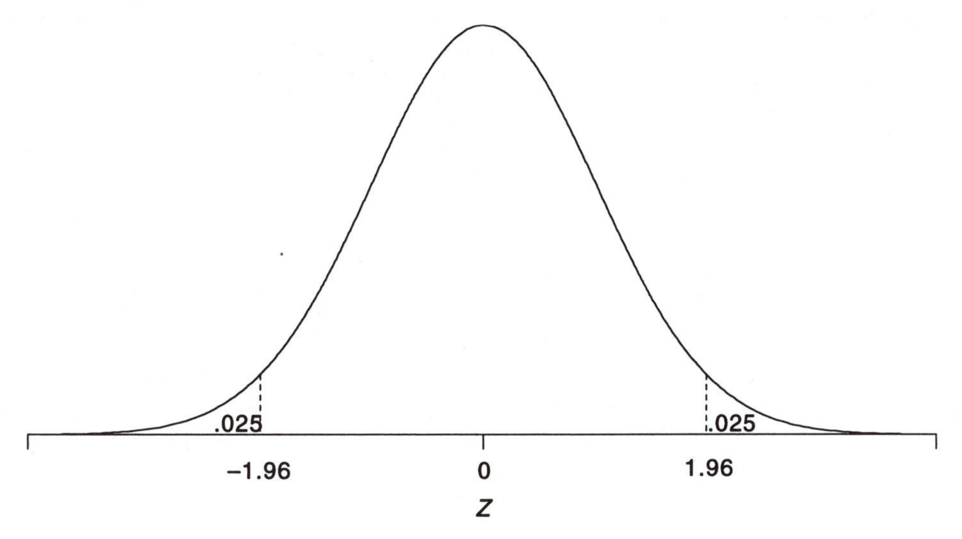

the null hypothesis that the true slope is zero is only .05. This probability is called the significance of the observed sample slope.

If you are doing formal hypothesis testing, the α level would be specified (e.g., .05), and a critical value of t for that level would be looked up in a t table. To determine the critical value, the degrees of freedom (df) for t would be needed, which is $df = n - 2$ for a bivariate slope. If $n = 102$, for example, $df = 102 - 2 = 100$. The critical value of t for a two-tailed test with $\alpha = .05$ and with $df = 100$ is given in a t table as 1.984 (see Table A.2 in the Appendix). This is slightly larger than $Z = 1.96$, the comparable critical value for $\alpha = .05$ in a normal distribution. When $df = 200$, $t = 1.972$, which is quite close to the comparable value in a normal distribution (1.96). After determining the critical t, the value of t for the observed sample b would be calculated, and if it is greater than or equal to the critical t, the null hypothesis would be rejected.

The Siblings and Fertility Example

The bivariate regression example from Chapter 2 (Table 2.9) involving the effect of number of siblings on number of children born to ever-married women aged 35 or older is partially reproduced in Table 4.1. The sample regression slope equals .1116. It is possible, however, that this slope might be zero in the U.S. population of all ever-married women aged 35 or older from which this sample of 493 women was selected; that is, the positive sample value .1116 might have resulted from random sampling error when the true value is zero ($\beta_{YX} = 0$). For

TABLE 4.1 Number of Children Regressed on Number of Siblings

	Mean	Std. Dev.
SIBS	4.286	3.166
CHILDS	2.736	1.935
n	493	

$$\hat{Y} = 2.258 + .1116X$$

	Sum of Squares
Regression	61.48959
Residual	1784.23049
Total	1845.72008

$$\hat{s}_{Y-\hat{Y}} = \sqrt{\frac{\sum (Y - \hat{Y})^2}{n - 2}} = \sqrt{\frac{1784.230492}{491}} = \sqrt{3.63387} = 1.906$$

$$s_b = \frac{\hat{s}_{Y-\hat{Y}}}{s_X\sqrt{n}} = \frac{1.906}{3.166\sqrt{493}} = .0271$$

$$t = \frac{b_{YX}}{s_b} = \frac{.1116}{.0271} = 4.114$$

a formal test of the null hypothesis that $\beta_{YX} = 0$ against a two-tailed alternative hypothesis that $\beta_{YX} \neq 0$, we first specify the α level (probability of falsely rejecting the null hypothesis) that we wish to use; in this case the conventional value of $\alpha = .05$ will be used (this value, of course, would be specified before looking at the value of the sample slope). For a two-tailed $\alpha = .05$ and $df = n - 2 = 493 - 2 = 491$, the critical value of t is approximately 1.97 (see Table A.2 in the Appendix). This means that for samples of size 493, 5% of the samples that might be selected will have regression slopes that are ± 1.97 standard errors or more from the value of the population slope, which is $\beta_{YX} = 0$ according to our null hypothesis. If our sample slope is at least 1.97 standard errors above or below $\beta_{YX} = 0$, then we will reject the null hypothesis, knowing that there is no more than a .05 probability of being in error.

Using Equation 4.6, Table 4.1 shows that the estimated standard error of the sample slope for children regressed on siblings equals .0271; that is, it is estimated that the standard deviation of the values of the sample slopes for all samples of size 493 that might be selected is .0271. The sample slope divided by the standard error gives us $t = 4.114$. This means that our sample slope is estimated to be 4.1 standard deviations (standard errors) above the hypothesized value $\beta = 0$. Since this is greater than the critical $t = 1.97$, we would reject the null hypothesis and conclude that the number of siblings has a positive effect on the number of children that a woman has. A table of t values (or the printout of an SPSS regression analysis or another standard regression program) would also tell us that the two-tailed probability associated with the observed $t = 4.114$ is $p < .0001$; a sample slope as large as the absolute value of the slope that we observed (.1116) would occur by chance less than one in ten thousand times if the population slope were equal to zero. This probability is the observed *level of significance*. The extremely small level of significance in this case means that it is very unlikely that $\beta = 0$; that is, we can be very confident that $\beta \neq 0$.

Let us imagine, however, that our sample size is shrunken. *What if* $n = 102$ but all of our other sample statistics (b, r, s_X, and s_Y) are exactly the same as in the real case of $n = 493$? We can use Equation 4.7 to calculate the value of the standard error in this imaginary case:

$$s_b = \frac{s_Y\sqrt{1 - r^2}}{s_X\sqrt{n - 2}} = \frac{1.935\sqrt{1 - .0333}}{3.166\sqrt{102 - 2}} = .0601$$

Now the value of the standard error of the sampling distribution is more than twice as great as when $n = 493$. With the smaller sample size we can expect greater variance from sample to sample, and thus any particular sample value has a greater probability of deviating from the true population parameter. Consequently,

$$t = \frac{b_{YX}}{s_b} = \frac{.1116}{.0601} = 1.857$$

For $n = 102$ and a two-tailed $\alpha = .05$, the critical $t = 1.984$. Our observed t is now less than the critical value, and thus we cannot reject the null hypothesis. The failure to reject the null hypothesis in this hypothetical case is entirely due to the relatively small sample size; everything else was equal to the case of $n = 493$. Or, on the other hand, we might say that the rejection of the null hypothesis in the case of $n = 493$ was entirely due to the relatively large sample size. This illustrates the extremely important role of sample size in hypothesis testing.

Finally, let us now engage in one last "what if." What if we had been so confident that siblings would have a positive effect on the number of children that our alternative to the null hypothesis had been $\beta > 0$? If so, we would be conducting a one-tailed t test. If we had chosen a one-tailed $\alpha = .05$, our critical t would be approximately 1.65 in the case of $n = 493$. Since the observed $t = 4.05$, this would not have affected our decision; we would have rejected the null hypothesis whether we conducted a one-tailed test or a two-tailed test, in the case of $n = 493$. But in the hypothetical case of $n = 102$, our $t = 1.86$ is greater than the one-tailed critical value of t (the critical $t = 1.66$ for $df = 100$), and thus we would have rejected the null hypothesis $\beta = 0$ in favor of $\beta > 0$.

In the case of $n = 102$, the choice between a one-tailed test and a two-tailed test makes all the difference in our decision. In this regard, we need to be clear about two points. First, we should choose between the one- and two-tailed tests before we calculate our sample slope(s). Second, there is a potential cost to one-tailed tests, just as there is a disadvantage to two-tailed tests. If our sample slope turns out to be opposite in sign from that predicted (e.g., if we find that number of siblings had a negative effect on number of own children, perhaps because women from large families didn't like competing with their brothers and sisters for their parents' attention and because those from small families had always longed for some/more brothers and sisters with which to play), it is not possible to reject the null hypothesis no matter how large the effect is observed to be in the sample. We have placed the entire α in one tail of the sampling distribution in order to increase our chances of supporting our prediction, and thus we have forsaken the chance of obtaining a "surprise" finding.

Violations of Assumptions

The characteristics of the sampling distribution of b specified in Equations 4.2, 4.3, and 4.4 are based on four assumptions about ε. Some effects of violating these assumptions have been discussed: if ε and X are correlated, b is not an unbiased estimator of β (Equation 4.2 is not valid); if ε is not homoskedastic or if the observations of ε are not independent of one another, the formula for the standard error of b is not correct (Equation 4.3 is not valid) and b is not the best linear unbiased estimator (BLUE) of β (the Gauss-Markov theorem is not valid); and if ε is not normally distributed, b is not normally distributed in small samples (Equation 4.4 is not valid), although the distribution of b becomes more and more similar to the normal distribution as sample size increases. We will now discuss the effects of assumptions violations on hypothesis testing.

If ε and X are correlated, the fact that b is biased is probably more of a concern than any consequences for hypothesis testing. The violation of this assumption, however, does have implications for testing the null hypothesis. Violation of this assumption means that there are unmeasured causes of Y that should be included in the regression equation but are presently subsumed under ε. We can infer the consequences of omitting these causes by examining the formula for the t statistic,

$$t = \frac{b}{s_b} = \frac{b}{\hat{s}_{Y-\hat{Y}}/(s_X\sqrt{n})} = \frac{b(s_X\sqrt{n})}{\hat{s}_{Y-\hat{Y}}}$$

Since the omitted causes are correlated with X, the failure to control them means that b is too small or too large (depending on the nature of the redundancy or suppression that exists between X and the omitted causes), and thus

t will tend to be too large or too small. Furthermore, the standard error of estimate $\hat{s}_{Y-\hat{Y}}$ is biased upward by the omission of these causes; if they were added to the equation, they would increase the explained variance and thus reduce $\hat{s}_{Y-\hat{Y}}$. Since $\hat{s}_{Y-\hat{Y}}$ is in the denominator of t, omitting these causes tends to reduce t. The net effect of omitting these causes depends on whether the bias in b (which may be upward or downward) is offset or reinforced by the upward bias in $\hat{s}_{Y-\hat{Y}}$.[7] Thus, although it is impossible to reach a general conclusion about the effect of violating the assumption that ε and X are uncorrelated, there is a good chance of some effect on t. This effect on t could either lead us to reject H_0 when we should not do so or to fail to reject H_0 when we should.

If the homoskedasticity assumption is violated, hypothesis tests will be biased. First, Equation 4.6 for the estimated standard error of b will not be valid. As was the case for the true standard error of b (Equation 4.3), if X and the variance of ε are positively correlated (the variance of ε increases as X increases), Equation 4.6 will tend to underestimate the standard error of b. This underestimate will in turn inflate the value of $t = b/s_b$. Since the computed value of t will be too high, the null hypothesis $\beta = 0$ may be incorrectly rejected. If X and the variance of ε are negatively correlated, Equation 4.6 will tend to overestimate the standard error of b. Too high a value of s_b will lead to a value of t that is too small. Thus, the null hypothesis may not be rejected when it should be. In sum, heteroskedasticity may lead to incorrect decisions about the null hypothesis. Even if it doesn't lead to incorrect decisions (e.g., if the biased t value and the correct t value were both greater than the critical t value for the specified level of α), the reported level of significance for the biased t value will be incorrect.

The second consequence of heteroskedasticity for hypothesis testing stems from the fact that the least-squares estimate of β will not be the most efficient estimator, i.e., b will not be BLUE. There will be other estimators of β that have smaller standard errors than the least-squares slope. Thus, even if we correct the biased estimate of the standard error of b given by Equation 4.6, there are other methods of estimating b, such as *weighted least-squares*, that are more efficient. Consequently, the correct value of t for the least-squares b might be too small for us to reject the null hypothesis. That is, the correct decision for the least-squares estimate of β may be not to reject H_0. However, if a more efficient estimator of β were used, t would tend to be larger because we would be dividing the estimate of β by a smaller standard error. Thus, it is quite possible that we could reject the null hypothesis by using a more efficient estimator in cases where the least-squares estimator resulted in a failure to reject the null hypothesis. In sum, a potential consequence of using least-squares in the pres-

7. There is another term in the formula for the standard error of b when there is more than one X in the equation. This term, which is called the *variance inflation factor*, will be covered in the next section. The omission of a correlated cause always creates a downward bias in the variance inflation factor, which in turn tends to inflate the value of t.

ence of heteroskedasticity is to fail to reject a null hypothesis that would be rejected by more efficient estimators.

The consequences of violating the assumption of independent ε's in our observations (i.e., the presence of autocorrelation) are the same in principle as those for heteroskedasticity. If we choose the usual formula for the estimated standard error, we may incorrectly reject the null hypothesis in the case of positive autocorrelation (the more common case) and incorrectly fail to reject the null hypothesis in the case of negative autocorrelation. If we correct our estimate of the least-squares standard error, on the other hand, we may fail to reject a null hypothesis that a more efficient estimator would reject.

Hypothesis tests may also be affected if ε is not normally distributed as assumed. Although b and the usual formula for s_b will not be biased in such cases, b/s_b will not be exactly distributed as t because b will not be exactly normally distributed. Thus, the level of significance associated with the computed "t" will be incorrect. It may either be too high, leading to an incorrect failure to reject the null hypothesis, or too low, resulting in an incorrect rejection of the null hypothesis, depending upon the nature of the distribution of ε and thus of b. The degree of error in the level of significance will depend upon sample size. Since the distribution of b resembles a normal distribution more and more as sample size increases, the greatest errors in the level of significance will occur in small samples. In large samples the distribution of b/s_b will be very close to the t distribution, so there should be little error in the reported level of significance.

A second problematic consequence of a nonnormal distribution of ε is a potential loss of efficiency for the least-squares b. The least-squares b is still the BLUE. There may be other estimators, however, with smaller standard errors, i.e., greater efficiency. One of these is the least absolute residual (LAR) estimator. The LAR estimator chooses estimates of α and β that minimize the sum of absolute residuals,

$$\sum |Y - \hat{Y}| = \sum |Y - a'' - b''X|$$

where a'' and b'' are the LAR estimates. The sampling distributions of b'' becomes closer and closer to being centered on β (i.e., unbiased) and more and more similar to a normal distribution as sample size increases. That is, for large samples the distribution of b'' is nearly unbiased and normal. When ε has a normal distribution, the OLS estimates are much more efficient than the LAR estimates. But for a large class of nonnormal distributions with fat tails (more extreme values than a normal distribution), the LAR estimates of β are much more efficient than the OLS estimates *in large samples* (Kamenta 1986, p. 264). The greater efficiency of the LAR estimates means that they may lead to the rejection of the null hypothesis in cases where the OLS estimators do not. Thus, the supremacy of estimates of α and β that minimize the sum of squared residuals (OLS) over those that minimize the sum of absolute residuals (LAR)—which was mentioned in the discussion of the Gauss-Markov theorem—can only be guaranteed when ε is normal or approximately normal.

Two Independent Variables

The standard errors of the partial slopes for an equation with two independent variables are

$$s_{b_1} = \frac{\hat{S}_{Y-\hat{Y}_{12}}}{s_1\sqrt{n}\,\sqrt{1 - r_{12}^2}} \tag{4.10}$$

$$s_{b_2} = \frac{\hat{S}_{Y-\hat{Y}_{12}}}{s_2\sqrt{n}\,\sqrt{1 - r_{12}^2}} \tag{4.11}$$

As with the standard error of the bivariate slope, the standard error of a partial slope involves dividing the standard error of estimate by the product of the standard deviation of the independent variable times the square root of the sample size.

The additional term in the denominators of Equations 4.10 and 4.11 is the square root of 1 minus the squared r between the two independent variables. This term has a very important influence on the standard error. In particular, as r_{12}^2 increases, the standard error of the slope increases. You should examine the equations closely to see why this is so. A greater standard error means that any given sample b will be a less reliable estimate of the population partial β. As the standard error increases, too, the t value of the observed partial slope decreases, as can be seen from the formula of t for X_1,

$$t = \frac{b_1}{s_{b_1}} \tag{4.12}$$

This means that as the standard error increases because of increased correlation between the X's, it will be more difficult to reject the null hypothesis. The more highly the independent variables are correlated, the more difficult it is to separate their independent effects on Y reliably. In the case where the independent variables are not correlated, we will get the most reliable estimates of the partial b's. That is, the significance level or probability of making a Type 1 error will be minimized, all other things being equal.

Equations 4.10 and 4.11 also indicate why there is an advantage to using multiple regression even if the independent variables are uncorrelated. Remember that in this case the bivariate slopes will be equal to the partial slopes. The numerator of the formula for the standard error, however, indicates that as the standard deviation of the residuals decreases, the standard error decreases, and thus the value of the t statistic increases. When we have two uncorrelated X's in the equation we will get smaller residuals than when only one predictor is used, assuming that each X has some effect on Y. Thus, the use of multiple regression even when the X's are uncorrelated is an advantage because the significance level (p) of the partial slopes will be less than the level of significance for the bivariate slopes.

The t test for a partial slope is conducted in the same way that it is conducted for a bivariate slope. The null hypothesis states that the partial slope is equal to zero ($\beta_i = 0$) and the two-tailed alternative hypothesis is that it does not equal zero ($\beta_i \neq 0$). After choosing an α level, the critical value of t can be looked up in a t table. *The degrees of freedom for a partial slope in an equation with k independent variables is $df = n - k - 1$.* Finally, the t test for the partial slope is also the test for B_i, sr_i, and pr_i. If the partial $b_i = 0$, then $B_i = sr_i = pr_i = 0$ must also be true. Thus, the test of the null hypothesis that one of these partial coefficients is equal to zero is a test for all of these measures.

Earnings Example. Chapter 3 (Table 3.9) presented the following example of a multiple regression equation with two independent variables:

$$\hat{Y} = 10.72 + 6.798X_1 + .954X_2$$

where Y is the 1977 annual salary (in hundreds of dollars) of 547 men, X_1 is years of schooling completed, and X_2 is years of work experience. The correlation between experience and education is $r_{12} = -.323$ (Table 3.9). Using Equations 4.10 and 4.11 and the values of the standard error of estimate and the standard deviations of experience and education from Table 3.9, the standard errors of the partial slopes for education and experience are

$$s_{b_1} = \frac{47.229}{(3.361)\sqrt{547}\sqrt{1 - (-.323)^2}} = .635$$

$$s_{b_2} = \frac{47.229}{(10.924)\sqrt{547}\sqrt{1 - (-.323)^2}} = .195$$

Using these standard errors and the values of the slopes given above, the t values are

$$t_1 = \frac{6.789}{.635} = 10.70$$

$$t_2 = \frac{.954}{.195} = 4.88$$

The critical $t \approx 1.96$ for a two-tailed α with $df = 547 - 2 - 1 = 544$. Since both of the calculated t's are much greater than the critical value, we can confidently reject the null hypotheses that age and education have no effects on earnings. The actual level of significance for both variables is $p < .0001$.

s_b for Residualized Variables. In Chapter 3 it was shown that the slope for X_1 in an equation with two independent variables is equal to the bivariate regression of Y on the residualized X_1, i.e., $X_1 - \hat{X}_{1\cdot 2}$. Recall Equation 3.4, which stated that

$$b_{Y1\cdot 2} = b_{Y(X_1 - \hat{X}_{1\cdot 2})} = \frac{\sum (X_1 - \hat{X}_{1\cdot 2})(Y - \overline{Y})}{\sum (X_1 - \hat{X}_{1\cdot 2})^2}$$

In addition, it was noted that $b_{Y1\cdot 2}$ is also equal to the regression of the residualized Y on the residualized X_1:

$$b_{Y1\cdot 2} = b_{(Y - \hat{Y}_2)(X_1 - \hat{X}_{1\cdot 2})} = \frac{\sum (X_1 - \hat{X}_{1\cdot 2})(Y - \hat{Y}_2)}{\sum (X_1 - \hat{X}_{1\cdot 2})^2}$$

We can now address the question of whether these bivariate regressions also give the correct standard error for $b_{Y1\cdot 2}$. We will use the academic-salary data from Chapter 3 to illustrate the answer to this question. Taking the standard deviation of X_1 from Table 3.1, the standard error of estimate from Table 3.3, and r_{12} from Table 3.5, Equation 4.10 gives the correct value for the standard error of b_1,

$$s_{b_1} = \frac{\hat{s}_{Y - \hat{Y}_{12}}}{s_1\sqrt{n}\sqrt{1 - r_{12}^2}} = \frac{2.6458}{3.6871\sqrt{5}\sqrt{1 - .8060^2}} = .5422$$

Using Equation 4.6 as a model, a basic conceptual formula for the standard error of a bivariate regression slope is

$$\text{Standard Error of } b = \frac{\text{Standard Error of Estimate}}{(\text{Standard Deviation of Independent Variable})\sqrt{n}}$$

$$= \frac{\sqrt{(\text{Residual Sum of Squares})/(n - 2)}}{(\text{Standard Deviation of Independent Variable})\sqrt{n}}$$

Whether the variables are raw scores, residual scores, or one of each, the above equation will equal the value of the standard error of b that a computer program would calculate for an equation with one independent variable.

We will first examine the case in which the dependent variable is the raw Y and the independent variable is the residualized X_1. The value of the standard error of $b_{Y(X_1 - \hat{X}_{1\cdot 2})}$ that a least-squares program would compute will equal

$$s_{bY(X_1-\hat{X}_{1\cdot 2})} = \frac{\sqrt{\dfrac{(1 - sr_1^2)\sum(Y - \overline{Y})^2}{n-2}}}{s_{(X_1-\hat{X}_{1\cdot 2})}\sqrt{n}} = \frac{\sqrt{\dfrac{(1 - .0507)\cdot 470}{3}}}{2.1827\sqrt{5}}$$

$$= \frac{\sqrt{\dfrac{446.17}{3}}}{4.8807} = \frac{12.1952}{4.8807} = 2.4987$$

This value is far larger than the true value, .5422, which was given above. The reason for the discrepancy is in the numerator of the equation. Because the squared semipartial correlation sr^2 equals the proportion of the variance of Y that is uniquely explained by X_1, $(1 - sr_1^2)\,\Sigma\,(Y - \overline{Y})^2 = 446.17$ equals the amount of the sum of squares of Y that cannot be explained by the residualized X_1. The residual sum of squares that should be used for the standard error of $b_{Y1\cdot 2}$ equals $(1 - R^2_{Y\cdot 12})\,\Sigma\,(Y - \overline{Y})^2 = 14$ (Table 3.3). Because $sr_1^2 = .0507$ (Table 3.6) is much smaller than $R^2_{Y\cdot 12} = .9702$ (Table 3.3), the residual sum of squares, and consequently the standard error of estimate as well, is much greater according to the bivariate formula than it should be. Consequently, the bivariate formula for the standard error of $b_{Y(X_1-\hat{X}_{1\cdot 2})}$ is not a valid measure of the standard error of $b_{Y1\cdot 2}$. It is important to note, however, that the denominator $s_{(X_1-\hat{X}_{1\cdot 2})}\sqrt{n}$ is correct:

$$s_{(X_1-\hat{X}_{1\cdot 2})}\sqrt{n} = \sqrt{s_1^2\,(1 - r_{12}^2)}\sqrt{n} = s_1\sqrt{(1 - r_{12}^2)}\sqrt{n}$$

This shows that the denominator is equivalent to the denominator in Equation 4.10, the correct formula for the standard error. Thus, when s_1 is multiplied by $\sqrt{(1 - r_{12}^2)}$ in Equation 4.10, we get the standard deviation of the residualized X_1. This shows that a strong r_{12} will reduce the amount of unique variance in X_1 and consequently increase the standard error of the slope.

If we regress the residualized Y on the residualized X_1 we have better luck in getting the correct standard error of $b_{Y1\cdot 2}$. The following formula shows the value that a least-squares program would compute for $Y - \hat{Y}_2$ regressed on $X_1 - \hat{X}_{1\cdot 2}$:

$$s_{b(Y-\hat{Y}_2)(X_1-\hat{X}_{1\cdot 2})} = \frac{\sqrt{\dfrac{(1 - pr_1^2)\sum(Y - \hat{Y}_2)^2}{n-2}}}{s_{(X_1-\hat{X}_{1\cdot 2})}\sqrt{n}}$$

$$= \frac{\sqrt{\dfrac{(1 - .6298)\cdot 37.82}{3}}}{2.1827\sqrt{5}} = \frac{\sqrt{\dfrac{14}{3}}}{4.8807} = \frac{2.1602}{4.8807} = .4426$$

The denominator is the same and thus is also correct. The numerator indicates that the residual sum of squares equals $(1 - pr_1^2)\,\Sigma(Y - \hat{Y}_2)^2 = 14$. Since pr_1^2 equals the proportion of the variance or sum of squares in Y that is not explained by X_2 and is explained by X_1, this part of the numerator must equal the sum of squares of Y that is not explained by either X_1 or X_2. This is verified in Table 3.3. The standard error of .4426 is slightly smaller than the correct figure of .5422 because the degrees of freedom in the numerator is incorrect. Instead of $n - 2$, which is the residual degrees of freedom for a bivariate regression, the degrees of freedom should be $n - k - 1 = n - 2 - 1 = n - 3$ because X_2 was used to residualize both Y and X_1. Thus, the correct formula is

$$s_{bY1\cdot 2} = s_{b(Y-\hat{Y}_2)(X_1-\hat{X}_{1\cdot 2})} = \frac{\sqrt{\dfrac{(1 - pr_1^2)\sum(Y - \hat{Y}_2)^2}{n - 3}}}{s_{(X_1-\hat{X}_{1\cdot 2})}\sqrt{n}}$$

To summarize, the bivariate regression of $Y - \hat{Y}_2$ on $X_1 - \hat{X}_{1\cdot 2}$ will give the correct value for $b_{Y1\cdot 2}$. It will also give the correct residual sum of squares. The standard error of this bivariate slope, however, is not equal to the standard error of $b_{Y1\cdot 2}$ because the degrees of freedom by which the residual sum of squares is divided is $n - 2$ instead of the correct $n - 3$ (Weisberg 1985, p. 53). In large samples this difference will be trivial and the regression program will give a value for the standard error that is quite close to being correct. But it is still mathematically incorrect, and the bias is to underestimate the standard error of $b_{Y1\cdot 2}$.

Three or More Independent Variables

When there are three or more independent variables, the standard error of the partial slope is

$$s_{b_i} = \frac{\hat{s}_{Y-\hat{Y}_{12\cdots k}}}{s_i\sqrt{n}\sqrt{1 - R_i^2}} \tag{4.13}$$

This is a general fromula for k independent variables where the subscript i indicates the *ith* X, with $i = 1, 2, \ldots, k$. The only difference between this formula and that for two independent variables is that Equation 4.13 contains $\sqrt{1 - R_i^2}$ in the denominator. R_i^2 is the squared multiple correlation between X_i and all of the other *independent* variables; if there were three independent variables, R_i^2 for X_1, for example, would be $R_{1\cdot 23}^2$. It replaces r_{12}^2 for the case of two independent variables.

Multicollinearity. It is important to stress the fact that R_i^2 in Equation 4.13 is not a squared multiple correlation for Y; only the X's are included in this multiple correlation. It is also different for each X. Some X's may be correlated more strongly with the set of all other independent variables than others. X's that are

TABLE 4.2 Variance Inflation Factor and the Standard Error of the Slope

R_i	$VIF_i = 1/(1 - R_i^2)$	$\sqrt{VIF_i}$
.00	1.0	1.0
.20	1.04	1.02
.40	1.19	1.09
.60	1.56	1.25
.80	2.78	1.67
.90	5.26	2.29
.95	10.30	3.30
.99	50.30	7.09
1.00	∞	∞

correlated more strongly with the other independent variables are said to be more *multicollinear*. R_i^2 is often treated as a measure of the multicollinearity of X_i. Remember that R^2 ranges from 0 to 1. Thus, the larger the R_i^2, the greater is the multicollinearity of X_i. Most importantly, the greater the value of R_i^2, the greater will be the standard error of the slope, as can be seen by carefully examining Equation 4.13.

The effect of R_i^2 on the standard error of the slope can be illustrated more effectively by rewriting Equation 4.13 as

$$s_{b_i} = \frac{\hat{S}_{Y-\hat{Y}12\cdots k}}{s_i\sqrt{n}} \times \frac{1}{\sqrt{1 - R_i^2}} = \frac{\hat{S}_{Y-\hat{Y}12\cdots k}}{s_i\sqrt{n}} \times \sqrt{\text{VIF}_i}$$

where VIF_i is known as the *variance inflation factor*. VIF_i is always greater than or equal to 1, and thus it indicates how many times R_i^2 increases the sampling variance $s_{b_i}^2$. The square root of VIF_i indicates the effect of R_i^2 on the standard error of the slope. Fox (1991) illustrates this effect with Table 4.2. When $R_i = R_i^2 = 0.0$, i.e., multicollinearity is totally absent, the effect on the standard error is $\sqrt{\text{VIF}_i} = 1$; there is no variance inflation or effect on the standard error. When $R_i = .60$, the effect of VIF on the standard error equals 1.25 (a 25% increase). When there is a high degree of multicollinearity, e.g., $R_i = .90$, the standard error will be 2.29 times larger (129% inflation) than when there is no multicollinearity. Finally, if X_i were perfectly correlated with the other independent variables as a set ($R_i = 1$), it would not be possible to calculate the standard error.

Thus, all other things being equal, the X_i having the highest multiple correlation with the other variables will have the largest standard error and consequently the lowest level of significance. The same principle holds here as with the case of two independent variables; it is difficult to estimate reliably the slopes of variables that are highly correlated with the other independent variables. The nature of the multicollinearity problem will be examined further in the next section.

Anomia and Income Example

The use of *t* tests for multiple regression equations with three or more independent variables will be illustrated with an analysis of data from the 1986 Akron Area Survey, a random-digit-dialed telephone survey of residents of Summit County, Ohio, conducted by the Department of Sociology at the University of Akron. The dependent variable in this example is *anomia* (Srole 1956), a sense of normlessness or alienation from society. Anomia is measured by the four statements shown in Figure 4.8a. The respondents' anomia scores equal the sum of their responses to these four statements (*strongly agree* = 4; *somewhat agree* = 3; *somewhat disagree* = 2; *strongly disagree* = 1) and thus range from 4 to 16. The independent variables are years of education (1–20), family income (*less than $5,000* = 1; *$5,000–9,999* = 2; *$10,000–14,999* = 3; *$15,000–19,999* = 4; *$20,000–24,999* = 5; *$25,000–34,999* = 6; *$35,000–49,000* = 7; *$50,000 or more* = 8), and three measures of the respondents' subjective assessment of the level of their incomes (Figure 4.8b). Notice that for each of the subjective income variables a high score indicates a positive assessment of income.

It is well known that people with lower socioeconomic status (SES) are more alienated. In this analysis we attempt to determine whether the SES-anomia relationship is caused by subjective feelings of deprivation or by the objective level of deprivation; e.g., does poverty itself cause alienation, or does feeling

FIGURE 4.8 Anomia Statements and Subjective Income Questions

(a) Anomia Statements

Nowadays a person has to live pretty much for today and let tomorrow take care of itself.

It's hardly fair to bring children into the world with the way things look for the future.

These days a person doesn't really know whom he can count on.

There's little use writing to public officials because they often aren't interested in the problems of the average person.

(b) Subjective Income Questions

SHORTINC: How seriously do you feel a personal shortage of money these days—a great deal, quite a bit, some, or little, or none? (*great deal* = 1; *quite a bit* = 2; *some* = 3; *little or none* = 4)

REWRDINC: Which number from 1 to 7 best describes how disappointing or rewarding is the standard of living provided by your family income? (*disappointing* = 1; ⅔ *rewarding* = 7)

SATINC: How dissatisfied or satisfied are you with your standard of living? (*dissatisfied* = 1; ⅔ *satisfied* = 7) impoverished cause alienation? Let's assume at first that we have only one measure of subjective income (SHORTINC). Our first regression equation will contain anomia as the dependent variable and education, income, and whether the respondent feels a shortage of money (SHORTINC) as the independent variables (Equation 1, Table 4.3). The partial regression slopes (b's) for each of these variables are negative, indicating that the higher the respondent's level of education is, the higher will be the level of income, and the less the respondent feels a shortage of money, the lower will be the level of anomia. These are sample slopes, however, and might occur even if the true relationships in the population are zero. We will use the t test for the slopes to address this issue.

The R_i^2 for each X indicates the proportion of each independent variable's variance that it shares with the other two independent variables. Thus, about .15 of the variance of EDUC is shared with INCOME and SHORTINC, .29 of the variance of INCOME is shared with EDUC and SHORTINC, and .26 of the variance of SHORTINC is shared with EDUC and INCOME. These squared multiple correlations indicate the degree of multicollinearity (i.e., intercorrelatedness) for each independent variable; thus, education has a lower level of multicollinearity than the two income variables. The standard errors of the slopes are given in the column following the R_i^2's. Using Equation 4.13 and the values of the appropriate statistics from Table 4.3, the standard errors of the slopes are calculated as follows:

$$s_{b_1} = \frac{\hat{s}_{Y-\hat{Y}_{123}}}{s_1\sqrt{n}\sqrt{1 - R^2_{1\cdot 23}}} = \frac{2.6679}{2.732\sqrt{513}\sqrt{1 - .1548}} = .0469$$

$$s_{b_2} = \frac{\hat{s}_{Y-\hat{Y}_{123}}}{s_2\sqrt{n}\sqrt{1 - R^2_{2\cdot 13}}} = \frac{2.6679}{2.059\sqrt{513}\sqrt{1 - .2895}} = .0679$$

$$s_{b_3} = \frac{\hat{s}_{Y-\hat{Y}_{123}}}{s_3\sqrt{n}\sqrt{1 - R^2_{3\cdot 12}}} = \frac{2.6679}{1.334\sqrt{513}\sqrt{1 - .2596}} = .1027$$

The slopes in Table 4.3 are divided by the above standard errors to give the following t values for each independent variable:

$$t_1 = \frac{-.1361}{.0469} = -2.898$$

$$t_2 = \frac{-.2635}{.0679} = -3.879$$

$$t_3 = \frac{-.2128}{.1027} = -2.072$$

For $df = 513 - 3 - 1 = 509$, the critical value of t for a two-tailed test at α

TABLE 4.3 Regression of Anomia on Education, Income, and Subjective Income

		Mean	*Std. Dev.*
Y	ANOMIA	8.297	2.820
X_1	EDUC	13.415	2.732
X_2	INCOME	5.197	2.059
X_3	SHORTINC	3.150	1.334
X_4	SATINC	5.230	1.649
X_5	REWRDINC	5.237	1.569
	n	513	

Correlations

		Y	X_1	X_2	X_3	X_4	X_5
Y	ANOMIA	—	−.233	−.290	−.236	−.161	−.176
X_1	EDUC	−.233	—	.364	.309	.138	.169
X_2	INCOME	−.290	.364	—	.490	.416	.519
X_3	SHORTINC	−.236	.309	.490	—	.546	.586
X_4	SATINC	−.161	.138	.416	.546	—	.813
X_5	REWRDINC	−.176	.169	.519	.586	.813	—

Equation 1

		b	B	R^2_i	s_b	t	p
X_1	EDUC	−.1361	−.1318	.1548	.0469	−2.898	.0039
X_2	INCOME	−.2635	−.1924	.2895	.0679	−3.879	.0001
X_3	SHORTINC	−.2128	−.1007	.2596	.1027	−2.072	.0388
	a	12.1632					
	Std. Error	2.6679					
	R^2	.1102					

Equation 2

		b	B	R^2_i	s_b	t	p
X_1	EDUC	−.1358	−.1315	.1656	.0473	−2.868	.0043
X_2	INCOME	−.2694	−.1967	.3750	.0725	−3.713	.0002
X_3	SHORTINC	−.2127	−.1006	.4280	.1170	−1.817	.0697
X_4	SATINC	−.0613	−.0358	.6694	.1245	−.492	.6229
X_5	REWRDINC	.0661	.0368	.7129	.1405	.470	.6384
	a	12.1615					
	Std. Error	2.6724					
	R^2	.1106					

= .05 is approximately 1.96. Since all of the above t values are larger in absolute value than 1.96 (only slightly, however, in the case of SHORTINC), we conclude that each of the independent variables has a statistically significant negative effect on anomia. Thus, it appears that both objective circumstances of low income and subjective feelings of deprivation contribute to alienation. The standardized slopes (B_i's) indicate that the effect of actual income may be greater than that of subjective income and education. The three SES variables together explain about 11% of the variance in anomia.

Now, let us assume that instead of having just the single measure of subjective income (SHORTINC), we have and use each of the three subjective measures shown in Figure 4.8b. The three variables (SHORTINC, SATINC, REWRDINC) may represent three somewhat distinct and independent types of feelings that an individual may have about his or her income and standard of living. The correlations in Table 4.3, however, show that these variables are not entirely independent; SHORTINC correlates .546 and .586 with SATINC and REWRDINC, respectively, and the latter two variables correlate .813. As implied by these r's, when all three subjective measures are included in the regression equation along with education and income, SATINC and REWRDINC have the greatest amount of multicollinearity; SATINC shares .67 of its variance with the other four independent variables and REWRDINC shares .71 of its variance with the other four variables (see R_i^2 in Table 4.3, Equation 2). Although SHORTINC has substantially less multicollinearity than the other two subjective income variables, it now shares .43 of its variance with the other three independent variables, as compared to $R^2 = .26$ when it was the only subjective measure in the equation. The effect of the increased multicollinearity on the standard error of SHORTINC can be seen below:

$$s_{b3} = \frac{\hat{s}_{Y-\hat{Y}12345}}{s_3\sqrt{n}\sqrt{1 - R^2_{3\cdot1245}}} = \frac{2.6724}{1.334\sqrt{513}\sqrt{1 - .4280}} = .1170$$

The standard error of estimate in the numerator is virtually identical to that in the equation with only three independent variables. That is because adding the two additional subjective income measures produced only a very slight increase in R^2 (from .1102 to .1106). Of course, the standard deviation of SHORTINC and the sample size did not change at all. Thus, the only appreciable change in the statistics that determine the standard error of the slope was in R_3^2, which increased from .2596 to .4280. As a consequence, s_b increased from .1027 to .1170, an increase of about 14%. How did this affect our decision regarding the null hypothesis that $\beta = 0$ for SHORTINC?

$$t_3 = \frac{b_{Y3\cdot1245}}{s_{b3}} = \frac{-.2127}{.1170} = -1.817$$

Notice that the slope did not change as a result of adding the two additional variables to the equation. Thus, the value of t decreased from -2.07 to -1.817 solely as a result of the increased value of the standard error of the slope. Since the observed t is now less than the critical t of 1.96, we would *not* reject the null hypothesis. The additional multicollinearity created by the two additional measures of subjective income was entirely responsible for the failure to reject the null hypothesis. That is, higher multicollinearity makes it more difficult to separate reliably the effects of the independent variables.

Notice that the other two subjective income variables are also not significant ($p = .6229$ and .6384 for SATINC and REWRDINC, respectively). Thus, our t tests do not allow us to conclude that any of the measures of subjective income affect anomia. This is entirely the result, apparently, of the degree of intercorrelatedness among these variables.

There are at least two possible solutions to this problem of multicollinearity. First, if the three variables were considered to be multiple indicators of a single underlying subjective income factor, we might create a single measure of subjective income by combining the three variables into an index. This is a measurement problem, however, and as such it is beyond the scope of this book (see, for example, Zeller and Carmines 1980). Second, we might use one of the F tests described in the next section to determine whether the subjective income variables as a set (or perhaps some subset of them) have a significant effect on anomia. The consequence of multicollinearity, however, is that we cannot yet disentangle the effects, if any, of the three correlated measures of subjective income. We could, of course, potentially reduce our standard errors by drawing a larger sample size. All other things remaining constant, if we could afford to collect a sufficiently larger sample of observations, we might then reject the null hypothesis at the $\alpha = .05$ level of significance.

F Tests

An alternate test statistic for the partial slopes is the F test. It can also be used to test the combined effects of all of the X's (a test of R^2) and to test the combined effects of some subset of independent variables. A general formula for this family of F tests is

$$F = \frac{[(R^2 \text{ for All } X\text{'s}) - (R^2 \text{ for } X\text{'s Held Constant})]/df_1}{(1 - R^2 \text{ for All } X\text{'s})/df_2}$$

$$df_1 = \text{Number of } X\text{'s Being Tested}$$

$$df_2 = n - k - 1 \qquad \textbf{(4.14)}$$

The numerator of Equation 4.14 is the increase in the explained variance created by the variable or set of variables being tested, divided by the number of variables being tested (df_1). The denominator of F is the proportion of the variance in Y that cannot be explained by all of the independent variables together (the residual variance), divided by the degrees of freedom of the residuals (df_2).

F for Partial Slopes

The general F formula will first be used as an alternate test for the partial slope. In the case of three independent variables, F for the partial slope of X_1 is

$$F_1 = \frac{(R^2_{Y \cdot 123} - R^2_{Y \cdot 23})/1}{(1 - R^2_{Y \cdot 123})/(n - 3 - 1)} \tag{4.15}$$

The first term in the numerator is R^2 for all the independent variables, as in Equation 4.14. The second term is R^2 for the variables being held constant, namely, X_2 and X_3. Since one variable is being tested, the numerator $df = 1$. In the case of an F test for a single variable, the numerator is sr^2 for the variable being tested; it represents the amount of variance that X_1 explains over and above that which is explained by the other independent variables. The denominator is the proportion of the variance that is not explained by the set of all independent variables, i.e., the proportion of variance that is residual variance, divided by the df of the residuals $(n - k - 1)$. In sum, F is equal to the proportion of variance uniquely explained by the variable being tested, divided by the proportion of variance that is unexplained, each divided by their degrees of freedom. Equation 4.15 is sometimes written in a somewhat simpler form as

$$F_1 = \frac{(R^2_{Y \cdot 123} - R^2_{Y \cdot 23})(n - 3 - 1)}{(1 - R^2_{Y \cdot 123})}$$

This formula, however, does not indicate as clearly that $n - 3 - 1$ is the denominator degrees of freedom. The F formulas for testing the other two independent variables are

$$F_2 = \frac{(R^2_{Y \cdot 123} - R^2_{Y \cdot 13})/1}{(1 - R^2_{Y \cdot 123})/(n - 3 - 1)}$$

$$F_3 = \frac{(R^2_{Y \cdot 123} - R^2_{Y \cdot 12})/1}{(1 - R^2_{Y \cdot 123})/(n - 3 - 1)}$$

Since the numerator $df = 1$ in these cases, it would be appropriate to omit it from the formula. F_3, for example, would then be

$$F_3 = \frac{(R^2_{Y \cdot 123} - R^2_{Y \cdot 12})}{(1 - R^2_{Y \cdot 123})/(n - 3 - 1)}$$

The significance level of F or the partial coefficient that is being tested is determined by looking up the critical value of F for the α level being tested in an F table. F tables give critical values of F for different combinations of the numerator and denominator degrees of freedom, at selected α levels (usually .05, .01, and .001, as in Tables A.3a–A.3c in the Appendix). If the calculated F is greater than or equal to the critical F for the specified α level, the null hypothesis ($\beta_i = 0$) is rejected. As with the t test for the partial slope, the F test for

TABLE 4.4 Some Regression Equations for Anomia

		Equations				
		1	2	3	4	5
		b	b	b	b	b
X_1	EDUC	—	−.1770	−.1518	−.1361	−.1358
X_2	INCOME	−.3143	—	−.3234	−.2635	−.2694
X_3	SHORTINC	−.2612	−.3833	—	−.2128	−.2127
X_4	SATINC	—	—	—	—	−.0613
X_5	REWRDINC	—	—	—	—	.0661
	a	10.7530	11.9549	12.0152	12.1632	12.1615
	R^2	.09547	.08385	.10265	.11015	.11063
	n	513				

a single independent variable is simultaneously a test that all of the partial coefficients are zero ($b_i = B_i = sr_i = pr_i = 0$). It should also be noted that F can never be negative. Adding an additional variable to the regression equation can never cause R^2 to decrease.[8] Therefore, the numerator will always be positive. The increase in explained variance represented by the numerator, therefore, does not take into account whether the independent variable has a positive or negative effect on Y. Therefore, the alternate hypothesis in the F test is simply that the variable has a nonzero effect, i.e., it is the same as a two-tailed t test.

The t and F tests for partial slopes are equivalent because when $df_1 = 1$ for the F test,

$$t^2 = F \tag{4.16}$$

Thus, either test can be used. Some investigators use t, and others use F. Computer programs may report one or the other, or both. SPSS, for example, prints t by default, but F may be requested. F is more versatile, however, because it can be used to test all of the variables simultaneously or subsets of variables.

The F tests for the slopes will be illustrated for the anomia regression equation with three independent variables (Table 4.3). This equation is repeated in Table 4.4 as Equation 4. In order to calculate the F values for the three variables in Equation 4, the R^2's from Equations 1, 2, and 3 in Table 4.4 are needed. These are the R^2's for the equations that omit one by one the variables to be tested with F. Thus, Equation 1 omits X_1 to get $R^2_{Y \cdot 23}$, which is needed to compute the F statistic for X_1; Equation 2 omits X_2 to get $R^2_{Y \cdot 13}$, which is needed in the F statistic for X_2; and Equation 3 omits X_3 in order to get $R^2_{Y \cdot 12}$, which is used in the F statistic for X_3.

8. The shrunken or adjusted R^2 can decrease, however.

Using the R^2's from Equations 1–3, the F values are as follows:

$$F_1 = \frac{(R^2_{\hat{Y}\cdot 123} - R^2_{\hat{Y}\cdot 23})}{(1 - R^2_{\hat{Y}\cdot 123})/(513 - 3 - 1)} = \frac{.11015 - .09547}{(1 - .11015)/509} = 8.397$$

$$F_2 = \frac{(R^2_{\hat{Y}\cdot 123} - R^2_{\hat{Y}\cdot 13})}{(1 - R^2_{\hat{Y}\cdot 123})/(513 - 3 - 1)} = \frac{.11015 - .08385}{(1 - .11015)/509} = 15.044$$

$$F_3 = \frac{(R^2_{\hat{Y}\cdot 123} - R^2_{\hat{Y}\cdot 12})}{(1 - R^2_{\hat{Y}\cdot 123})/(513 - 3 - 1)} = \frac{.11015 - .10265}{(1 - .11015)/509} = 4.290$$

The critical value of F for $\alpha = .05$, $df_1 = 1$, and $df_2 = 509$ is 3.86. Thus, the null hypothesis would be rejected for each of the independent variables. That this is the same result that we got with the t tests can be seen by taking the square root of F to get t, as follows:

$$t_1 = \sqrt{F_1} = \sqrt{8.397} = 2.898$$
$$t_2 = \sqrt{F_2} = \sqrt{15.044} = 3.879$$
$$t_3 = \sqrt{F_3} = \sqrt{4.290} = 2.071$$

These values equal the t values given in Table 4.3, except that the true t values are all negative. Since F is always positive, the square root of F will equal the absolute value of t, but not necessarily the signed value of t. t^2, however, will always equal F.

F for R^2

The F test for R^2 in an equation with three independent variables is

$$F = \frac{R^2_{\hat{Y}\cdot 123}/3}{(1 - R^2_{\hat{Y}\cdot 123})/(n - 3 - 1)} \qquad \textbf{(4.17)}$$

It is simply the variance explained by all the variables divided by the unexplained variance, each divided by their degrees of freedom. There is only one R^2 in the numerator because all the variables are being tested. Since three variables are being tested, the numerator degrees of freedom is 3. This F test has no equivalent t test because the numerator degrees of freedom is greater than 1. A general form of Equation 4.17 for k independent variables is

$$F = \frac{R^2_{\hat{Y}\cdot 12\cdots k}/k}{(1 - R^2_{\hat{Y}\cdot 12\cdots k})/(n - k - 1)} \qquad \textbf{(4.18)}$$

The multiple F also may be defined in terms of the regression sum of squares and the residual sum of squares as follows:

$$F = \frac{\sum (\hat{Y}_{12\cdots k} - \overline{Y})^2/k}{\sum (Y - \hat{Y}_{12\cdots k})^2/(n - k - 1)}$$

The numerator includes the variation of the predicted scores, i.e., the regression sum of squares. When the regression sum of squares is divided by the number of independent variables k, we have the **regression mean square** as the numerator of F. The denominator includes the variation of the residuals, or residual sum of squares. After dividing the residual sum of squares by $n - k - 1$, we have the **residual mean square** in the denominator. These statistics are provided as output by many computer programs, such as SPSS. Since the regression sum of squares is equal to $R^2\Sigma (Y - \overline{Y})^2$ and the residual sum of squares equals $(1 - R^2)\Sigma (Y - \overline{Y})^2$, F can be expressed as

$$F = \frac{\text{Regression Mean Square}}{\text{Residual Mean Square}} = \frac{R^2\Sigma(Y-\overline{Y})^2/k}{(1-R^2)\Sigma(Y-\overline{Y})^2/n-k-1}$$

$$= \frac{R^2/k}{(1-R^2)/(n-k-1)}$$

In the above formula, $\Sigma (Y - \overline{Y})^2$ (the variation of Y), which is contained in both the numerator and denominator, cancels out to give Equation 4.18. Equation 4.18 and the other F tests that contain R^2 are emphasized because of their greater conceptual and notational simplicity.

The null hypothesis for the multiple F test is that $R^2 = 0$, and the alternate hypothesis is that $R^2 > 0$. It should be recognized, however, that R^2 can only be equal to zero if the slopes for all the independent variables are zero and that it only takes one nonzero slope for R^2 to be greater than zero. Thus, the hypotheses may also be stated as follows:

$$H_0: \beta_1 = \beta_2 = \beta_3 = 0$$

$$H_1: \text{one or more } \beta_i \neq 0$$

Thus, the F test for the entire equation is a global test to determine if *any* of the independent variables has an effect on Y. Finding the multiple F to be significant, thus, does not tell the researcher how many variables have effects on Y. The tests for the individual partial slopes must be examined if we are to determine which variables have significant effects.

The F statistics for the anomia equations with three independent variables and five independent variables (Tables 4.3 and 4.4) are

$$F_{Y\cdot 123} = \frac{R^2_{Y\cdot 123}/3}{(1 - R^2_{Y\cdot 123})/(n - 3 - 1)} = \frac{.11015/3}{(1 - .11015)/509} = 21.003$$

$$F_{Y\cdot 12345} = \frac{R^2_{Y\cdot 12345}/5}{(1 - R^2_{Y\cdot 12345})/(n - 5 - 1)} = \frac{.11063/5}{(1 - .11063)/507} = 12.613$$

The critical value of F for $\alpha = .05$, $df_1 = 3$, and $df_2 = 509$ is 2.62 (Table A.3a in the Appendix). Since the first F above is greater than 2.62, we can reject the null hypothesis that $R^2 = 0$ at the $\alpha = .05$ level. Actually, for $\alpha = .001$ the critical

$F = 5.51$ for the above numerator and denominator degrees of freedom (Table A.3c in the Appendix), so the level of significance is $p < .001$. For the equation with five independent variables, the critical F for $\alpha = .001$, $df_1 = 5$, and $df_2 = 507$ is 4.18. Since the second F above is greater than 4.18, the level of significance for R^2 in the equation with five independent variables is also $p < .001$.

It is possible for the multiple F to be statistically significant but for none of the individual F/t tests to be significant. This case would indicate that some variable(s) affects Y, but the individual tests are not sensitive enough to reveal reliably which variable(s) accounts for the significant multiple effect. This might result, for example, if the X's are too highly correlated to separate their effects.

A second inconsistency between the multiple and individual tests is that the multiple F may not be significant but one or more of the individual tests may be significant. Some writers argue that in this case you should not draw any conclusions about the individual variables when the multiple test is not significant. The reasoning here is that when a number of individual tests are conducted, the probability increases that at least one of the individual variables will be "significant" by chance. If, for example, 20 X's were tested individually at the $\alpha = .05$ level, then approximately .05 of the X's (or about one X) might be found to be significant by chance. Using the multiple F as a screening test before examining the individual F's provides some protection against incorrectly rejecting the null hypothesis for one or more of the X's. If the multiple F is significant, it means that at least one of the variables has an effect on Y. It would then be appropriate to use the individual tests to determine which one(s) has such an effect. However, some writers, including this one, do not feel that this rule should be applied universally. Although it is not possible to resolve this issue here, you should be sensitive to the problem.

F for Subsets of *X*'s

Finally, the F test may be used to determine whether any variable within some subset of the X's has an effect on Y. In the case of three independent variables, for example, we might wish to test the subset consisting of X_1 and X_2 while holding X_3 constant. The F test here is

$$F = \frac{(R^2_{Y \cdot 123} - r^2_{Y3})/2}{(1 - R^2_{Y \cdot 123})/(n - 3 - 1)} \tag{4.19}$$

This is just another specific application of the general form for F given in Equation 4.14. The second term in the numerator is the variance explained by the variables that are being held constant (X_3). Since only one variable is being controlled in this case, the second term in the numerator is a bivariate r^2 instead of a multiple R^2. The numerator $df = 2$ because two variables are being tested. For this test we have the following hypotheses:

$$H_0: \beta_1 = \beta_2 = 0$$

$$H_1: \text{one or both } \beta_i \neq 0$$

If F is significant, it would be concluded that at least one of the variables has an effect on Y.

This type of test might be used in a situation where none of the variables in some conceptually similar subset has a significant partial slope. The failure to find significant individual effects for any of the variables in the subset might be caused by relatively high correlations among the variables in the subset, i.e., **high multicollinearity** within the subset of X's. Remember that the higher the R_i^2 for a variable (see Equation 4.13), the greater will be its standard error, and thus the less likely is it to be significant. For example, if two measures of socioeconomic status, such as education and occupational prestige, were included in an equation and neither had a significant effect, they might have a relatively high r between them. If the above F test were conducted for the subset of variables, it might be significant. If so, we could conclude that at least one of these variables is affecting Y, although it could not be reliably ascertained which one(s) was responsible. This would be better, however, than concluding on the basis of the individual tests for each variable that neither has an effect on Y.

The case of the three subjective income measures in the anomia example illustrates this issue and F test. When all three variables (SHORTINC, SATINC, REWRDINC) are in the equation, none are significant at $\alpha = .05$. But if we add the three subjective measures to an equation containing education and income, will the three variables significantly increase the explained variance? Using the R^2's from Equations 3 and 5 in Table 4.4, the F statistic for this test is

$$F = \frac{(R^2_{Y\cdot12345} - R^2_{Y\cdot12})/3}{(1 - R^2_{Y\cdot12345})/(n - 5 - 1)} = \frac{(.11063 - .10265)/3}{(1 - .11063)/507} = 1.516$$

The critical $F = 2.62$ for $\alpha = .05$, so we cannot reject the null hypothesis. Thus, we cannot conclude that subjective income, i.e., feeling income deprivation, increases alienation over and above the level of anomia that is caused by low income itself.

This is a somewhat ambiguous situation; when we used only the single measure of subjective income, i.e., SHORTINC, it was significant at $p = .04$. It might be added that if we had used either of the other two variables (SATINC or REWRDINC) alone, neither of them would have been statistically significant (not shown in Table 4.4). Therefore, it might be tempting to conclude that SHORTINC is *the* subjective aspect of income that affects alienation. The counterargument to this reasoning is that we need to control each of the other two variables when testing any single one of them; otherwise, we cannot be sure that one of them is exerting an effect *independently* of the other two. Thus, since

SHORTINC was not significant when the other two were controlled, and since the set of three together does not significantly increase the explained variance, we should not conclude that subjective income has an effect on anomia.

Detection and Correction of Heteroskedasticity

The efficiency of the least-squares b as an estimate of β in $Y = \alpha + \beta X + \varepsilon$ depends in part on the assumption that σ_ε^2 is constant across all values of X (i.e., ε is homoskedastic). The consequences of violating this assumption have been discussed in general: although b will still be an unbiased estimate of β, b will no longer be the most efficient unbiased estimator and the conventional formula for the standard error of b will itself be incorrect. It is therefore important to consider how to determine whether the homoskedasticity assumption is valid, and if heteroskedasticity exists, how to go about correcting our estimates. But given that ε is an unobserved variable, how can we diagnose whether this assumption is true? We can use the least-squares residuals $e = Y - \hat{Y}$ as estimates of ε and see if the variance of e differs significantly across X. Both graphical and statistical diagnostics for heteroskedasticity will be considered. If we conclude that the homoskedasticity assumption is false, we will first learn how to correct the standard error of the ordinary least-squares b. We will also learn how to use *weighted least-squares* to get a more efficient estimate of β.

Plots for Heteroskedasticity

Plotting the least-squares residuals $e = Y - \hat{Y}$ against the independent variable X is the most elementary way to attempt to discern visually the presence of nonconstant σ_ε^2. Remember that the mean of the residuals is zero. Additionally, if there is a truly linear relationship between X and Y, the conditional means of the residuals for each value of X should also be approximately equal to zero. Therefore, we should not be looking for changes in the mean value of e. Instead, we should be examining the plot for evidence that the spread or variance of e around zero is not constant across X. The most common pattern of heteroskedasticity when it exists is for the variance of ε, and thus the variance of e, to increase as X increases.

An alternative to plotting e against X is to plot either $|e|$ or e^2 against X. These transformations remove the negative signs from e. As a consequence, instead of looking to see if the variance of e is nonconstant we will be looking to see if the mean of the transformed e varies across X. Remember that the variance of any variable is equal to the mean of its squared deviation values. Thus, $\text{Var}(e) = (\Sigma e^2)/n = \overline{e^2}$ because e is a deviation score. Thus, when the mean of e^2 is not constant across X, it means that the variance of e is changing. When the mean of $|e|$ changes, it means that the average deviation is not constant; thus, we can also say that the dispersion of e is changing. I prefer to plot the squared residuals rather than the absolute values of the residuals because the homoskedasticity assumption refers to the constancy of the variance of the residuals, which equals the mean squared residual.[9]

When there are two or more independent variables, one may conduct an overall examination for heteroskedasticity by plotting either e, |e|, or e^2 against $\hat{Y}$. Since $\hat{Y}$ is a linear function of the X's, any relationships between the variance of the residuals and the X's should often produce observable relationships between the variance of the residuals and $\hat{Y}$. In order to determine which independent variables are related to the nonconstant error variance it would be necessary to plot the residuals against each X. In this case, residualized X's (e.g., $X_1 - \hat{X}_{1 \cdot 23}$, $X_2 - \hat{X}_{2 \cdot 13}$, and $X_3 - \hat{X}_{3 \cdot 12}$ for $k = 3$) should be plotted instead of raw X's in order to observe the unique relationship between the variance of e and each X.

Siblings and Fertility Example. Figure 4.9 shows the slightly jittered scatterplot of the Y residuals against X for the example data in which number of children is regressed on number of siblings (Table 4.1). The parallel strands of data points running downward from left to right are produced by the fact that Y is a discrete variable that takes on only whole numbers between 0 and 8. Therefore, there are nine parallel strands of points in Figure 4.9, one for each of the nine values of Y. The strands slope downward because the residuals for any given value of Y get smaller as X increases, due to the fact that the predicted values of Y increase linearly as X increases.

When inspecting Figure 4.9, we should be looking to see if there are any differences in the spread or variance of the residuals for different numbers of siblings. Although there appears to be the same range of residuals for values of X up to about ten siblings (where the number of cases becomes much smaller), you cannot judge the variance of the conditional distributions by looking only at the maximum and minimum values. Also, it is quite difficult to judge visually whether the residuals are becoming more concentrated or less concentrated around the middle of the conditional distributions as X varies.

Figure 4.10 shows the scatterplot of the squared residuals against the number of siblings. This plot does not help much to clarify the analysis. If anything, the downward trend of the top two strands of data points in Figure 4.10 suggests that the mean squared residual, and thus the variance of the residuals, may be declining as X increases. This would be in contrast to most cases of heteroskedasticity, where the variance is more likely to increase as X increases.

---

9. Ironically, the residuals may be heteroskedastic even when ε is homoskedastic (Fox 1991, p. 25). High-leverage observations may have smaller residuals because they are able to force the regression surface to be close to them. Because of this, Fox (1991) recommends plotting the absolute value or the squared value of the *studentized residuals* against X. Studentized residuals are corrected for the heteroskedasticity created by leverage and standardized to a t distribution. Studentized residuals, however, may be highly correlated with the raw residuals, especially in large samples.

FIGURE 4.9 Jittered Plot of Children Residuals and Siblings

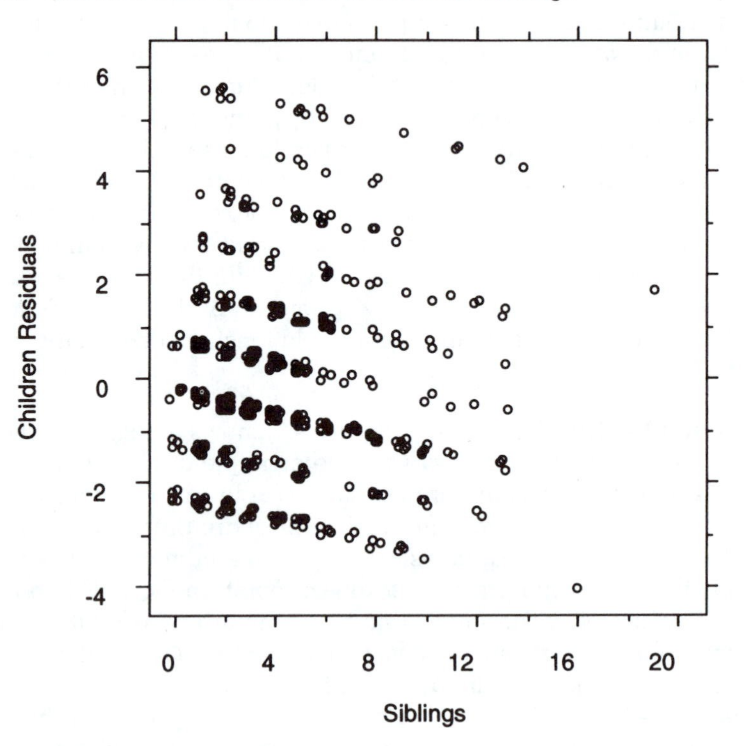

Earnings, Education, and Experience Example. We will look at the residuals resulting from the regression of earnings on experience and education; both of the independent variables had positive effects on earnings (Table 3.9; Figure 3.7). The first scatterplot (Figure 4.11) shows the earnings residuals plotted against the predicted values of earnings. The greater the number of years of education and experience, the greater will be the predicted earnings (e.g., 16 years of school and 40 years of experience lead to predicted earnings of 157.68 or $15,768). The scatterplot shows a fairly clear funnel-shaped pattern of earnings residuals, with the dispersion of earnings residuals increasing as predicted earnings increases. This means that heteroskedasticity appears to have a positive correlation with predicted earnings.

Although the picture provided by Figure 4.11 is fairly clear, we will also examine the squared residuals plotted against predicted earnings (Figure 4.12) to see how it displays the apparent heteroskedasticity. We would expect to see the mean squared residual increase as predicted earnings rise. Figure 4.12a shows five cases with squared residuals that are greater than 20,000 while the other 542 cases (n = 547) have squared residuals that are less than 20,000. As

FIGURE 4.10 Jittered Plot of Squared Children Residuals and Siblings

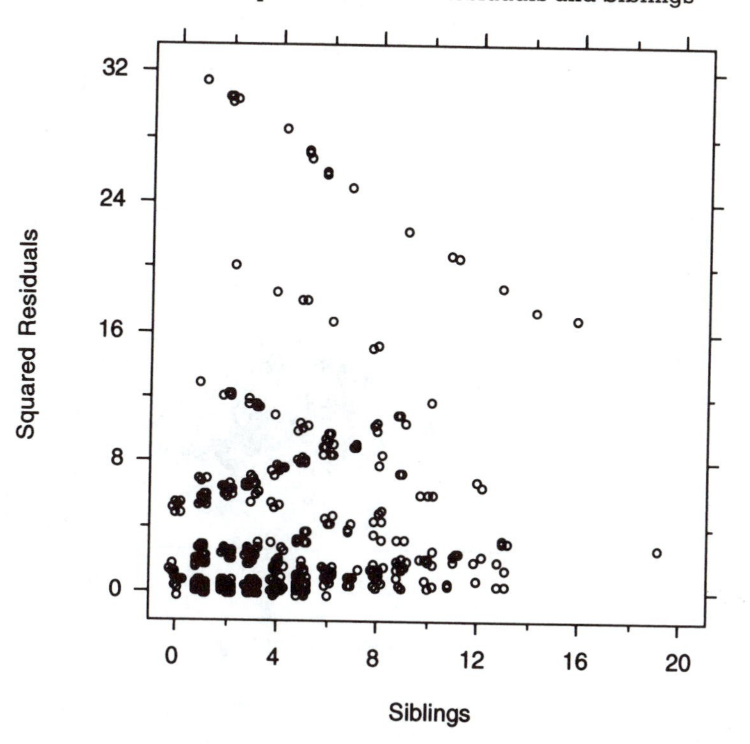

a result, 6/7 of the vertical dimension in the plot is allocated for the five largest residuals. Consequently, the remaining 542 residuals are compressed into a relatively small space, making it difficult to discern any pattern in the variance of the residuals for the vast majority of cases. Figure 4.12b rescales the vertical axis to a range of 0 to 20,000 and thus removes the five largest residuals. The rescaled plot spreads out the data points vertically and as a consequence more clearly displays the differences in squared residuals that exist for these 542 cases. The plot clearly shows that the squared residuals tend to increase, on the average, as predicted earnings increase; for example, as the predicted values increase there appears to be a larger and larger proportion of cases with squared residuals greater than 2,500. Thus, Figure 4.12b leads to the same visual interpretation as Figure 4.11; as predicted earnings increase the variance of e increases.

Since predicted earnings are a positive function of both experience and education, the heteroskedasticity might be attributed to experience, education, or both. Figure 4.13 plots the earnings residuals against the education residuals ($X_1 - \hat{X}_{1 \cdot 2}$) and against the experience residuals ($X_2 - \hat{X}_{2 \cdot 1}$). The funnel-shaped

FIGURE 4.11 Earnings Residuals and Earnings Predicted by Experience and Education

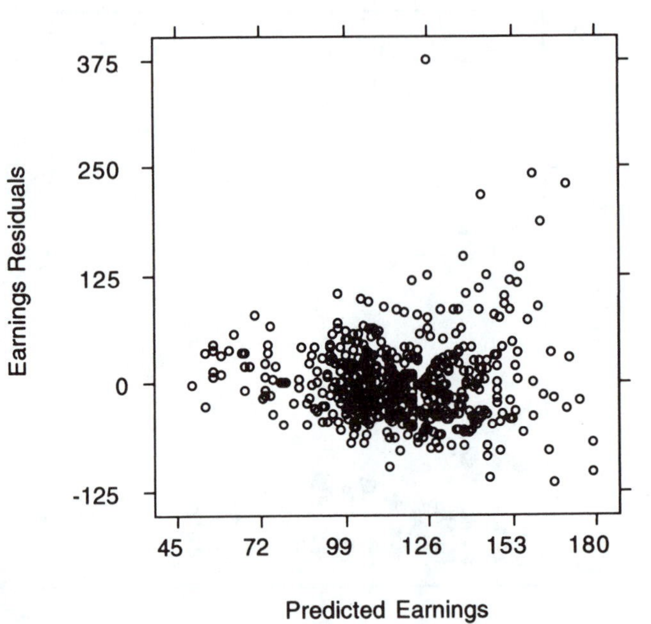

pattern of the dispersion of the residuals in relationship to education (Figure 4.13a) looks quite similar to its relationship to predicted earnings; as years of education increase, the variance of the residuals increases. The relationship between the dispersion of the residuals and years of experience (Figure 4.13b), however, does not look the same as for education. Although there does appear to be some increase in the variance of the residuals as experience increases, the relationship looks weaker and somewhat different in form (at least to the author's eyes). The statistical tests for heteroskedasticity to be discussed next will be helpful in this regard.

To summarize, the plots for diagnosing heteroskedasticity did not provide any conclusive basis for questioning the homoskedasticity assumption in the case of children regressed on siblings. For earnings regressed on experience and education, however, the plots showed much stronger evidence of heteroskedasticity, especially in relationship to education. We now turn to statistical tests for heteroskedasticity.

White's Test for Heteroskedasticity

A number of different statistical tests for violation of the homoskedasticity assumption have been proposed [see Gujarati (1988) and Kamenta (1986) for re-

FIGURE 4.12 (a & b) Squared Earnings Residuals and Predicted Earnings

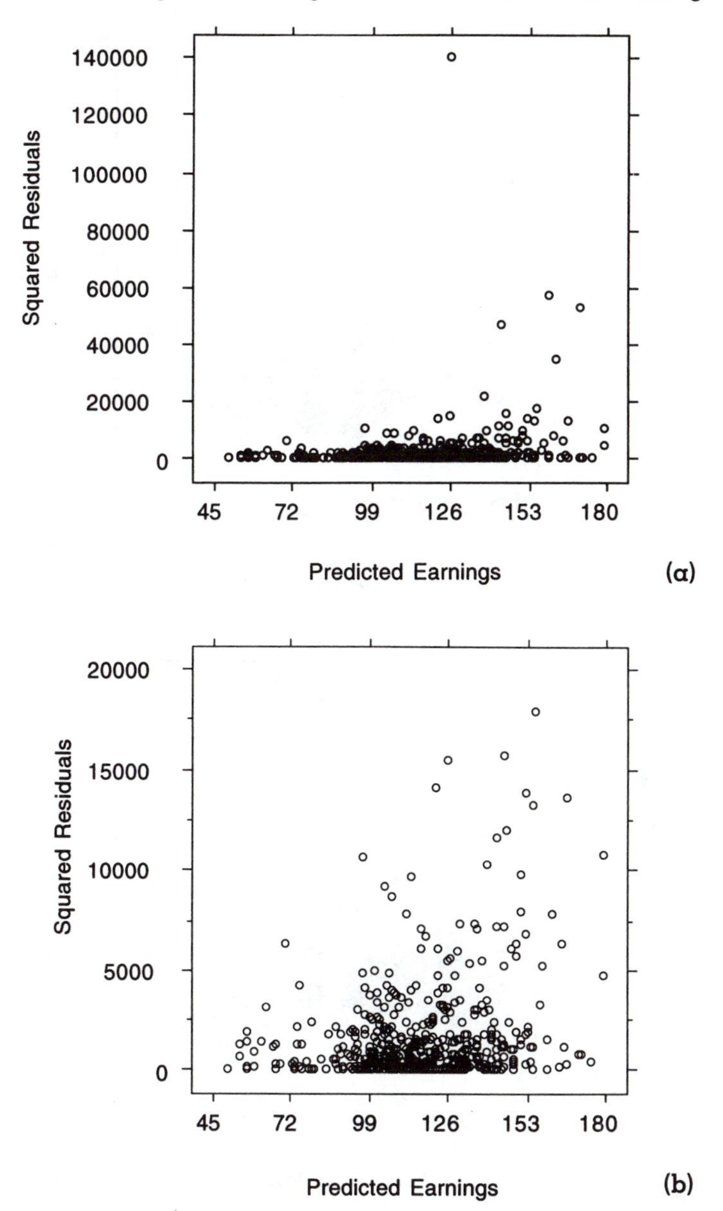

FIGURE 4.13 (a & b) Earnings Residuals Against Education Residuals and Experience Residuals

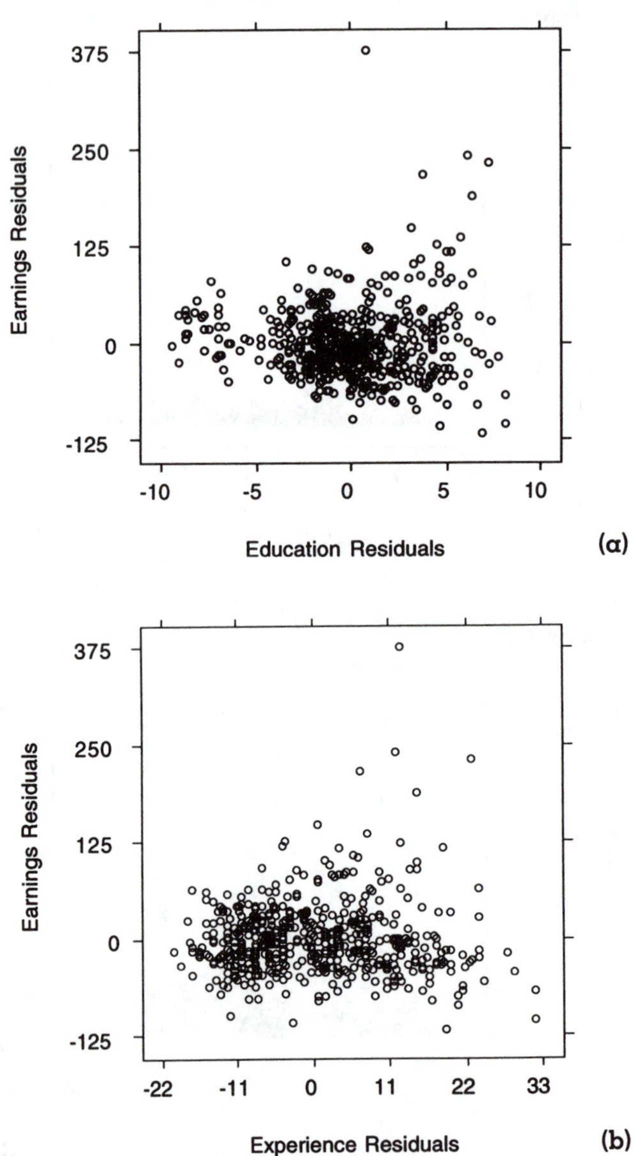

views]. Among these are three similar tests that are relatively simple to execute and understand (Breusch and Pagan 1979; White 1980; Cook and Weisberg 1983). White's test will be described here because it is not as dependent on the assumption of normality of e as are the other two methods (Kamenta 1986, p. 295).

White's test involves estimating a regression equation in which the squared residuals are specified to be a function of a set of X's such as

$$\widehat{e^2} = a' + b_1' X_1 + b_2' X_2 + \cdots + b_h' X_h \tag{4.20}$$

The X's are usually the variables included in the regression equation for Y, but they could include other X's as well. Thus, the h independent variables in Equation 4.20 may be different from the k independent variables in the regression equation for Y. The symbols a' and b' are used for the intercept and slopes in Equation 4.20 to distinguish them from a and b in the equation for Y. Equation 4.20 specifies only linear and additive effects of the X's. After nonlinear relationships (Chapter 6) and nonadditive relationships (Chapter 7) have been covered, you should be able to modify Equation 4.20 to allow for more complex effects.[10]

White's test involves the following steps to test the *null hypothesis of homoskedasticity*:

1. Run the regression equation $\hat{Y} = a + b_1 X_1 + b_2 X_2 + \cdots + b_k X_k$.
2. Save the residuals $e = Y - \hat{Y}$ from Step 1.
3. Run Equation 4.20.
4. Using R^2 from Step 3, calculate $\chi^2 = nR^2$ with $df = h$.
5. If the calculated χ^2 is greater than the critical χ^2 with $df = h$ at the selected α level, reject the null hypothesis.

This chi-square test is a general test for any heteroskedasticity that may be linearly related to at least one of the X's. The chi-square statistic is used instead of the F test for R^2 because the least-squares residuals e may be heteroskedastic even when ε is homoskedastic (Fox 1991, p. 25). High-leverage observations may have smaller residuals because they are able to force the regression surface to be close to them. Thus, the assumptions justifying the F test do not hold when e^2 is the dependent variable, as in Equation 4.20.

Siblings and Fertility Example. The results of regressing number of children on number of siblings were given in Table 4.1. We will use White's test to see if any heteroskedasticity is linearly related to X = SIBS. In order to conduct

10. White (1980) recommends $\widehat{e^2} = a' + b_1' X_1 + b_2' X_2 + b_3' X_1^2 + b_4' X_2^2 + b_5' X_1 X_2$, in the case of two independent variables, as sufficiently flexible to fit most kinds of nonlinearity and nonadditivity that might be encountered. Thus, White's test does not require the investigator to have a theory about the form of the relationship between e^2 and the X's.

TABLE 4.5 Squared Residuals of Children Regressed on Siblings

****MULTIPLE REGRESSION****

Dependent Variable ... RESIDSQ					
Multiple R	.13539	Analysis of Variance			
R Square	.01833		DF	Sum of Sqs	Mean Sq
Adjusted R Square	.01633	Regression	1	278.3318	278.3318
Standard Error	5.50933	Residual	491	14903.1875	30.3527
		F = 9.1699		Signif F = .0026	

------------------ Variables in the Equation ------------------

Variable	B	SE B	Beta	T	Sig T
SIBS	.23735	.07838	.1354	3.0282	.0026
(Constant)	2.60185				

White's test for heteroskedasticity, the children residuals must be saved, squared, and then regressed on siblings. Example SPSS commands for carrying out these steps are given below:

```
REGRESSION VARIABLES = CHILDS SIBS
  /DEPENDENT = CHILDS
  /ENTER SIBS
  /SAVE = RESID
COMPUTE RESIDSQ = RESID**2
REGRESSION VARIABLES = RESIDSQ SIBS
  /DEPENDENT = RESIDSQ
  /ENTER SIBS
  /SAVE = PRED
```

RESID is the name of a temporary variable that equals $Y - \hat{Y}$. The residuals from the first regression run are saved for use in subsequent procedures by use of the SAVE subcommand. The saved residuals are then squared by the COMPUTE command and placed in a new variable RESIDSQ, which is e^2. RESIDSQ is then regressed on SIBS in a second regression run to calculate the necessary statistics for the test of the null hypothesis of homoskedasticity. The output of the regression of e^2 on X is given in Table 4.5.

The results in Table 4.5 show the following ordinary least-squares estimate of the linear relationship between the variance of the residuals of children and the independent variable siblings:

$$\widehat{e^2} = 2.60185 + .23735X$$

This equation has a positive slope and thus indicates that as the number of siblings increases there is an increase in the variance of the residuals of chil-

FIGURE 4.14 Regression Line and Scatterplot

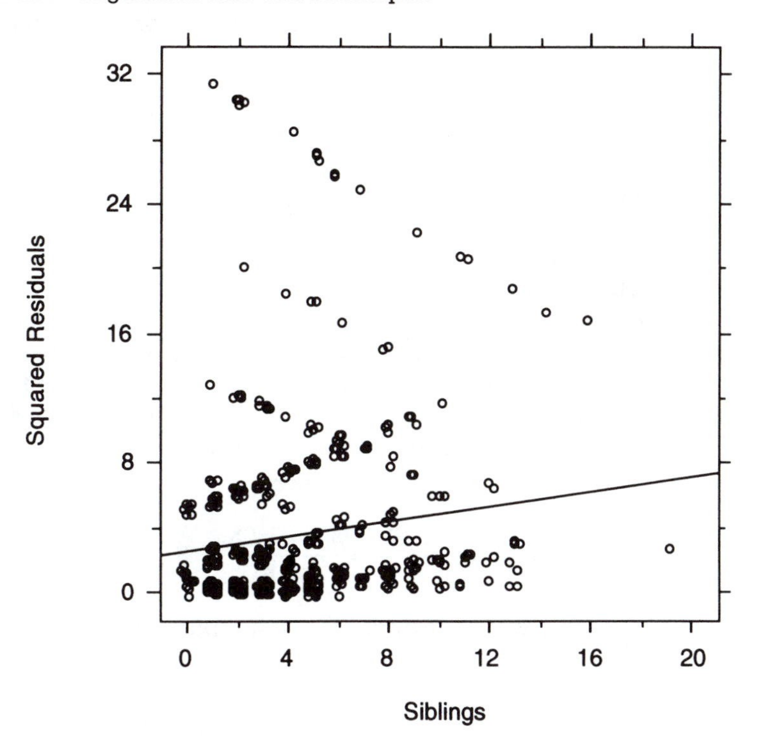

dren. The heteroskedastic variance of the residuals in the sample data might have occurred by chance when the population is characterized by homoskedasticity. Although the t and F tests in Table 4.5 indicate that this is not very likely, these tests are biased. The chi-square test described above is the appropriate procedure. Using R^2 from Table 4.5, we get

$$\chi^2 = nR^2 = 493(.01833) = 9.037$$

With $df = 1$, the critical value of χ^2 at $\alpha = .05$ equals 3.841. Therefore, we reject the null hypothesis of homoskedasticity.

The regression analysis has detected a pattern of heteroskedasticity that was not readily apparent to the eye in the plot of the squared residuals against siblings (Figure 4.10). Figure 4.14 shows the regression line $\hat{e^2} = 2.60185 + .23735X$ plotted on the scatterplot of squared residuals and siblings.

How much heteroskedasticity does this regression line indicate, and how serious are the consequences? The changes in $\hat{e^2}$ represent the estimated heteroskedasticity, that is, the changes in the estimated variance of ε. A summary description of this heteroskedasticity is provided by the variance or standard deviation of $\hat{e}^2$, which can be calculated from the regression sum of squares in Table 4.5:

$$\text{VAR}(\hat{e}^2) = \frac{\text{Reg SS}}{n-2} = \frac{\sum(\hat{e}^2 - \overline{e^2})^2}{n-2} = \frac{278.3318}{491} = .5669$$

$$s_{\hat{e}^2} = \sqrt{.5669} = .7529$$

Thus, we are estimating that the standard deviation of the error variance equals .7529, which is the estimated amount of heteroskedasticity.

The coefficient of variability (CV) is often used to help evaluate the magnitude of a standard deviation. The coefficient of variability equals the standard deviation of a variable divided by the mean of the variable. This statistic is based on the fact that variables with larger means tend to have larger standard deviations. Using the results given in Table 4.1, we can find the mean of $\hat{e}^2$ as follows:

$$\overline{\hat{e}^2} = \overline{e^2} = s_e^2 = \frac{\sum e^2}{n-2} = \frac{1784.2305}{491} = 3.6339$$

This is the estimated error variance under the assumption of homoskedasticity. The coefficient of variability is

$$\text{CV} = \frac{s_{\hat{e}^2}}{\overline{\hat{e}^2}} = \frac{s_{\hat{e}^2}}{s_e^2} = \frac{.7529}{3.6339} = .2072 \qquad \textbf{(4.21)}$$

The coefficient of variability indicates that the heteroskedasticity of the error variance is about .2 as large as the constant error variance would be under homoskedasticity.

Although there is no upper limit on the CV, it does not appear to indicate a large amount of heteroskedasticity. We can also compare the CV for children with the CV for earnings in the following example to determine which is estimated to have the greatest heteroskedasticity. The best indication of the seriousness of the heteroskedasticity can be seen after we correct the usual formula for the estimated standard error of the slope, which is biased under homoskedasticity, and after we construct a weighted least-squares b in order to get a more efficient estimate of β. If the heteroskedasticity is really large, we should notice a sizeable difference between the usual OLS standard error and the corrected OLS standard error and we should notice a substantial decrease in the WLS standard error in comparison with the corrected OLS standard error. The point is that although the unequal error variances are statistically significant, it does not automatically follow that correcting for heteroskedasticity will necessarily offer a substantial improvement.

TABLE 4.6 Squared Residuals of Earnings Regressed on Experience and Education

****MULTIPLE REGRESSION****

Dependent Variable ... RESIDSQ					
Multiple R	.208	Analysis of Variance			
R Square	.043		DF	Sum of Sqs	Mean Sq
Adjusted R Square	.040	Regression	2	.137448e+10	.687242e+09
Standard Error	7468.374	Residual	544	303425e+11	.557766e+08
		F = 12.321		Signif F = .000	

--------------- Variables in the Equation ---------------

Variable	B	SE B	Beta	T	Sig T
EXPER	117.251	30.914	.168	3.793	.000
EDUC	427.605	100.475	.189	4.256	.000
(Constant)	−5684.839				

Earnings, Education, and Experience Example. The results of regressing earnings on experience and education were given in Table 3.9. We will conduct White's test by using both experience and education to predict the earnings residuals. The form of the SPSS commands is identical to those that were shown for children and siblings. The results of regressing the squared residuals on experience and education are given in Table 4.6.

Both education and experience are positively related to the squared residuals: $\widehat{e^2} = -5684.839 + 117.251\text{EXPER} + 427.605\text{EDUC}$.[11] This result was rather clearly shown in the scatterplots, although the relationship appeared stronger for education than for experience. The standardized regression coefficients (BETA's in Table 4.6), however, indicate that the two independent variables have about equal effects on e^2. The chi-square statistic with $df = 2$ is

$$\chi^2 = nR^2 = 547(.043) = 23.521$$

This is greater than the critical value of 5.991 for $\alpha = .05$ and $df = 2$. We therefore reject the null hypothesis of homoskedasticity.

The quantities for calculating the CV according to Equation 4.21 are found in Tables 3.9 and 4.6.

11. Using the nonlinear and nonadditive function recommended by White in the case of two independent variables gives $\widehat{e^2} = 7884.1 - 1310.8X_1 - 148.4X_2 + 50.0X_1^2 - 0.45X_2^2 + 22.4X_1X_2$, with $R^2 = .058$. In this equation, X_1 is education and X_2 is experience. It shows a nonlinear effect of education, with heteroskedasticity increasing more rapidly as education increases. It also shows an interaction between education and experience such that education has a stronger positive effect on heteroskedasticity when experience is high. There are also no predictions of negative error variances by this equation, unlike the linear function (notice the negative intercept in the linear function).

$$CV = \frac{s_{\hat{e}^2}}{\overline{\hat{e}^2}} = \frac{s_{\hat{e}^2}}{s_e^2} = \frac{\sqrt{(.137448E + 10)/544}}{1{,}213{,}433.937/544} = .7126$$

This result indicates that the standard deviation of the error variance is approximately .7 as great as the overall error variance itself. The CV for earnings indicates considerably more heteroskedasticity than was indicated by CV = .21 for children. This suggests that correcting for heteroskedasticity of the earnings residuals may have a more substantial effect on the standard errors in the earnings equation than correcting for heteroskedasticity will have in the case of the equation for children.

Correcting the OLS Standard Error

When heteroskedasticity exists, the correct formula for the standard error of the least-squares slope is

$$\sigma_b' = \frac{\sqrt{\sum (X_i - \overline{X})^2 \sigma_{\varepsilon_i}^2/n}}{s_X^2\sqrt{n}} \tag{4.22}$$

When there is heteroskedasticity, $\sigma_{\varepsilon_i}^2$ is not constant across values of X; thus, the subscript i is used to indicate that the variance of ε is a variable. In Equation 4.22, each $\sigma_{\varepsilon_i}^2$ is weighted by $(X_i - \overline{X})^2$. Error variances associated with values of X that are far above or below the mean will count more than error variances that are associated with values of X that are close to the mean. Thus, if the pattern of heteroskedasticity is such that large error variances tend to go with large squared deviations of X (large ε's occur more frequently for extreme X's), the numerator of Equation 4.22 will be large, and thus the standard error of b will be large. If large error variances tend to go with small squared deviations of X, the standard error will be small.

When homoskedasticity exists, $\sigma_{\varepsilon_i}^2$ will have the same value for all values of X; that is, $\sigma_{\varepsilon_i}^2 = \sigma_\varepsilon^2$ for all X. Then Equation 4.22 becomes

$$\sigma_b' = \frac{\sqrt{\sum (X_i - \overline{X})^2 \sigma_{\varepsilon_i}^2/n}}{s_X^2\sqrt{n}} = \frac{\sigma_\varepsilon s_X}{s_X^2\sqrt{n}} = \frac{\sigma_\varepsilon}{s_X\sqrt{n}}$$

That is, Equation 4.22 simplifies to Equation 4.3, the formula for the standard error of b under the assumption of homoskedasticity.

Because Equation 4.22 contains $\sigma_{\varepsilon_i}^2$, an unknown, it cannot be used to correct σ_b when heteroskedasticity exists. White (1980), however, has proposed a method of estimating the standard error of the slope when heteroskedasticity exits. White's method is to use e_i^2 as an estimate of $\sigma_{\varepsilon_i}^2$ in Equation 4.22.

$$\sigma_b' = \frac{\sqrt{\sum (X_i - \overline{X})^2 e_i^2/n}}{s_X^2\sqrt{n}} \tag{4.23}$$

It makes sense to use e_i^2 as an estimate of $\sigma_{\varepsilon_i}^2$ because $\sigma_{\varepsilon_i}^2$ is the mean ε^2 for all cases with the same value of X. Equation 4.23 indicates that the squared residuals for each case are multiplied by each case's squared deviation score on X. Compare Equation 4.23 to the usual formula for the standard error (Equation 4.6):

$$s_b = \frac{\hat{s}_{Y-\hat{Y}}}{s_X\sqrt{n}} = \frac{\sqrt{(\sum e^2)/n - 2}}{s_X\sqrt{n}}$$

The main difference is that e^2 is not multiplied by $(X_i - \overline{X})^2$ in the usual formula.

Under very general conditions, the mean of the sampling distribution of White's corrected estimator (Equation 4.23) gets closer and closer to the true sampling error (Equation 4.22) as sample size increases. In other words, White's estimator is, roughly speaking, unbiased in large samples. There is some evidence, however, that in samples of small and moderate size, the estimator is downwardly biased (Kamenta 1986, p. 293).

Siblings and Fertility Example. White's test for heteroskedasticity showed a significant positive relationship between siblings and the variance of the residuals for children (Table 4.5). We will therefore use White's estimator (Equation 4.23) to correct the standard error of the slope. Some statistical packages (e.g., SAS) compute Equation 4.23 as an optional output. In other packages you will have to supply the necessary instructions to compute the corrected standard error. The new quantity that is needed to compute Equation 4.23 is $\Sigma (X_i - \overline{X})^2 e_i^2$. The following SPSS commands illustrate one method of computing Equation 4.23:

```
COMPUTE XDEVIATE = X - 4.286
COMPUTE XSQESQ = (XDEVIATE**2)*(RESID**2)
DESCRIPTIVES VARIABLES = XSQESQ
  /STATISTICS = SUM MEAN
```

The mean of siblings (4.286) that is needed to compute $X - \overline{X}$ (XDEVIATE) in the first command comes from the output of a previous procedure (Table 4.1). The second statement computes $(X - \overline{X}_i)^2 e_i^2$, using the RESID variable that can be saved from a previous REGRESSION procedure, and gives it the variable name XSQESQ. The third command specifies that a procedure named DESCRIPTIVES is to compute the sum and mean of XSQESQ, which equal 20,934.719 and 42.464, respectively. Either the sum or mean of $(X - \overline{X}_i)^2 e_i^2$ may be inserted in the numerator of Equation 4.23, along with n and the standard deviation of X (from Table 4.1) in the denominator, to get the corrected standard error.

$$\sigma_b' = \frac{\sqrt{\sum (X_i - \overline{X})^2 e_i^2/n}}{s_X^2\sqrt{n}} = \frac{\sqrt{20934.719/493}}{(3.166)^2\sqrt{493}} = \frac{\sqrt{42.464}}{(3.166)^2\sqrt{493}} = .0293$$

Because there is a positive correlation between X and e^2, we would expect the corrected standard error to get larger. The usual standard error equals .0271 (Table 4.1). Thus, the corrected standard error is slightly larger, as expected. The increase, however, is rather slight. The least-squares regression slope $b = .1116$ (Table 4.1) does not need to be corrected because the least-squares slope is still unbiased when the homoskedasticity assumption is violated. The next step is to compute the corrected t statistic as follows:

$$t' = \frac{b}{s_b'} = \frac{.1116}{.0293} = 3.809$$

The corrected t is only slightly smaller than the uncorrected $t = 4.114$ (Table 4.1). Although the biased s_b underestimated the OLS standard error, and thus might have led us to reject incorrectly the null hypothesis, the correction was so slight in this case that we must clearly still reject the null hypothesis. This illustrates the point that statistically significant heteroskedasticity does not necessarily seriously bias our hypothesis tests.

Earnings, Education, and Experience Example. White's test indicated significant heteroskedasticity of the earnings residuals, and Table 4.6 showed that this heteroskedasticity is positively related to both education and experience. Furthermore, the coefficient of variability indicated that this income heteroskedasticity is greater than the heteroskedasticity for the siblings and fertility example. We should therefore expect to see a bigger change in the corrected standard errors for education and experience than we observed for siblings.

Because we have two independent variables, we will have to modify Equation 4.23 slightly to correct the standard errors of education and experience.[12] The modification consists of substituting a residualized X for the raw X in Equation 4.23. In the case of X_1 = EDUC, we substitute $X_1 - \hat{X}_{1\cdot2}$ for $X_1 - \overline{X}_1$ and $s_{(X_1-\hat{X}_{1\cdot2})}$ for s_X to get

$$s_{b_1}' = \frac{\sqrt{\sum (X_1 - \hat{X}_{1\cdot2})^2 e^2/n}}{s^2_{(X_1-\hat{X}_{1\cdot2})}\sqrt{n}} \qquad \textbf{(4.24)}$$

In order to calculate Equation 4.24 we would regress X_1 on X_2 and save the residuals $X_1 - \hat{X}_{1\cdot2}$. We could also use the output of this regression to compute $s^2_{(X_1-\hat{X}_{1\cdot2})} = s_1^2(1 - r_{12}^2)$. Then we would use the saved residuals $X_1 - \hat{X}_{1\cdot2}$ and e (from the regression of Y on X_1 and X_2) to compute the product $(X_1 - \hat{X}_{1\cdot2})^2 e^2$. Finally, the mean of $(X_1 - \hat{X}_{1\cdot2})^2 e^2$ would be acquired from a descriptive statis-

12. The matrix algebra formula for White's corrected standard error is $\mathbf{S} = (\mathbf{X'X})^{-1}\mathbf{X'\hat{\Sigma}X(X'X)}^{-1}$, where $\hat{\mathbf{\Sigma}}$ is an $n \times n$ diagonal matrix with $e_1^2, \ldots, e_n^2$ on the diagonal (Fox 1991).

tics procedure. When these quantities are computed and entered into Equation 4.24, we get

$$s'_{b_1} = \frac{\sqrt{33660.24949}}{10.10071\sqrt{547}} = .7766$$

In an analogous procedure, the corrected standard error for experience is found to be

$$s'_{b_2} = \frac{\sqrt{\sum (X_2 - \hat{X}_{2\cdot 1})^2 e^2/n}}{s^2_{(X_2 - \hat{X}_{2\cdot 1})}\sqrt{n}} = \frac{\sqrt{319{,}822.7}}{106.6955\sqrt{547}} = .2266$$

The ratios of the corrected standard errors to the usual standard errors calculated earlier in this chapter are $s'_{b_1}/s_{b_1} = .777/.635 = 1.224$ and $s'_{b_2}/s_{b_2} = .227/.196 = 1.158$ for education and experience respectively. That is, the standard error of the slope for education increased by 22 percent and the standard error of the slope for experience increased by 16 percent. These increases are at least twice as great as the increase in the standard error of the slope for children regressed on siblings, as was expected. The corrected t values are

$$t'_1 = \frac{b_1}{s'_{b_1}} = \frac{6.789}{.777} = 8.737$$

$$t'_2 = \frac{b_2}{s'_{b_2}} = \frac{.954}{.227} = 4.20$$

Both of these values are still highly significant. Thus, although the usual formulas for the OLS standard errors lead to appreciable underestimates of the standard errors, the effects of education and experience on earnings are so strong that the existing heteroskedasticity was again insufficiently strong to lead to an incorrect rejection of the null hypothesis. We will now examine an alternative to ordinary least-squares regression that should be able to provide even more efficient estimates of the effects of the independent variables.

Weighted Least-Squares Regression

When heteroskedasticity exists, the OLS estimates of α and β are no longer BLUE; that is, they are no longer the most efficient unbiased estimators. Even though the OLS estimates are unbiased, and even though the OLS estimated standard errors may have been corrected, there are other unbiased estimators that have smaller standard errors. *Weighted least-squares* (WLS) is one such method. Whereas OLS regression selects values of a and b that minimize SSE $= \Sigma (Y - \hat{Y})^2 = \Sigma (Y - a - bX)^2$, WLS regression computes a regression equation $\hat{Y} = a^* + b^*X$ in which a^* and b^* are the values of the intercept and slope that minimize

$$\text{SSE}^* = \sum w_i(Y_i - \hat{Y}_i)^2 = \sum w_i(Y_i - a^* - b^*X_i)^2 \tag{4.25}$$

In Equation 4.25, w_i is a weight that is assigned to each case; the greater the w_i, the more case i influences the values of a^* and b^*. In OLS regression $w_i = 1$ for each case; i.e., each case is weighted equally.

When heteroskedasticity exists, the objective of weighted least-squares regression is to give cases with greater error variances ($\sigma^2_{\varepsilon i}$) less weight in the regression solution. Remember that according to Equation 4.22, the greater $\sigma^2_{\varepsilon i}$, the more case i will tend to inflate the standard error of the slope. Thus, ideally we would like to make $w_i = 1/\sigma^2_{\varepsilon i}$ and minimize

$$\text{SSE}^* = \sum \frac{1}{\sigma^2_{\varepsilon i}}(Y_i - \hat{Y}_i)^2 = \sum \frac{1}{\sigma^2_{\varepsilon i}}(Y_i - a^* - b^*X_i)^2 \tag{4.26}$$

In Equation 4.26 the greater the $\sigma^2_{\varepsilon i}$ associated with X_i, the less weight case i will have in determining the values of a^* and b^*. Of course, we don't know $\sigma^2_{\varepsilon i}$ and thus we will have to use estimates of $\sigma^2_{\varepsilon i}$ in order to weight the data.

The formulas for b^* and a^* that minimize Equation 4.25 are[13]

$$b^* = \frac{\sum w_i(X_i - \tilde{X})(Y_i - \tilde{Y})}{\sum w_i(X_i - \tilde{X})^2} \tag{4.27}$$

and

$$a^* = \tilde{Y} - b^*\tilde{X} \tag{4.28}$$

where

$$\tilde{X} = \frac{\sum w_i X_i}{\sum w_i} \quad \text{and} \quad \tilde{Y} = \frac{\sum w_i Y_i}{\sum w_i}$$

In Equations 4.27 and 4.28, $\tilde{X}$ and $\tilde{Y}$ are the weighted means of X and Y (if $w_i = 1$, then $\tilde{X} = \Sigma X/n = \overline{X}$ and $\tilde{Y} = \Sigma Y/n = \overline{Y}$). Thus, Equations 4.27 and 4.28 are similar to the usual OLS formulas for a and b; b^* is equal to the weighted covariation between X and Y (or weighted sum of cross-products) divided by the weighted variation in X (or weighted sum of squares). Once estimates of $\sigma^2_{\varepsilon i}$ have been chosen to construct the w_i, the quantities needed for b^* can be computed by first computing the weighted means, then computing the deviations from the weighted means, and then computing the weighted sum of cross-products and the weighted sum of squares.

An easier method of computing WLS estimates of α and β involves using the weights to transform X and Y and then applying OLS regression to the

13. The matrix algebra formula for the WLS estimates is $\mathbf{b}^* = (\mathbf{X}'\mathbf{W}\mathbf{X})^{-1}\mathbf{X}'\mathbf{W}\mathbf{Y}$, where $\mathbf{W}$ is a diagonal $n \times n$ matrix with $w_1, \ldots, w_n$ on the diagonal.

transformed variables. The logic of this approach can be shown by first rewriting Equation 4.25 as

$$\begin{aligned} SSE^* &= \sum w_i(Y_i - a^* - b^*X_i)^2 \\ &= \sum (\sqrt{w_i}\,Y_i - a^*\sqrt{w_i} - b^*\sqrt{w_i}\,X_i)^2 \\ &= \sum (Y_i^* - a^*w_i^* - b^*X_i^*)^2 \end{aligned} \quad \textbf{(4.29)}$$

where $Y_i^* = \sqrt{w_i}Y_i$, $w_i^* = \sqrt{w_i}$, and $X_i^* = \sqrt{w_i}X_i$. Equation 4.29 shows that minimizing the sum of the weighted squared residuals (the WLS solution) is equivalent to minimizing the sum of the squared residuals from the following regression equation:

$$\hat{Y}_i^* = a^*w_i^* + b^*X_i^* \quad \textbf{(4.30)}$$

Equation 4.30 has two independent variables, the transformed weights and the transformed X. The slope for the transformed weight variable is a^* and serves as our WLS estimate of α. The slope for the transformed X is b^* and serves as our WLS estimate of β. You have probably noticed that Equation 4.30 does not have an intercept term. Regression equations without an intercept are referred to as *regression through the origin*. In such cases the regression surface is constrained to pass through the origin, e.g., $X_1 = X_2 = Y = 0$. In other words, when the independent variables are all equal to zero, the equation is constrained to predict $\hat{Y} = 0$. Formulas for computing bivariate and partial b's under the constraint of regression through the origin may be found in Gujarati (1988, p. 137) and Kamenta (1986, p. 477). These formulas will not be utilized here because most computer programs have an option for regression through the origin.

As indicated above, we would like to conduct a WLS analysis in which cases with greater error variances would receive less weight. Thus, we will have to estimate $\sigma^2_{\varepsilon_i}$ and use $w_i = 1/\hat{\sigma}^2_{\varepsilon_i}$. In White's test for heteroskedasticity, the error variances were estimated with Equation 4.20, $\hat{e}^2 = a' + b_1' X_1 + b_2'X_2 + \cdots + b_h' X_h$. We will therefore use $\hat{e}^2$ as our estimate of $\sigma^2_{\varepsilon_i}$ and thus our weights will be $w_i = 1/\hat{e}_i^2$. Substituting $\hat{e}^2$ for $\sigma^2_{\varepsilon_i}$ in Equation 4.26, our WLS solution will minimize the following equation:

$$SSE^* = \sum \frac{1}{\hat{e}_i^2}(Y_i - \hat{Y}_i)^2 = \sum \frac{1}{\hat{e}_i^2}(Y_i - a^* - b^*X_i)^2 \quad \textbf{(4.31)}$$

The a^* and b^* that minimize Equation 4.31 will be the WLS estimates of α and β. The standard error of the slope provided by the WLS solution (s_{b^*}) is the primary objective of the analysis. The s_{b^*} should be smaller than the OLS s_b.

OLS regression can be used to compute a^* and b^* in a fashion specified by Equations 4.29 and 4.30. In order to use Equation 4.30 to compute a^* and b^* we will need to create the following new variables:

$$Y_i^* = \sqrt{w_i}\, Y_i = \sqrt{\frac{1}{\hat{e}_i^2}}\, Y_i = \frac{Y_i}{\sqrt{\hat{e}_i^2}}$$

$$X_i^* = \sqrt{w_i}\, X_i = \sqrt{\frac{1}{\hat{e}_i^2}}\, X_i = \frac{X_i}{\sqrt{\hat{e}_i^2}}$$

$$w_i^* = \sqrt{w_i} = \frac{1}{\sqrt{\hat{e}_i^2}}$$

With these transformed variables we will be able to compute the WLS a^* and b^* with the OLS regression $\hat{Y}_i^* = a^* w_i^* + b^* X_i^*$. If we have a multiple regression equation, we will need to compute a new X^* variable for each X by dividing each X by $\sqrt{\hat{e}_i^2}$.

Siblings and Fertility Example. A weighted least-squares solution for number of children regressed on number of siblings will be computed using the procedures outlined above. Some of the SPSS commands that are needed in this method were given previously in the section on White's (1980) test for heteroskedasticity. They are given again as follows, along with the additional necessary commands.

```
REGRESSION VARIABLES = CHILDS SIBS
  /DEPENDENT = CHILDS/ENTER SIBS
  /SAVE = RESID
COMPUTE RESIDSQ = RESID**2
REGRESSION VARIABLES = RESIDSQ SIBS
  /DEPENDENT = RESIDSQ/ENTER SIBS
  /SAVE = PRED
COMPUTE WTCHILDS = CHILDS/SQRT(PRED)
COMPUTE WTSIBS = SIBS/SQRT(PRED)
COMPUTE WTSQROOT = 1/SQRT(PRED)
REGRESSION VARIABLES=WTCHILDS WTSIBS WTSQROOT
  /ORIGIN
  /DEPENDENT = WTCHILDS
  /ENTER WTSIBS WTSQROOT
```

The steps involved in this procedure are as follows:

1. Run an OLS regression of children (Y) on siblings (X), and save the residuals $e = Y - a - bX$ (named RESID in the SPSS commands).
2. Square the residuals to obtain a variable that will be used to estimate σ_ε^2 (named RESIDSQ in the SPSS commands).

TABLE 4.7 Weighted Least-Squares Regression of Children on Siblings

****MULTIPLE REGRESSION****

Dependent Variable . . . WTCHILDS

- - - - - - - - - - - - - - - - -Variables in the Equation- -

Variable	B	SE B	T	Sig T
WTSQROOT	2.21538	.13723	16.143	.000
WTSIBS	.12154	.02915	4.169	.000

3. Run an OLS regression of e^2 on siblings, and save the predicted squared residuals $\widehat{e^2}$ (named PRED in the SPSS commands).
4. Transform children and siblings by dividing each by $\sqrt{\widehat{e^2}}$, and compute $1/\sqrt{\widehat{e^2}}$ (named WTCHILDS, WTSIBS, and WTSQROOT, respectively, in the SPSS commands).
5. Run an OLS regression through the origin procedure with the weighted children variable regressed on the weighted siblings variable and $1/\sqrt{\widehat{e^2}}$, (the regression through the origin option is invoked by the SPSS subcommand ORIGIN).

The output of Step 5 gives a^*, b^*, and s_{b^*}, which are the WLS estimates of α, β, and σ_{b^*}.

Table 4.7 gives part of the results of regressing the weighted children variable on the weighted siblings variable and the square root of the weights variable. Table 4.7 gives only part of the output because statistics such as the residual sum of squares, the regression sum of squares, and R^2 are either not accurate or not relevant due to the facts that the regression involves transformed variables and does not have an intercept (Pindyck and Rubinfield 1981; Gujarati 1988). The slopes in Table 4.7 provide the WLS estimates for

$$\widehat{\text{CHILDS}} = a^* + b^*\text{SIBS} = 2.2154 + .1215\text{SIBS}$$

The WLS intercept and slope are very similar to $a = 2.2582$ and $b = .1116$ from the OLS solution (Table 4.1). The standard error of the WLS slope equals .02915, which is only a tiny fraction smaller (i.e., more efficient) than the corrected OLS standard error, which equals .0293. Thus, in the case of children and siblings, despite the detection of statistically significant heteroskedasticity, the uncorrected OLS solution has not been improved upon.

One further technical point needs to be mentioned, however. The above WLS equation can be used to get WLS predictions of Y that in turn can be used to get the WLS residuals $Y - \hat{Y}$. These can then be used to calculate r^2 (or R^2) in the usual manner:

$$r^2 = 1 - \frac{\sum (Y - \hat{Y}_{WLS})^2}{\sum (Y - \bar{Y})^2} = 1 - \frac{1784.722290}{1845.72\ 8} = .03305$$

This value is very close to the OLS $r^2 = .03331$.

It should also be mentioned that many regression procedures have options for computing WLS solutions that eliminate the need for computing the transformed variables that we have used to obtain a WLS solution. After saving the predicted squared residuals, the following SPSS commands can be used to invoke the WLS option:

```
COMPUTE WT = 1/PRED
REGRESSION VARIABLES = CHILDS SIBS
  /REGWGT = WT
  /DEPENDENT = CHILDS
  /ENTER SIBS
```

The subcommand REGWGT invokes the WLS solution, which uses WT $= 1/\hat{e}^2$ as the weight for minimizing Equation 4.25. The results are identical to those in Table 4.7.

Earnings, Education, and Experience Example. In order to compute a WLS solution for earnings (Y) regressed on education (X_1) and experience (X_2) with the method that was used for children and siblings, the following OLS regression is computed:

$$\hat{Y}^* = a^*w^* + b_1^*X_1^* + b_2^*X_2^*$$

which represents

$$\frac{\hat{Y}}{\sqrt{\hat{e}^2}} = a^* \frac{1}{\sqrt{\hat{e}^2}} + b_1^* \frac{X_1}{\sqrt{\hat{e}^2}} + b_2^* \frac{X_2}{\sqrt{\hat{e}^2}}$$

Because subscripts are needed for the two X variables, the subscript i has been dropped to reduce notational clutter. Using SPSS commands analogous to those outlined in detail for the siblings and fertility example, the output shown in Table 4.8 can be obtained.

The WLS solution from Table 4.8 is only slightly different from the OLS solution given in Table 3.9, as can be seen below:

$$\widehat{\text{EARN}} = 14.793 + 6.203\text{EDUC} + 1.107\text{EXPER} \quad \text{(WLS)}$$

$$\widehat{\text{EARN}} = 10.720 + 6.798\text{EDUC} + 0.954\text{EXPER} \quad \text{(OLS)}$$

The OLS, corrected OLS, and WLS standard errors are listed below to facilitate comparisons.

TABLE 4.8 **WLS Regression of Earnings on Education and Experience**

* * * * M U L T I P L E   R E G R E S S I O N * * * *

Dependent Variable . . . WTEARNS

- - - - - - - - - - - - - - - - - - -Variables in the Equation- -

Variable	B	SE B	T	Sig T
WTSQROOT	14.79297	5.74042	2.57698	.0102
WTEDUC	6.20253	.38974	15.91453	.0000
WTEXPER	1.10718	.16935	6.53765	.0000

	s_b	s'_b	s^*_b
EDUC	.635	.777	.389
EXPER	.196	.227	.169

The expected patterns are displayed: the corrected OLS s'_b is somewhat larger than the usual OLS for both education and experience, which we would expect from the positive correlation of the error variance of earnings with each independent variable; the WLS s^*_b is smaller than the corrected OLS s'_b for both education and experience, as we would also expect. But it is also apparent that the differences in standard errors are much greater for education than for experience. Especially noticeable is the fact that the WLS standard error is only half as large as the corrected OLS standard error for education. The WLS solution produced a much more efficient estimate of the effect of education on earnings in comparison to the OLS solution (even though the null hypothesis is decisively rejected in both solutions).

Summary

In order to understand better the subject of testing hypotheses about population parameters, we began this chapter with a discussion of two basic sampling concepts, the sampling distribution and the standard error of the least-squares slope. The distribution of the values of b for all distinct combinations of elements that might occur in a random sample of size n is called the sampling distribution of b. The standard error of b is simply the standard deviation of the sampling distribution. The theoretical sampling distribution of b was described in terms of three important properties: b is normally distributed; the mean of the distribution equals β (i.e., b is unbiased); and the standard error of b is a direct function of the standard error of estimate and an inverse function of the standard deviation of X and the sample size n. When there is more than one X, the standard error of b is also affected by a variance inflation factor (VIF); the higher the multiple correlation of one X with the other X's (an indicator of multicollinearity), the greater will be the VIF and thus the larger will be the standard error.

Knowledge about the properties of the sampling distribution of b allows us to test hypotheses about β. A common null hypothesis is that β equals zero. This hypothesis can be tested with a t statistic, which equals the sample b divided by the estimated standard error of b. The t statistic indicates how many standard deviations b is from the mean of the sampling distribution, under the assumption that the null hypothesis is true. Knowledge about the form of the t distribution allows us to calculate the probability that the null hypothesis is true, given the sample t. If that probability is very small (conventionally, less than or equal to .05), then the null hypothesis is unlikely to be true and is thus rejected in favor of the alternative hypothesis (e.g., the two-tailed hypothesis that b is not equal to 0).

A general formula for a very flexible F statistic that can be used in hypothesis testing was also presented. The F statistic can be used as an alternative to the t statistic for testing the hypothesis that the population slope equals zero. In this case (when the numerator degrees of freedom equals unity), $F = t^2$. The F statistic also allows us to test hypotheses about sets of variables, such as whether the squared multiple correlation is zero and whether a subset of variables has any relationship to the dependent variable.

In order to justify least-squares hypothesis testing, four assumptions are made about the error term in the population equation: ε is not correlated with X (thus, b is unbiased); the variance of ε is constant for all values of X, and different ε's are not correlated with one another (thus, the formula for the standard error of b is correct); and ε is normally distributed (thus, b is normally distributed). If these assumptions are correct, the Gauss-Markov theorem says that b is the most efficient unbiased estimator of β. If ε is correlated with X, then b will be biased. If ε does not have constant variance (heteroskedasticity), or if the ε's are correlated with one another, then the usual formula for the standard error of b is incorrect (and thus, the t test is biased) and the least-squares b is not the most efficient estimator of β. If ε is not normally distributed, then t will be biased in small samples, and there are biased estimators of β that are more efficient than the least-squares b in large samples.

Finally, we considered how to detect and correct for the violation of the assumption that the error variance is homoskedastic. Plotting residuals and squared residuals against X, or against the predicted Y, may provide a visual detection of heteroskedasticity. White's statistical test for heteroskedasticity was described and illustrated. Next, two methods of correcting for significant heteroskedasticity were described. The first involves White's formula for correcting the standard error of the least-squares slope so that an unbiased t test can be conducted. The second method involves using weighted least-squares instead of ordinary least-squares to estimate the population slope. Weighted least-squares gives less weight to cases that have greater variances of ε. Weighted least-squares can provide a more efficient estimate of the population slope when heteroskedasticity is present than can ordinary least-squares, even when the standard error of the ordinary least-squares b has been corrected.

References

Berry, William D., and Stanley Feldman. 1985. *Multiple Regression in Practice*. Beverly Hills: Sage.

Breusch, T. S., and A. R. Pagan. 1979. "A Simple Test for Heteroskedasticity and Random Coefficient Variation." *Econometrica* 47:1287–1294.

Cook, R. Dennis, and Sanford Weisberg. 1983. "Diagnostics for Heteroskedasticity in Regression." *Biometrika* 70:1–10.

Fox, John. 1991. *Regression Diagnostics*. Newbury Park: Sage.

Greene, William H. 1990. *Econometric Analysis*. New York: Macmillan.

Gujarati, Damodar N. 1988. *Basic Econometrics* (2nd edition). New York: McGraw-Hill.

Judd, Charles M., and Gary H. McClelland. 1989. *Data Analysis: A Model-Comparison Approach*. San Diego: Harcourt Brace Jovanovich.

Kamenta, Jan. 1986. *Elements of Econometrics*. New York: Macmillan.

Pindyck, R., and D. Rubinfield. 1981. *Econometric Models and Economic Forecasts*. New York: McGraw-Hill.

Srole, Leo. 1956. "Social Integration and Certain Corollaries: An Exploratory Study." *American Sociological Review* 30 (October):709–716.

Weisberg, Sanford. 1985. *Applied Linear Regression*. New York: Wiley.

White, Halbert. 1980. "A Heteroskedasticity-Consistent Covariance Matrix Estimator and a Direct Test for Heteroskedasticity." *Econometrica* 48:817–838.

Zeller, Richard A., and Edward Carmines. 1980. *Measurement in the Social Sciences*. New York: Cambridge University Press.

5 Nominal Independent Variables

In the preceding chapters, we have assumed that the variables used in a regression equation are measured at least at the *interval* level or, if *ordinal* variables are used, that scores can be assigned to the ordered categories that approximate the intervals between the categories (see Chapter 1 for definitions of the levels of measurement). In order to compute sums, deviation scores, and the amount by which Y changes per unit increase in X, it is necessary to have measures of the variables that reflect the amount by which one case differs from another case on each variable. *Nominal* variables do not provide information about the intervals between the categories of a variable nor do they rank-order the categories. Nominal variables are simply classification schemes that group units into mutually exclusive categories, where each category is defined by some attribute shared by all members of the category. Examples of nominal variables are sex (male or female), race (e.g., black, white, American Indian, or Asian), marital status (married, separated, divorced, widowed, or never-married), and employment status (employed full-time, employed part-time, not employed, or retired). For variables such as these, there is no measurement system that specifies how high or low members of one category are in comparison to members of other categories. Although it is common to assign numeric codes to the categories of nominal variables (e.g., full-time = 1, part-time = 2, not employed = 3, retired = 4), these numbers are purely arbitrary and are used only for purposes of record-keeping and data processing.

Nevertheless, nominal variables may still contain information that is useful for explaining variation in Y. The fact that these variables do not provide ordinal and interval information does not mean that there are no meaningful differences between the categories in terms of the dependent variable under investigation. Before we can use regression analysis to extract such potentially valuable information, however, it is necessary to create one or more appropriately coded variables from the nominal variable. These new variables are then included in the regression equation to represent the nominal variable. Although not just any coding scheme for these variables will work, there are several alternative schemes that will provide meaningful interpretations and tests of the effects of the nominal variable on Y. We will examine *dummy variable coding* (perhaps the most widely used coding) as well as *effects coding* and *contrast coding*.

Dummy Variable Coding

Dichotomous Nominal Variables

We will begin with the simplest type of nominal variable, one that contains only two categories (i.e., a dichotomous variable). We will also begin with bivariate regression before moving on to the multivariate case. When using regression to analyze a dependent variable and a nominal independent variable, the focus is on whether the categories of the nominal variable (whether there are two or more than two) have different means on Y. For example, do blacks and whites (the nominal variable) have different income means ($\bar{Y}$'s), or do Protestants, Catholics, Jews, and atheists (the nominal variable) have different mean frequencies of church or temple attendance ($\bar{Y}$'s)? When there are only two groups/categories, a t test for the difference of means may be used to decide if the mean difference is significant; when there are two or more categories of the nominal variable, the F test provided by a one-way analysis of variance (ANOVA) may be used to determine whether there are any mean differences in Y between the groups defined by the nominal variable. Regression analysis can also be used to extract the same information and more.

In order to understand how regression analysis can be used to test for differences in means, it is useful to remember that the linear regression of Y on an interval level X is also a test for differences in means. Figure 5.1 shows a regression line fit to a scatterplot of a perfectly linear relationship between X and Y. For each value of X, there is a distribution of Y scores (a conditional distribution of Y). For each value of X, the regression line passes through the conditional mean of Y ($\bar{Y}_1$, $\bar{Y}_2$, $\bar{Y}_3$, $\bar{Y}_4$), that is, through the mean of Y for all cases with the same value of X. The discussion of the sum of squares in Chapter 2 pointed out that *the mean is a least-squares statistic*; that is, the sum of squared deviations around the mean of a variable is smaller than the sum of squared deviations around any other value of that variable. As a result, the prediction

FIGURE 5.1 Scatterplot and Linear Regression Line for X and Y

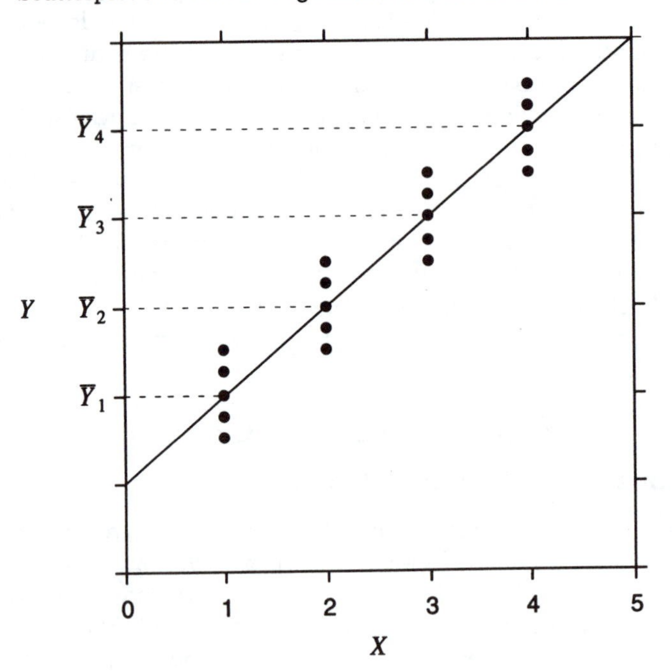

of Y that minimizes the sum of squared residuals (the least-squares criterion) for all cases with the same value of X is the mean of Y for those cases (i.e., the conditional mean of Y). Thus, a perfectly linear relationship between X and Y is one in which the conditional means of Y all fall on a straight line. The test for the regression slope is actually a test for a linear trend in the conditional means of Y.

When X is a dichotomous nominal variable, there will be a conditional distribution of Y for each of the two groups defined by X. When using X to predict Y, the least-squares criterion would lead to the prediction of the conditional mean of Y for each group/category defined by X. This is shown in Figure 5.2 for the two categories of the nominal variable, which are labeled A and B. The prediction of $\overline{Y}_A$ for all cases in group A would minimize the sum of squared residuals for group A, i.e., $\Sigma\,(Y_A - \overline{Y}_A)^2$, and the prediction of $\overline{Y}_B$ for all cases in group B would minimize $\Sigma\,(Y_B - \overline{Y}_B)^2$. Thus, the least-squares regression line would pass through the two conditional means in order to minimize the total sum of squared residuals, as shown in Figure 5.2. Notice that no numeric values are shown for groups A and B. In terms of finding the best-fitting line, it is arbitrary how we score A and B. Whatever scores we assign to A and B (which

FIGURE 5.2 Least-Squares Line for Y and a Dichotomous Nominal X

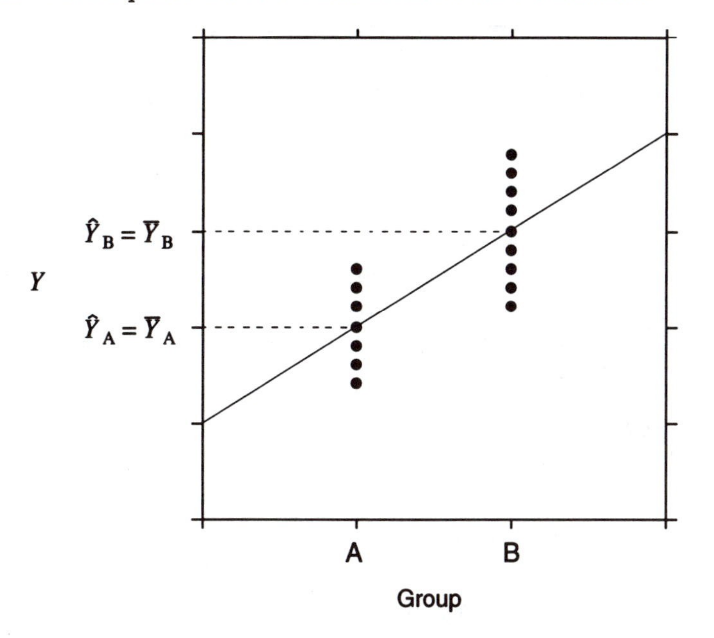

we can refer to as X_A and X_B), the least-squares line will pass through the coordinates $(X_B, \bar{Y}_B)$ and $(X_A, \bar{Y}_A)$. The slope of the line will be

$$b_{YX} = \frac{\bar{Y}_B - \bar{Y}_A}{X_B - X_A} = \frac{\bar{Y}_A - \bar{Y}_B}{X_A - X_B} \quad \textbf{(5.1)}$$

Equation 5.1 indicates that regardless of which group is given the higher score and what the difference in the scores is, we can find a line that gives the best predictions of Y simply by dividing the difference between the means of Y for groups A and B by the difference between whatever scores we assign to groups A and B. We will see, however, that some scores that we can assign to X will give particularly meaningful interpretations of the slope.

To illustrate the above points we will examine some data from the 1982 Akron Area Survey, a random-digit-dialed telephone survey of residents of Summit County, Ohio. The dependent variable Y is a life satisfaction scale ranging from one to seven, with $Y = 1$ meaning *dissatisfied* and $Y = 7$ meaning *satisfied*. The independent variable is race, with $X = 1$ being the arbitrary code assigned to blacks and $X = 2$ being the code assigned to whites by the Akron Area Survey. Table 5.1 summarizes the values of Y for each race.

Nonsense Coding. To illustrate the arbitrariness of the coding of the race variable, *nonsense coding* will be used instead of the codes provided by the

TABLE 5.1 Life Satisfaction (Y) by Race (X)

Label	*n*	*Mean*	*St. Dev.*	*Variance*
Black	63	4.7619	2.1153	4.4747
White	656	5.5168	1.3546	1.8348
Total	719	5.4506	1.4512	2.1058

Akron Area Survey data set, which were RACE = 1 for blacks and RACE = 2 for whites. The nonsense codes that have been assigned are 74 for blacks and 19 for whites. The SPSS commands for transforming the race variable to the nonsense codes and using it in a regression run are shown below.

```
RECODE RACE (1 = 74) (2 = 19)
REGRESSION VARS = SATLIFE RACE/
  DEP = SATLIFE/ ENTER RACE
```

The output of the regression run will give a value for the least-squares slope that will be equal to the following value computed by Equation 5.1:

$$b_{YX} = \frac{\overline{Y}_B - \overline{Y}_W}{X_B - X_W} = \frac{4.7619 - 5.5168}{74 - 19} = \frac{-.7549}{55} = -.01373$$

The intercept for the equation is computed according to the usual formula. First, the mean for the nonsense-coded X must be calculated:

$$\overline{X} = \frac{n_W X_W + n_B X_B}{n} = \frac{656(19) + 63(74)}{719} = 23.8192$$

The intercept can now be computed as follows:

$$a = \overline{Y} - b\overline{X} = 5.4506 - (-.01373)23.8192 = 5.7776$$

The least-squares regression equation is thus

$$\hat{Y} = 5.7776 - .01373X$$

This equation has a negative slope, indicating that race has a negative effect on life satisfaction. The negative slope has meaning only when we remember that blacks were arbitrarily given a higher score on X than whites. Thus, the negative b reflects the fact that the group with the higher score on RACE (blacks) has lower life satisfaction. If we interpret the slope literally as the change in Y per unit increase in X, we would say that as race increases by one unit, satisfaction decreases by .01373 units (on a seven-point scale). This literal interpretation doesn't make any sense in terms of the groups that are represented by X because a one-unit difference on X doesn't correspond to the difference between the scores of blacks and whites on X; the difference between blacks and whites equals $74 - 19 = 55$, not 1. Thus, the nonsense coding leads to a nonsense interpretation of the slope. Furthermore, the intercept is meaningless because no group has a score of 0 on X.

The equation, however, does lead to meaningul predictions of Y:

$$\hat{Y}_W = 5.7776 - .01373(19) = 5.5167$$

$$\hat{Y}_B = 5.7776 - .01373(74) = 4.7616$$

The predicted score for whites equals the mean for whites, and the predicted score for blacks equals the mean for blacks (see Table 5.1). These predictions can be used to compute the coefficient of determination. In order to compute this statistic we need the sum of squared residuals for the predicted values of Y. Since the predicted score for each race equals its mean on Y, the variance of Y for each race shown in Table 5.1 equals the variance of the residuals for each race. If we multiply the variance for each race by the number of respondents of each race minus 1, we will get the sum of squared residuals for each race:

$$\sum (Y_W - \hat{Y}_W)^2 = (n_W - 1)s^2_{\hat{Y}_W} = 655(1.8348) = 1201.794$$

$$\sum (Y_B - \hat{Y}_B)^2 = (n_B - 1)s^2_{\hat{Y}_B} = 62(4.4747) = 277.4314$$

Adding the above equations together gives the total sum of squared residuals for all respondents:

$$\sum (Y - \hat{Y})^2 = 1201.7940 + 277.4314 = 1479.2254$$

The sum of squares for Y can be computed by multiplying $n - 1$ times the variance of Y shown in Table 5.1 for all respondents:

$$\sum (Y - \overline{Y})^2 = (n - 1)s^2_Y = 718(2.1058) = 1511.9644$$

The coefficient of determination, or r^2, equals

$$r^2_{YX} = \frac{\sum (Y - \overline{Y})^2 - \sum (Y - \hat{Y})^2}{\sum (Y - \overline{Y})^2} = \frac{1511.9644 - 1479.2254}{1511.9644} = .02165$$

This indicates that race explains about 2 percent of the variance in life satisfaction. This isn't a very high proportion, which indicates that there is a lot of variance in satisfaction within each race. The racial difference, however, is not trivial; blacks and whites differ by about .75 points on a seven-point scale. This difference is equivalent to about half of a standard deviation in Y. The reason r^2 is low is that blacks comprise less than ten percent of the sample; thus, there is very little variance in X with which to explain Y. This fact, however, does not mean that the racial difference is unimportant. This is a case in which you must pay attention to factors in addition to measures of the strength of association.

Nonsense coding will also give valid significance tests (t and/or F) for the effect of the dichotomous X on Y. We can use r^2 to calculate the F statistic for the null hypothesis that $b_{YX} = 0$:

$$F = \frac{r^2_{YX}}{(1 - r^2_{YX})/n - 2} = \frac{.02165}{(1 - .02165)/717} = 15.8666$$

For $F = 15.8666$, $df_1 = 1$, and $df_2 = 718$, $p < .001$. Thus, we would reject the null hypothesis that blacks and whites are equally satisfied with life and conclude that life satisfaction is higher for whites. Since $t^2 = F$ for b, $t = \sqrt{F} = \sqrt{15.8666} = -3.9832$. (Note that although the square root of a number is always positive, the proper sign of t is negative because the slope is negative.)

We have seen that nonsense coding can be validly used with a dichotomous nominal variable; any codes will work. The coding will not affect the measures of strength of association, such as r, r^2, and B; this will be illustrated later. The regression slope will have the correct sign, which is easily interpreted if you keep in mind which group is given the highest score on X. You can make sense of the value for b if you use the difference between the scores of the two groups on X; if you multiply the slope times the difference in scores you will get the difference between the means for the two groups, i.e., $\overline{Y}_B - \overline{Y}_W = b_{YX}(X_B - X_W) = -.01373(74 - 19) = -.7552$. A final important point that will be illustrated below is that no matter what scores you give to the two groups, your decision will not affect the values of t and F in the test for the significance of the slope; t and F are invariant with respect to the scoring of X.

Dummy Coding. *Dummy variable* coding is often used as a method that provides more readily interpretable slopes and intercepts for nominal independent variables. In dummy coding, one group is given a score of 1 and the other group is given a score of 0. In this example, blacks will be given a score of 1 and whites a score of 0.

The regression slope can again be calculated with Equation 5.1, since the best prediction for each group will still be each group's mean on Y:

$$b_{YX} = \frac{\overline{Y}_B - \overline{Y}_W}{X_B - X_W} = \frac{4.7619 - 5.5168}{1 - 0} = \frac{-.7549}{1} = -.7549$$

As can be seen, *the slope for a dummy variable will equal the mean of Y for the group coded 1 on X minus the mean of Y for the group coded 0 on X.* This results because the difference between their scores on X is unity. The sign of b indicates whether the group scored 1 has a higher mean (+ slope) or a lower mean (− slope) than the group scored 0. The slope for race indicates that the mean life satisfaction for blacks is .7549 less than that for whites. The slope can also be interpreted in the usual manner; when there is a positive unit difference in X (which corresponds to the racial difference in values of X) there will be a difference in Y equal to b (i.e., −.7549).

Since the least-squares equation predicts the conditional mean of Y for each of the two groups represented by a dummy variable X, the line will pass through the coordinates $(0, \bar{Y}_W)$ and $(1, \bar{Y}_B)$. The first coordinate corresponds to $X = 0$; thus, the predicted value when $X = 0$ is $\bar{Y}_W$, which by definition is the Y-intercept. *The intercept in a dummy variable regression is therefore the mean of the group coded 0 on the dummy variable* (in this case, whites). Thus, the intercept equals 5.5168. This makes the value of the intercept a meaningful quantity. Having determined the slope and intercept, we can write the dummy variable equation for race and life satisfaction as

$$\hat{Y} = 5.5168 - .7549X$$

Figure 5.3 shows the least-squares line for life satisfaction regressed on the dummy race variable. The line crosses the Y-axis at a value equal to the mean for whites, the group coded 0 on X. The figure also shows that the slope of the line equals the mean of the group scored 1 (blacks) minus the mean of the group scored 0 (whites).

FIGURE 5.3 Life Satisfaction (Y) Regressed on Dummy-Coded Race (X)

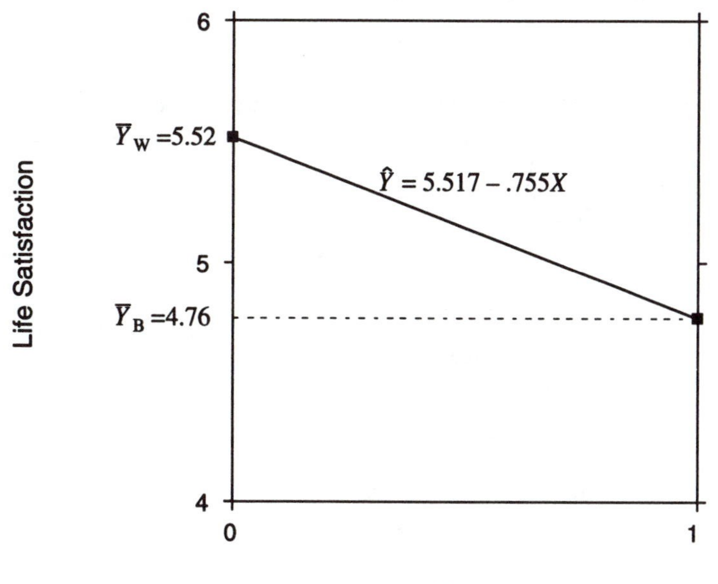

We can verify that the intercept equals the mean of the group scored 0 by using the usual formula for the intercept. First, we need to compute the mean of X, which is

$$\overline{X} = \frac{\sum X}{n} = \frac{1 + 1 + \cdots + 1 + 0 + 0 + \cdots + 0}{n} = \frac{n_B}{n} = \frac{63}{719} = .0876 = p$$

Since the scores on X are either 1's or 0's, the sum of X will equal the number of cases scored 1 on X. In this case, that will be the number of blacks. When the sum is divided by n, it is shown that the mean equals a proportion, i.e., the proportion of all cases with a score of 1. In the above formula the mean is .0876, which indicates that .0876 of the respondents are black. In general, the mean of a dummy variable equals the proportion of cases that are in the group coded 1. Without proof, the standard deviation of a dummy variable can be computed with the following formula:

$$s_X = \sqrt{p(1 - p)} = \sqrt{(.0876)(.9124)} = \sqrt{.0799} = .2827$$

Thus, the standard deviation of the dummy race variable is .2827, and the variance (the number under the radical) is .0799; the variance equals $p(1 - p)$. The intercept of the least-squares line can now be computed as

$$a = \overline{Y} - b\overline{X} = 5.4506 - (-.7549)(.0876) = 5.5167$$

This verifies that the intercept equals the mean of the group scored 0. The same value for the intercept is reported in the SPSS output shown in Table 5.2.

TABLE 5.2 Regression of Life Satisfaction on Dummy-Coded Race

```
RECODE RACE (1 = 1) (2 = 0)
REGRESSION VARS = SATLIFE RACE/
  DEP = SATLIFE/ ENTER RACE
```

* * * * MULTIPLE REGRESSION * * * *

Multiple R	.14718	Analysis of Variance			
R Square	.02166		DF	Sum of Squares	Mean Square
Adjusted R Square	.02030	Regression	1	32.75310	32.75310
Standard Error	1.43635	Residual	717	1479.24412	2.06310

F = 15.87566 Signif F = .0001

------------------------------Variables in the Equation------------------------------

Variable	B	SE B	Beta	T	Sig T
RACE	−.75486	.18945	−.14718	−3.984	.0001
(Constant)	5.51677	.05608		98.373	.0000

Table 5.2 gives the SPSS statements that may be used to run a dummy variable regression of life satisfaction on race, along with the output of the SPSS run. The values for the bivariate slope and the intercept shown in Table 5.2 are virtually identical to those calculated previously.

To state in a more general form the characteristics of dummy variable regression that have been covered thus far, let

$$Y_0 = Y \text{ scores when } X = 0$$

$$Y_1 = Y \text{ scores when } X = 1$$

The least-squares solution for a dummy-coded X always gives the following values for the slope and intercept:

$$b_{YX} = \overline{Y}_1 - \overline{Y}_0 \quad \textbf{(5.2)}$$

$$a_{YX} = \overline{Y}_0 \quad \textbf{(5.3)}$$

As was implied in the discussion of nonsense coding, dummy variable coding will give the same values of t, F, r, and r^2 as nonsense coding or any other coding. This is illustrated in the SPSS regression analysis shown in Table 5.2, where we can see that F and r^2 are nearly identical to the values that we calculated earlier for nonsense coding. The dummy variable solution differs from the solutions using other coding only with respect to the slope and intercept of the regression equation.

It should also be noted that the t test for b when X is dummy-coded is equivalent to a t test for the difference of means. This is because the slope for a dummy X equals the difference between the two groups' means on Y (as shown above). Thus, the following hypotheses apply for a bivariate regression with a dummy X:

$$H_0: b = 0 \text{ or } \overline{Y}_1 = \overline{Y}_0$$

$$H_1: b \neq 0 \text{ or } \overline{Y}_1 \neq \overline{Y}_0$$

The equivalence between dummy regression for a dichotomous X and the t test for a difference of two group means is illustrated in Appendix 5A, where the output of the SPSS procedure T–TEST is discussed.

Polychotomous Nominal Variables

When a nominal variable has more than two categories, we can no longer use a single dummy variable to represent all of the categories. We cannot use a single variable coded with three or more scores, either, because a regression program would treat the numeric codes as if they represented an interval variable. The data in Table 5.3 will be used to illustrate the analysis of nominal variables with three or more categories; these data are also from the 1982 Akron Area Survey. The dependent variable is again life satisfaction, and the nominal

TABLE 5.3 Life Satisfaction by Marital Status

Code	*Status*	*n*	*Mean*
1	Married	411	5.6837
2	Divorced/Separated	106	4.9057
3	Widowed	69	5.4348
4	Never-Married	150	5.1600
	Total	736	5.4410

independent variable is marital status (MARITAL). In Table 5.3, the separated and divorced categories in the original data have been merged because of the small number of separated respondents.

If we were to use a single variable coded as in Table 5.3, regression analysis would assume that never-married are one unit higher than widowed, that widowed are one unit higher than divorced, and that divorced are one unit higher than married. This would be an inappropriate scoring since the variable is nominal. The solution to the scoring problem is to use dummy variables. It is not possible, however, to represent the marital variable with just one dummy variable. But how many dummy variables will be needed? There are $g = 4$ dummy variables that *could* be created, where g equals the number of categories in a nominal variable. One dummy could be created for each marital status category, where each category would have a score of 1 on its dummy variable and all the other categories would have a 0 score on that variable. The coding would look like this:

The $g = 4$ Possible Dummy Variables for Marital Status

Marital Status	X_1 DIVORCED	X_2 WIDOWED	X_3 NEVER	X_4 MARRIED
Divorced/Separated	1	0	0	0
Widowed	0	1	0	0
Never-Married	0	0	1	0
Married	0	0	0	1

Each person that is divorced would have a score of 1 on a dummy variable called DIVORCED; all other persons would have a score of 0 on DIVORCED. Thus, this variable would represent two groups, divorced and not divorced. The DIVORCED variable would make no distinction between the various categories that are not divorced, i.e., the widowed, never-married, and married. The widowed persons would have a score of 1 on WIDOWED, and all of the other groups would have a score of 0. Thus, WIDOWED would represent two groups, widowed and not widowed. The dummy variables NEVER and MARRIED would be interpreted in an analogous fashion.

Although there are g distinct dummy variables that could be created, all we need for the regression equation is $g - 1$ dummy variables. In fact, the regression coefficients cannot be computed for an equation that includes all of the g dummy variables because there will be perfect multicollinearity among the dummy variables in that case. This is because of the following identity:

$$\text{DIVORCED} + \text{WIDOWED} + \text{NEVER} + \text{MARRIED} = 1$$

Since each person has a score of 1 on one and only one of the dummies, and a score of 0 on all of the other dummies, the sum of the four dummies will equal 1 for each person. If we rearrange the above equation by leaving the MARRIED variable, for example, to the left of the equal sign and moving all of the other variables to the right, we have

$$\text{MARRIED} = 1 - \text{DIVORCED} - \text{WIDOWED} - \text{NEVER}$$

where

$$0 = 1 - 1 - 0 - 0 \quad \text{(for divorced persons)}$$
$$0 = 1 - 0 - 1 - 0 \quad \text{(for widowed persons)}$$
$$0 = 1 - 0 - 0 - 1 \quad \text{(for never-married)}$$
$$1 = 1 - 0 - 0 - 0 \quad \text{(for married persons)}$$

The above shows that the MARRIED dummy variable is a perfectly linear function of the other three dummy variables. Thus, $R^2 = 1$ for MARRIED as the dependent variable and the other three dummies as predictors. There would be perfect multicollinearity among the four variables, and thus it would not be possible to estimate their separate effects on Y.[1]

The above discussion leads to the conclusion that one of the g dummies must be left out of the equation. In general, only $g - 1$ dummy variables can be used. The group represented by the dummy that will be omitted will not, however, be left out of the analysis. It will serve as a reference group, as we shall see. The dummy variable to be omitted depends on which group should theoretically serve as a reference group. In the following analysis the married respondents have been chosen to be the reference group, and thus the MARRIED dummy will not be included in the equation. This decision is based on the premise that the married status has the greatest normative support, and thus it would be of interest to compare each of the deviant marital statuses with the married group.

The omission of the MARRIED dummy leaves the following three dummies for entry into the regression equation:

1. The cross-product matrix **x'x** used in the matrix solution for the multiple regression equation would be singular and thus could not be inverted.

The $g - 1 = 3$ Dummy Variables Selected for the Regression Equation

Marital Status	X_1 DIVORCED	X_2 WIDOWED	X_3 NEVER
Divorced/Separated	1	0	0
Widowed	0	1	0
Never-Married	0	0	1
Married	0	0	0

Notice that married respondents will have a 0 on each of the three dummy variables. The regression equation to be estimated is

$$\hat{Y} = a + b_1\text{DIVORCED} + b_2\text{WIDOWED} + b_3\text{NEVER}$$

Analogously to bivariate dummy regression, the solution that minimizes the sum of squared residuals of Y will be to select coefficients such that no matter which marital status a person is in, including the currently married group, the equation will predict the mean of Y for the group to which the person belongs. The following symbols will be used for these conditional means of Y:

$$\bar{Y}_M = \text{Mean for Married Group}$$
$$\bar{Y}_D = \text{Mean for Divorced Group}$$
$$\bar{Y}_W = \text{Mean for Widowed Group}$$
$$\bar{Y}_N = \text{Mean for Never-Married Group}$$

If the least-squares solution predicts the conditional mean of Y for the group to which each person belongs, then we can interpret each of the regression coefficients by substituting the scores for each group into the equation and simplifying as follows:

$$\hat{Y}_M = a + b_1(0) + b_2(0) + b_3(0) = a = \bar{Y}_M$$
$$\hat{Y}_D = a + b_1(1) + b_2(0) + b_3(0) = a + b_1 = \bar{Y}_D$$
$$\hat{Y}_W = a + b_1(0) + b_2(1) + b_3(0) = a + b_2 = \bar{Y}_W$$
$$\hat{Y}_N = a + b_1(0) + b_2(0) + b_3(1) = a + b_3 = \bar{Y}_N$$

Since the married group has a 0 score for each dummy variable, the above results show that when 0 is substituted for each variable, the predicted score for the married group is a, the intercept. Since the best prediction for the married group is the mean of Y for all married persons, we can see that the intercept equals the mean of the married group, which is the reference group. With respect to the divorced group, that has a score of 1 on the first dummy variable and 0 on the others, the predicted score is the intercept plus the slope for the

DIVORCE dummy variable; thus, $a + b_1$ is equal to the mean for the divorced group. The predicted scores for widowed and never-married are derived in analogous fashion.

We can use the above predicted scores and the means from Table 5.3 to specify the meaning of each regression coefficient, and their numeric values, as follows:

$$\overline{Y}_M = a = 5.6837 \quad \textbf{(5.4)}$$

$$\overline{Y}_D - \overline{Y}_M = (a + b_1) - a = b_1 = 4.9057 - 5.6837 = -.778 \quad \textbf{(5.5)}$$

$$\overline{Y}_W - \overline{Y}_M = (a + b_2) - a = b_2 = 5.4348 - 5.6837 = -.2489 \quad \textbf{(5.6)}$$

$$\overline{Y}_N - \overline{Y}_M = (a + b_3) - a = b_3 = 5.1600 - 5.6837 = -.5237 \quad \textbf{(5.7)}$$

Equation 5.5 shows that the slope for the DIVORCE variable (b_1) equals the difference between the mean for the divorced group and the mean for the married group. The value of the slope indicates that divorced persons are .778 less satisfied with their lives than married persons. Equation 5.6 indicates that the slope for WIDOWED (b_2) equals the widowed mean minus the married mean and indicates that widowed persons are .2489 less satisfied than married persons. Finally, according to Equation 5.7, the slope for NEVER (b_3) equals the never-married mean minus the married mean and indicates that never-married persons are .5237 less satisfied than married persons.

The least-squares regression equation has thus been shown to be equal to

$$\hat{Y} = 5.6837 - .778 \cdot \text{DIVORCED} - .2489 \cdot \text{WIDOWED} - .5237 \cdot \text{NEVER}$$

Although each dummy variable represents the group coded 1 versus all other groups that are coded 0, the slope for the dummy doesn't represent the mean difference between the group coded 1 and all persons coded 0 as it does in the bivariate case. These slopes are partial slopes, so they indicate the change in Y per unit change in the dummy variable, holding all other dummy variables constant. Thus, for example, when DIVORCE scores change by one unit but WIDOWED and NEVER are held constant, the unit change in DIVORCE represents the difference between married persons and divorced persons. When WIDOWED changes by one unit and DIVORCED and NEVER are held constant, the change in WIDOWED represents the difference between widowed and married persons. Thus, *each partial slope equals the mean of Y for the group coded 1 on that dummy variable minus the mean of Y for the group coded 0 on all of the dummies*; the latter group is referred to as the reference group, which in this case is the married group. *The intercept a equals the mean of Y for the reference group.*

In general, if the regression equation for a set of $g - 1$ dummy variables is $\hat{Y} = a + \Sigma b_i X_i$ $(i = 1, \ldots, g - 1)$, where X_i is the dummy variable for the ith

TABLE 5.4 Life Satisfaction Regressed on the Dummy-Coded Marital-Status Variables

* * * * MULTIPLE REGRESSION * * * *

Multiple R	.20640	Analysis of Variance			
R Square	.04260		DF	Sum of Squares	Mean Square
Adjusted R Square	.03868	Regression	3	66.43387	22.14462
Standard Error	1.42818	Residual	732	1493.05390	2.03969

F = 10.85685 Signif F = .0000

---------------Variables in the Equation---------------

Variable	B	SE B	Beta	T	Sig T
NEVER	−.52370	.13624	−.14493	−3.844	.0001
WIDOWED	−.24892	.18581	−.04984	−1.340	.1808
DIVORCED	−.77804	.15558	−.18767	−5.001	.0000
(Constant)	5.68370	.07040		80.681	.0000

group (i.e., the code of the *i*th group equals unity on X_i), then the intercept and slopes for the dummy variable regression equation are

$$a = \overline{Y}_0 \tag{5.8}$$

$$b_i = \overline{Y}_i - \overline{Y}_0 \tag{5.9}$$

where $\overline{Y}_0$ equals the mean of the reference group (i.e., the group coded 0 on each X_i).

The following SPSS commands may be used to transform the marital-status nominal variable, which has four categories, into the three dummy variables that have been chosen for entry into the regression equation.

```
IF MARITAL EQ 1 MARRIED = 1
IF MARITAL NE 1 MARRIED = 0
IF MARITAL EQ 2 DIVORCED = 1
IF MARITAL NE 2 DIVORCED = 0
IF MARITAL EQ 3 WIDOWED = 1
IF MARITAL NE 3 WIDOWED = 0
IF MARITAL EQ 4 NEVER = 1
IF MARITAL NE 4 NEVER = 0
REGRESSION VARS = SATLIFE DIVORCED WIDOWED NEVER/
  DEP = SATLIFE/ ENTER DIVORCED WIDOWED NEVER
```

The output from the SPSS regression run is given in Table 5.4. Notice that the partial slopes and the intercept given in Table 5.4 are exactly equal to those values computed from the group means using Equations 5.1 to 5.4.

The hypotheses that are tested by the *t* and *F* statistics in Table 5.4 are

F for R^2 $\quad$ H_0: $b_1 = b_2 = b_3 = 0$ or $\overline{Y}_M = \overline{Y}_D = \overline{Y}_W = \overline{Y}_N$

t for b_i $\quad$ H_0: $b_i = 0$ or $\overline{Y}_i = \overline{Y}_M$

The F value and associated level of significance for the test that all slopes are equal to 0 or that the means are equal for all groups is rejected at $p = .0000$. Therefore, at least one of the marital statuses has a different mean satisfaction level than the others. The squared multiple correlation indicates that we can account for about 4.3% of the variance in life satisfaction simply by knowing whether a person is married, divorced/separated, widowed, or never-married. The t tests for the individual slopes indicates that the never-married and the divorced are significantly less satisfied with their lives than the married group, but there is not a significant difference between the widowed and married persons.

When dummy variables are used to represent a nominal independent variable with three or more categories, the F test for R^2 in the regression equation is equivalent to the F test for group differences in means performed by a one-way analysis of variance (ANOVA). See Appendix 5B for a description of an equivalent analysis of variance.

Inclusion of Another *X* with the Dummy Variables

If we add another X to the equation containing the dummy variables, the interpretation changes somewhat from that given above. For purposes of this discussion, the new X might be an ordinal, interval, or ratio variable, or it might even be a dummy variable representing an additional nominal variable. In the new equation, the intercept equals the predicted value of Y when all the variables are equal to zero, including the new X that has been added. The value of a no longer equals the mean for the reference group; instead, it is the predicted value for members of the reference group who have a score of zero on the new X. The slopes no longer equal the difference between the mean of the group coded 1 and the mean of the reference group, since the new X is now being held constant when the groups are contrasted; the partial slope now equals the predicted difference between the group coded 1 and the reference group for persons who do not differ on the new X.

Table 5.5 shows the output of an SPSS run that adds the variable AGE to the dummy variables already in the equation. The intercept is not a meaningful value in this equation because it represents the predicted score for married persons who are zero years of age. The partial slopes for each dummy variable now represent the difference between a particular marital status and the married group, holding age constant. The partial slopes, however, still represent a comparison between the group coded 1 on the dummy variable and the ref-

TABLE 5.5 Life Satisfaction Regressed on the Dummy-Coded Marital Status Variables and Age

****MULTIPLE REGRESSION****

			DF	Sum of Squares	Mean Square
Multiple R	.23153	Analysis of Variance			
R Square	.05361				
Adjusted R Square	.04831	Regression	4	81.91767	20.47942
Standard Error	1.42319	Residual	714	1446.17691	2.02546

F = 10.11101 Signif F = .0000

Variables in the Equation

Variable	B	SE B	Beta	T	Sig T
NEVER	−.35075	.15074	−.09728	−2.327	.0203
WIDOWED	−.47926	.21054	−.09556	−2.276	.0231
DIVORCED	−.75238	.15611	−.18226	−4.820	.0000
AGE	.01197	.00406	.13499	2.949	.0033
(Constant)	5.18099	.18729		27.663	.0000

erence group. Thus, the slope for DIVORCED indicates that when any age difference between divorced/separated persons and married persons is held constant, divorce or separation reduces life satisfaction by about .75 points. Although the DIVORCED and NEVER dummy variables are still significant after age is controlled, the negative effect of having never been married is reduced from −.52 to −.35. Apparently part of the original difference between never-married persons and married persons was due to the fact that people who have never been married are younger, on the average, than married individuals. We also see that after controlling for age, WIDOWED now also has a significant negative effect on life satisfaction; the fact that the average age of widowed persons is greater than the age of married persons had been *suppressing* the true negative effect of widowhood on life satisfaction. In sum, after controlling for age we can now infer that if the three nonmarried groups did not differ in age from the married group, being divorced, widowed, or never-married would each reduce satisfaction with life.

It was noted above when a single nominal variable is converted to dummy variables for inclusion in a regression equation, the F test for the regression equation is equal to the F test in a one-way analysis of variance for the same nominal variable. We now note that the equivalence between regression analysis and analysis of variance continues when an interval X (or X's) is added to the regression equation containing the dummy variables; the F tests in the regression analysis are now equal to the F tests in an analysis of covariance (ANCOVA). An interval variable such as age is called a covariate in analysis of variance. The equivalence of regression analysis and analysis of covariance is shown in Appendix 5B.

Effects Coding

Dummy variable coding requires that one category of the nominal variable be specified as a reference group against which each of the other categories are tested. This is often a very satisfactory procedure for testing differences between groups. There may be times, however, when it is difficult to choose one group with which to compare each of the others. There might be two or more categories that would serve equally well from a theoretical standpoint as reference groups, or there may not be any theoretical reasons for specifying any of the groups as a reference group, as in an exploratory study, for example. In both of these cases the analyst may want to consider *effects coding*.

An example of effects coding for the marital status variable is given below.

Effects Coding

Marital Status	X_1 DIVORCED	X_2 WIDOWED	X_3 NEVER
Divorced/Separated	1	0	0
Widowed	0	1	0
Never-Married	0	0	1
Married	−1	−1	−1

As can be seen above, the only difference between effects coding and dummy variable coding is that one of the groups is coded −1 on each variable instead of 0. Just as with dummy coding, there will be $g - 1$ effects-coded variables.

Effects coding, however, changes the meaning of the regression coefficients. In effects coding the intercept will equal the mean of the group means on the dependent variable,

$$a = \overline{\overline{Y}} = \frac{\overline{Y}_1 + \overline{Y}_2 + \cdots + \overline{Y}_g}{g} \tag{5.10}$$

If there are g groups, the intercept will equal the mean of the g group means. This mean of means is indicated by $\overline{\overline{Y}}$ and is often called the *unweighted grand mean*. In the case of marital status and life satisfaction, the grand mean of the $g = 4$ group means given in Table 5.3 is

$$a = \overline{\overline{Y}} = \frac{\overline{Y}_M + \overline{Y}_D + \overline{Y}_W + \overline{Y}_N}{4} = \frac{5.6837 + 4.9057 + 5.4348 + 5.1589}{4} = 5.2958$$

Notice that the grand mean is smaller than the mean given in Table 5.3, which is the regular mean of Y computed as the arithmetic average of all 737 observations. The arithmetic mean of Y is influenced greatly by the higher mean of the married persons who comprise the majority of the respondents. The married persons, however, do not carry any more weight than the unmarried persons in the computation of the grand mean. Thus, the grand mean is lower because the smaller unmarried groups all have lower life satisfaction than the married group.

In effects coding, *the unstandardized regression slope for an effect-coded variable equals the difference between the mean of the group coded 1 on the variable and the grand mean of all groups.* Thus, the unstandardized slope for group i equals

$$b_i = \overline{Y}_i - \overline{\overline{Y}} \qquad \textbf{(5.11)}$$

Instead of comparing each group to a single reference group, effects-coded variables compare each group to the mean of all groups. In the case of marital status and satisfaction, we have

$$b_D = \overline{Y}_D - \overline{\overline{Y}} = 4.9057 - 5.2958 = -.3901 \quad \text{[divorced and separated]}$$
$$b_W = \overline{Y}_W - \overline{\overline{Y}} = 5.4348 - 5.2958 = .1390 \quad \text{[widowed]}$$
$$b_N = \overline{Y}_N - \overline{\overline{Y}} = 5.1589 - 5.2958 = -.1369 \quad \text{[never-married]}$$

In sum, the regression equation with effects-coded variables for each unmarried group is

$$\hat{Y} = 5.2958 - .3901\text{DIVORCED} + .1390\text{WIDOWED} - .1369\text{NEVER}$$

If an effects-coded variable had been included for married respondents in place of one of the unmarried groups, the slope would be

$$b_M = \overline{Y}_M - \overline{\overline{Y}} = 5.6837 - 5.2958 = .3879 \quad \text{[married]}$$

But just as with dummy-coded variables, only $g - 1$ variables can be included in the equation to represent the g categories of the nominal variable. The fact that one group is not coded 1 on any of the variables (and is coded -1 on all of the variables) does not mean it is not influencing the regression equation coefficients. The married group, for example, influences the grand mean and consequently influences the intercept and each of the unstandardized slopes.

The null hypothesis to be tested by either t or F for each effects-coded variable is

$$H_0\text{: } b_i = 0; \quad \text{or } \overline{Y}_i = \overline{\overline{Y}}$$

However, since the mean of group i is included in the grand mean, the only way that the null hypothesis can be true is for the mean of group i to be equal to the mean of all the other groups, excluding group i. Thus, the null hypothesis can also be stated as

$$H_0\text{: } \overline{Y}_i = \overline{\overline{Y}}_{(-i)}$$

The subscript $(-i)$ means that it is the mean of the means of all groups except

TABLE 5.6 Life Satisfaction Regressed on the Effects-Coded Marital Status Variables

****MULTIPLE REGRESSION****

Multiple R	.20640	Analysis of Variance			
R Square	.04260		DF	Sum of Squares	Mean Square
Adjusted R Square	.03868	Regression	3	66.43387	22.14462
Standard Error	1.42818	Residual	732	1493.05390	2.03967

F = 10.85685 Signif F = .0000

------------------Variables in the Equation------------------

Variable	B	SE B	Beta	T	Sig T
NEVER	−.13604	.10493	−.07456	−1.297	.1952
WIDOWED	.13875	.13781	.06296	1.007	.3144
DIVORCED	−.39038	.11761	−.19537	−3.319	.0009
(Constant)	5.29604	.06489		81.620	.0000

group *i*. Thus, for each effects-coded variable, we are testing whether the mean of one of the groups differs from the mean of the means of all other groups.

SPSS statements that may be used to create the effects-coded variables are shown below.

```
IF MARITAL EQ 2 DIVORCED = 1
IF MARITAL NE 2 DIVORCED = 0
IF MARITAL EQ 3 WIDOWED = 1
IF MARITAL NE 3 WIDOWED = 0
IF MARITAL EQ 4 NEVER = 1
IF MARITAL NE 4 NEVER = 0
IF MARITAL EQ 1 DIVORCED = -1
IF MARITAL EQ 1 WIDOWED = -1
IF MARITAL EQ 1 NEVER = -1
```

The output from the SPSS regression program is shown in Table 5.6. The results for the multiple correlation statistics, the standard error of estimate, and the analysis of variance for the entire equation are the same as for the dummy variable regression in Table 5.4. Whether dummy coding or effects coding is used, the predicted values of the dependent variable will be identical because in each case three variables are being used to represent the conditional means of the four marital status groups. Thus, the F test for the entire equation indicates that the null hypothesis that all of the conditional means are equal should be rejected.

The t tests for the three effects-coded independent variables are not identical to those for the dummy-coded variables. The test for NEVER in Table 5.4 indicated that the never-married respondents had significantly lower life satisfaction than the married respondents. But the effects-coded NEVER is not significant at $p < .05$; the mean for never-married respondents is not less than the mean of the means for the other three groups (the "ever" married respondents). Table 5.4 showed that the widowed respondents did not differ significantly from the married respondents in satisfaction with their lives, and the effects coding indicates that the widowed do not differ significantly from the mean of the means of the never-married, divorced, and married respondents. Both dummy and effects coding indicate that the slope for the DIVORCED variable is significantly different from zero; the divorced group has the lowest life satisfaction. If we had chosen to include an effects-coded variable for married respondents in place of any of the other effects-coded variables, it would probably be significant at the .05 level since the married respondents have the highest life satisfaction. This indicates that the number of variables for which the null hypothesis is rejected may depend on which group is used as the reference group. If the mean of the group coded −1 on each of the $g - 1$ variables is near the middle of the group means, there may be more significant variables than if the group coded −1 has one of the highest or lowest means. The number of significant results, of course, may also depend on which category is selected as the reference group in dummy variable coding.

Contrast Coding

Effects coding is atheoretical in that the investigator does not specify which groups are to be tested for differences of means but instead simply compares each group with all of the others together. Dummy coding gives the investigator more control over which groups are to be compared, but all tests for differences of means are between a single reference group and each of the other groups. The final coding scheme to be considered—*contrast coding*—provides the researcher with the most control and flexibility in specifying the group comparisons or contrasts that are to be tested.

As with effects coding and dummy coding, contrast coding involves creating $g - 1$ variables to represent the g categories of the nominal variable. For each of these variables the investigator specifies one subset of groups that is to be contrasted with a second subset of groups. Each subset may consist of only one group. Each of these variables is referred to as a *contrast*. Thus, for example, three contrasts would be specified or created for the four categories of marital status.

For each of the $g - 1$ contrast variables a set of codes c_i ($i = 1, 2, \ldots, g$) must be selected for the g groups. There are two sets of restrictions on the c_i. First, the codes for each contrast variable must sum to zero:

$$\sum_{i=1}^{g} c_i = 0$$

This means, of course, that some of the groups must have positive codes and some must have negative codes. Some may also be zero. Furthermore, the codes for the subset of groups with positive codes must all be equal, and the codes for the subset of groups with negative codes must all be equal.[2] The consequence of these restrictions, in combination with the second type of restriction to be specified below, is that *contrast coding implements a contrast between the subset of groups with positive codes and the subset of groups with negative codes.*

The second restriction is that

$$\sum_{i=1}^{g} c_{ij} c_{ik} = 0, \qquad j \neq k$$

This restriction means that the sum of products of the c_i for each pair of contrast-coded variables must equal zero. If the sum of products of the codes equals zero, the codes are said to be *orthogonal*, or uncorrelated. This does not mean that the coded variables themselves will be orthogonal; that is, when the correlation between the variables is calculated across all *n* observations, it need not be zero. Only when each group has an equal number of cases will orthogonal coding produce orthogonal variables.

An example of contrast coding for the marital status variable follows.

Contrast Coding

Marital Status	X_1 LOSS	X_2 WIDOWED	X_3 NEVER
Divorced/Separated	$+\frac{1}{3}$	$-\frac{1}{2}$	$-\frac{1}{4}$
Widowed	$+\frac{1}{3}$	$+\frac{1}{2}$	$-\frac{1}{4}$
Never-Married	0	0	$+\frac{3}{4}$
Married	$-\frac{2}{3}$	0	$-\frac{1}{4}$
Sum	0	0	0

As indicated above, the sum of the c_i for each variable equals zero. There are three pairs of variables, and the sums of cross-products of the codes for these variables are

2. This restriction applies only when contrasts are being specified for a purely nominal variable. When contrasts are specified for ordinal/interval categorical variables, sometimes referred to as *orthogonal polynomial contrasts*, only the restriction that the c's must sum to zero holds.

$$\sum_{i=1}^{4} c_{i1}c_{i2} = \left(\frac{1}{3}\right)\left(-\frac{1}{2}\right) + \left(\frac{1}{3}\right)\left(\frac{1}{2}\right) + (0)(0) + \left(-\frac{2}{3}\right)(0)$$

$$= -\frac{1}{6} + \frac{1}{6} + 0 + 0 = 0$$

$$\sum_{i=1}^{4} c_{i1}c_{i3} = \left(\frac{1}{3}\right)\left(-\frac{1}{4}\right) + \left(\frac{1}{3}\right)\left(-\frac{1}{4}\right) + (0)\left(\frac{3}{4}\right) + \left(-\frac{2}{3}\right)\left(-\frac{1}{4}\right)$$

$$= -\frac{1}{12} - \frac{1}{12} + 0 + \frac{2}{12} = 0$$

$$\sum_{i=1}^{4} c_{i2}c_{i3} = \left(-\frac{1}{2}\right)\left(-\frac{1}{4}\right) + \left(\frac{1}{2}\right)\left(-\frac{1}{4}\right) + (0)\left(\frac{3}{4}\right) + (0)\left(-\frac{1}{4}\right)$$

$$= \frac{1}{8} - \frac{1}{8} + 0 + 0 = 0$$

Each sum of cross-products equals zero, and thus the codes are orthogonal. Since the restrictions for contrast coding are satisfied, these variables are contrasts.

The first variable contrasts divorced and widowed persons (+ codes) with married persons. The variables have been named after the groups with positive codes. Since the two groups with positive codes on the first variable have both lost spouses, this variable has been named LOSS. Thus, this contrast is specified in order to estimate the effect of having lost a spouse. The second variable is WIDOWED, and it contrasts widowed persons (+ codes) with divorced persons (− codes) to examine the effect of the two different ways of losing a spouse. The third variable (NEVER) contrasts never-married persons (+ code) with married, widowed, and divorced persons (− codes) in order to estimate the effect of never having been married.

Let $g_{(+)}$ be the number of groups in the positive subset for a contrast variable, and let $g_{(-)}$ be the number of groups in the negative subset. The values of the positive codes c^+ and the negative codes c^- were chosen as follows:

$$c^+ = +\frac{g_{(-)}}{g_{(+)} + g_{(-)}}$$

$$c^- = -\frac{g_{(+)}}{g_{(+)} + g_{(-)}}$$

For example, for the NEVER variable the codes are

$$c^+ = +\frac{3}{1 + 3} = +\frac{3}{4}$$

$$c^{-} = -\frac{1}{1+3} = -\frac{1}{4}$$

These rules satisfy the restriction that the sum of the c_i's must equal zero. They also provide for the following interpretation of the unstandardized regression coefficients:

$$b_i = \overline{\overline{Y}}_{(+)} - \overline{\overline{Y}}_{(-)} \qquad \textbf{(5.12)}$$

Equation 5.12 indicates that the regression slope for each contrast-coded variable will equal the mean of the means for the groups in the positive subset minus the mean of the means for the groups in the negative subset:

$$b_1 = \frac{\overline{Y}_D + \overline{Y}_W}{g_{(+)}} - \overline{Y}_M = \frac{4.9057 + 5.4348}{2} - 5.6887 = 5.1703 - 5.6887 = -.5185$$

$$b_2 = \overline{Y}_W - \overline{Y}_D = 5.4348 - 4.9057 = .5291$$

$$b_3 = \overline{Y}_N - \frac{\overline{Y}_M + \overline{Y}_D + \overline{Y}_W}{g_{(-)}} = 5.1589 - \frac{5.6837 + 4.9057 + 5.4348}{3}$$

$$= 5.1589 - 5.3414 = -.1825$$

This interpretation of the slopes comes from the fact that the difference between the positive code and the negative code for each variable is unity. Thus, an increase of one unit on the independent variable is equivalent to changing from a level represented by the code of the negative subset to a level represented by the code of the positive subset. It is not necessary to assign codes in this manner (as long as the sum of the codes equals zero and the sum of cross-products of codes equals zero), but it provides an easy and meaningful interpretation of the slopes.

The regression intercept for a contrast-coded set of variables is equal to the unweighted mean of means, i.e., the grand mean, just as it was in effects coding:

$$a = \overline{\overline{Y}} = \frac{\sum_{i=1}^{g} \overline{Y}_i}{g} \qquad \textbf{(5.13)}$$

Thus, $a = 5.2958$, just as in effects coding. The regression equation for the above contrast-coded version of the marital status variable is

$$\hat{Y} = 5.2958 - .5185\text{LOSS} + .5291\text{WIDOWED} - .1825\text{NEVER}$$

The null hypothesis to be tested by the t or F test for the slope is

$$H_0\!: b_i = 0; \quad \text{or } \overline{\overline{Y}}_{(+)} = \overline{\overline{Y}}_{(-)}$$

The output from the SPSS regression program is shown in Table 5.7. The

TABLE 5.7 **Life Satisfaction Regressed on Contrast-Coded Marital Status Variables**

****MULTIPLE REGRESSION****

Multiple R	.20640	Analysis of Variance			
R Square	.04260		DF	Sum of Squares	Mean Square
Adjusted R Square	.03868	Regression	3	66.43387	22.14462
Standard Error	1.42818	Residual	732	1493.05390	2.03967

F = 10.85685 Signif F = .0000

------------------------Variables in the Equation------------------------

Variable	B	SE B	Beta	T	Sig T
NEVER	−.18138	.13990	−.05019	−1.297	.1952
WIDOWED	.52912	.22091	.08815	2.395	.0169
LOSS	−.51348	.13101	−.15325	−3.919	.0000
(Constant)	5.29604	.06489		81.620	.0000

multiple correlation statistics, the standard error of estimate, and the analysis of variance results for the entire equation are identical to those for dummy coding and effects coding. The tests for the individual contrasts show that the never-married persons are not significantly less satisfied with their lives than the ever-married persons (married, divorced, and widowed). Widowed respondents, however, are significantly more satisfied than divorced respondents, and those who have lost their spouses (divorced and widowed) are less satisfied with their lives than currently married individuals.

An example of an alternate set of contrast codes for marital status is given below.

Contrast Coding

Marital Status	X_1 MARRIED	X_2 NEVER	X_3 WIDOWED
Divorced/Separated	$-\frac{1}{4}$	$-\frac{1}{3}$	$-\frac{1}{2}$
Widowed	$-\frac{1}{4}$	$-\frac{1}{3}$	$+\frac{1}{2}$
Never-Married	$-\frac{1}{4}$	$+\frac{2}{3}$	0
Married	$+\frac{3}{4}$	0	0
Sum	0	0	0

Verify that the sum of cross-products of the codes for each pair of variables equals zero. The first variable (MARRIED) contrasts currently married persons with those who are not currently married (divorced, widowed, and never-married). The second variable (NEVER) contrasts those who have never been married with those who have lost a spouse (divorced and widowed). Finally, the third variable (WIDOWED) contrasts widowed respondents with divorced respondents, just as in the first set of contrast codes. Compute the slopes for the three variables in the above contrast coding of marital status.

Summary

In general, nominal variables cannot be entered directly into regression equations. New variables must be created, with appropriate coding, to represent the nominal variable in an regression equation. If the nominal variable has g categories, then $g - 1$ specially coded variables must be used to represent fully the nominal variable. Dummy variable coding, effects coding, and contrast coding are three useful ways of scoring the new variables.

In dummy coding the investigator must specify one category of the nominal variable as a reference group and create a dummy variable for each of the other groups. The slope for each dummy variable equals the mean for the group coded 1 on that dummy variable minus the mean of the reference group. The consequence of this coding is that the reference group is compared to each of the other groups. Thus, it is important to pick a reference group that provides the most interesting or important theoretical comparisons.

Effects coding appears more important for exploratory studies or for other situations where it is not desirable to single out one category as a reference group. In effects coding, $g - 1$ variables are created, just as in dummy coding. Thus, there will be an effects-coded variable for each group except one. The slope for each effects-coded variable represents the mean of the group represented by that variable minus the grand mean for all of the other groups (the unweighted mean of means).

Contrast coding is the most flexible type of coding, and thus it gives the investigator the most control over the various types of group comparisons that may be investigated. In contrast coding, each variable contrasts the mean of means of one subset of groups with the mean of means of another subset of groups. Each subset, however, may consist of only one group. Just as in dummy coding and effects coding, only $g - 1$ contrast variables can be used.

Finally, it is important to remember that each method of coding leads to the same predicted scores; that is, for each method, the predicted score for each group is the mean of that group on the dependent variable. The consequence is that the squared multiple correlation is the same for each type of coding; dummy, effects, and contrast coding all explain the same proportion of variance in the dependent variable. Thus, if the objective is to determine how much variance is explained by a nominal variable, or simply to control for the nominal variable while focusing on the effects of other variables, then one coding method is as good as another.

Reference

Blalock, Hubert M., Jr. 1979. *Social Statistics*. New York: McGraw Hill.

APPENDIX 5A

t Test for a Difference of Means

It was claimed in Chapter 5 that the *t* (or *F*) test for the slope of *Y* regressed on a single dummy-coded independent variable is equivalent to the *t* test for the difference in means between two independent samples (groups). This equivalency will be illustrated with output from the SPSS procedure T–TEST. The data again involve RACE as the dichotomous independent variable defining the two groups, blacks and whites. The dependent variable is satisfaction with life (SATLIFE). The dummy variable regression results from the SPSS procedure REGRESSION were presented in Table 5.2. The output from the T–TEST procedure is shown in Table 5A.1.

The means for blacks and whites (Table 5A.1) are identical to those in Table 5.1. There are two *t* tests in Table 5A.1. The first test—labeled the *Pooled Variance Estimate*—is identical to the *t* test for the slope in Table 5.2. Thus, the *t* test for the ordinary least-squares regression slope with a dummy-coded *X* is the same as the *t* test for a difference of means based on a *pooled variance estimate* of the standard error of the difference of means, i.e., $s_{\bar{Y}_B - \bar{Y}_W}$ [see any standard introductory statistics text (e.g., Blalock 1979) for the formula for this standard error].

However, before using the standard *pooled variance* method, an *F* test for the between-group difference in the variance of *Y* should be conducted. The *F* value ($F = 2.44$) and significance level ($p = .000$) for this test are shown immediately following the means and standard deviations for the groups in Table 5A.1. The results indicate that we can reject the null hypothesis that blacks and whites have equal variances in life satisfaction. In this case we would conclude that blacks have greater variance in *Y* than do whites. As a consequence of

TABLE 5A.1 SPSS *t* Test for Racial Differences in Mean Life Satisfaction

T–TEST GROUPS = RACE (1,2)/VARIABLES = SATLIFE

t tests for independent samples of RACE
GROUP 1—RACE EQ 1: BLACK
GROUP 2—RACE EQ 2: WHITE

							Pooled Var. Est.			Separate Var. Est.		
	N	Mean	S.D.	S.E.	F	P	t	DF	p	t	DF	P
GROUP 1	63	4.7619	2.115	.267	2.44	.000	−3.98	717	.000	−2.78	66.97	.007
GROUP 2	656	5.5168	1.355	.053								

the significant F test, we should not use the pooled variance estimate of the standard error of the difference of means. Instead, we should use the *separate variance estimate* [see Blalock (1979) for the appropriate formula for this estimate of the standard error]. The separate variance estimate results in $t = -2.78$ and $p = .007$. Thus, we would still reject the null hypothesis of no racial difference in mean life satisfaction and conclude that whites are more satisfied, on the average, than blacks.

Note, however, that we have to qualify the claim that the t test for the dummy regression slope and the t test for the difference of means are always identical. Only when we can validly assume (or fail to reject the hypothesis) that both of the groups have the same variance on Y are the two tests identical. That is, only the pooled variance estimate produces a t value equal to that for the regression slope. It must be emphasized, however, that the pooled variance estimate is based on one of the assumptions that justifies the use of the ordinary least-squares formula for the standard error of the slope (Formula 4.6). That assumption is that the variance of ε is the same for all values of X, i.e., the variance of ε is *homoskedastic* (see Chapter 4). In our race example, the variance of Y for blacks is the same as the estimated variance of ε for blacks, and the variance of Y for whites is the estimated variance of ε for whites. Thus, the finding in the T–TEST procedure that blacks have greater Y variance than whites means that the assumption of homoskedasticity has been rejected; blacks and whites are the two categories/values of X for our nominal independent variable, and the fact that these two groups do not have equal variances on Y means that heteroskedasticity is present. Thus, the example of race and life satisfaction has highlighted the importance of the regression assumptions that justify the use of the ordinary least-squares t test. Only when these assumptions are true (e.g., only when the error variance is homoskedastic) will the t test for the regression slope equal the t test for a two-sample (group) difference in means.

APPENDIX 5B

F Tests in Analysis of Variance

One-Way Analysis of Variance

The equivalence between a one-way analysis of variance and the dummy variable regression for a polychotomous nominal independent variable will be shown in this appendix. The SPSS statement for the analysis of variance procedure ANOVA and the output from that procedure are given in Table 5B.1, using the same life satisfaction and marital status data analyzed in Table 5.4.

TABLE 5B.1 **Analysis of Variance of Life Satisfaction by Marital Status**

ANOVA VARIABLES = SATLIFE BY MARITAL (1,4)/STATISTICS = ALL

*** CELL MEANS ***

TOTAL POPULATION

	5.44
	(736)

MARITAL

	1	2	3	4
	5.68	4.91	5.43	5.16
	(411)	(106)	(69)	(150)

*** ANALYSIS OF VARIANCE ***

Source of Variation	Sum of Squares	DF	Mean Square	F	Sig of F
Main Effects	66.434	3	22.145	10.857	.000
MARITAL	66.434	3	22.145	10.857	.000
Explained	66.434	3	22.145	10.857	.000
Residual	1493.054	732	2.040		
Total	1559.488	735	2.122		

Since this is a one-way analysis of variance, there is only one independent variable (MARITAL), and thus the Main Effects Sum of Squares is equal to the Sum of Squares for MARITAL, which in turn equals the total Explained Sum of Squares. The explained sum of squares equals 66.434, which is identical to the regression sum of squares of 66.43387 shown in Table 5.4. And the $F = 10.857$ for the analysis of variance is equal to $F = 10.85685$ for the regression equation. Thus, the one-way analysis of variance F test is equivalent to the F test for the regression equation.

Analysis of Covariance

The analysis of variance with the covariate age included is shown in Table 5B.2. The covariate age is included to adjust the main effects of marital status to take into account the age differences between the different marital status groups. Thus, age is controlled when the effect of marital status is estimated. The SPSS procedure ANOVA provides two alternatives for measuring and testing the effects of covariates; they may be tested before marital status (i.e., with no control for the nominal variable), or they may be tested with marital status controlled (which is the way multiple regression tests all variables in the equation). The SPSS subcommand COVARIATES = WITH in Table 5B.2 specifies

TABLE 5B.2 **Analysis of Covariance of Life Satisfaction by Marital Status with Age**

```
ANOVA VARIABLES = SATLIFE BY MARITAL (1,4) WITH AGE/
  COVARIATES = WITH/STATISTICS = ALL

                    *** CELL MEANS ***

TOTAL POPULATION
          5.45
         (719)

MARITAL
             1           2         3          4
          5.69        4.93      5.49       5.16
         (399)       (105)      (67)      (148)

              *** ANALYSIS OF VARIANCE ***
```

Source of Variation	Sum of Squares	DF	Mean Square	F	Sig of F
Main Effects	81.918	4	20.479	10.111	.000
MARITAL	56.037	3	18.679	9.222	.000
AGE (Covar)	17.611	1	17.611	8.695	.003
Explained	81.918	4	20.479	10.111	.000
Residual	1446.177	714	2.025		
Total	1528.095	718	2.128		

that the effect of AGE is to be measured *with* marital status controlled. The analysis of covariance gives F tests for each of the two main effects (independent variables) and an F test for the total explained variance. $F = 10.111$ for the total explained variance (Table 5B.2), which is identical to the F value for the test of the regression equation in Table 5.5 ($F = 10.11101$).

The tests for the main effects of MARITAL and AGE, with each controlled when testing the other, show that both have significant effects on life satisfaction. $F = 8.695$ for AGE, and thus $t = \sqrt{F} = \sqrt{8.695} = 2.949$. This is identical to $t = 2.949$ for AGE in the regression equation (Table 5.5). The test for the main effect of marital status gives $F = 9.222$. There is no comparable test statistic that is part of the standard regression output. Remember, in the regression analysis we do not have a single marital status variable, but instead, marital status is represented by three dummy variables. Thus, the regular regression output contains t tests (or F tests) for each marital status dummy, but when other X's, such as age, are in the equation, an overall test for marital status is not provided. We can, however, readily conduct such a test. We saw in Chapter 4 (e.g., Equation 4.19) how to conduct F tests for subsets of independent variables.

Using this procedure, we can use the regression results in Table 5.5, plus the bivariate $r^2 = .01694$ for life satisfaction and age (not shown in Table 5.5), to compute F for the three marital status dummy variables (letting X_1 represent AGE and X_2, X_3, and X_4 represent NEVER, DIVORCED, and WIDOWED, respectively), as follows:

$$F = \frac{(R^2_{Y.1234} - r^2_{Y1})/g - 1}{(1 - R^2_{Y.1234})/n - g - 1} = \frac{(.05361 - .01694)/3}{(1 - .05361)/714} = 9.2218$$

We can see that this value of F is equal to $F = 9.222$ for MARITAL in the analysis of covariance (Table 5B.2). Thus, all of the F tests provided by the analysis of variance with covariates are also provided or readily computed with dummy variable regression analysis.

6 Nonlinear Relationships

In Chapters 2 through 4, we examined linear relationships and tests of statistical significance for linear relationships. In this chapter, we will see how to describe various kinds of nonlinear relationships, how to conduct significance tests for nonlinearity, and how to transform nonlinear relationships into linear relationships when appropriate.

Unlike linear relationships, nonlinear relationships may take on many different functional forms. We will begin by outlining a general significance test for nonlinearity that does not require a specification of the form of nonlinearity. Next, we will examine a family of related nonlinear functions called power polynomials. These functions are very flexible in that they can describe a wide range of different shapes of nonlinearity. We will see how to test for the order of the polynomial function that best fits the data.

Both the general test and the method of power polynomials involve introducing additional variables into the linear regression equation in order to describe and test for nonlinearity. In certain types of nonlinear relationships, called monotonic nonlinear relationships, it is sometimes possible to transform the values of X (or Y) such that the values of the transformed variable are linearly related to Y. In such cases, it is not necessary to introduce additional variables into the regression equation. Before taking up these regression techniques for describing and testing nonlinear functions, however, it will be useful to take a careful look at what is meant by a nonlinear function.

FIGURE 6.1 *Y* as a Nonlinear Function of *X*

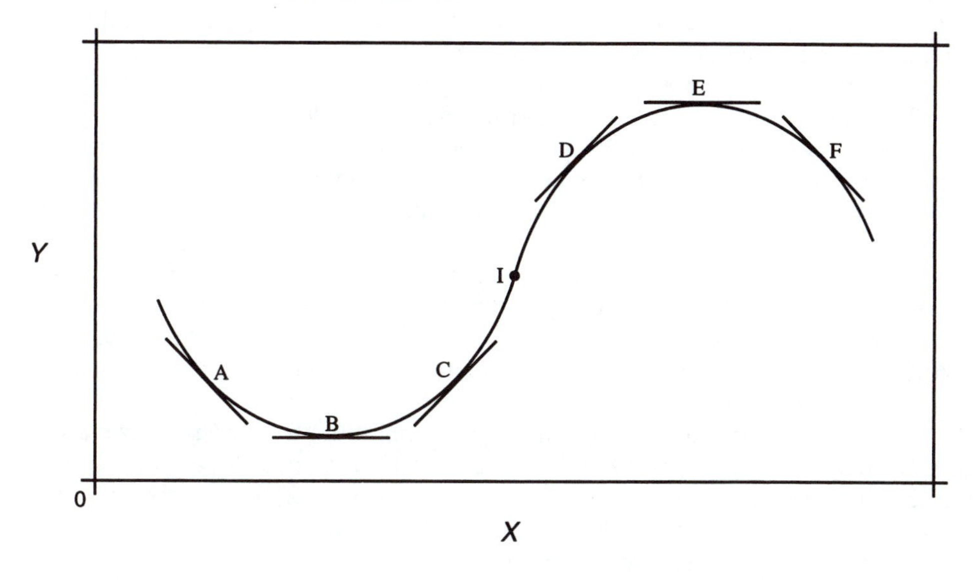

Characteristics of Nonlinear Functions

The linear relationship $Y = a + bX$ means that the effect of X on Y is the same at all levels of X; that is, the change in Y per unit increase in X is the same regardless of where in the range of X the unit increase occurs. For example, if X changes from 1 to 2, the change in Y will equal b, and if X changes from 99 to 100, the change in Y will again equal b.

A nonlinear relationship, on the other hand, means that the slope of the function that relates Y to X is not identical for all values of X. The effect of X may either increase or decrease as the values of X increase, and the effect may also change signs, either from positive to negative or vice versa. Figure 6.1 shows a curvilinear relationship between X and Y for which the effect of X, or its slope, is constantly changing throughout the range of X.

The lines touching the curve at points A through F are *tangents*; a tangent touches the curve at one, and only one, point. The slope of the tangent at a particular point is the slope of the curve at that point. Thus, the slope of the function at A is equal to the slope of the tangent at point A.[1] The slope of the tangent at point A is negative. As you move along the graph of the function

1. For those familiar with differential calculus, the slope may be determined by taking the first derivative of Y with respect to X. The numerical value of the slope equals the result achieved by entering some value of X (such as the value of X at point A) into the formula for the derivative. If $Y = a + b_1X + b_2X^2$, the derivative of Y with respect to X is $dY/dX = b_1 + 2b_2X$. Entering any value of X into the formula for the derivative gives the slope of the function at that value of X.

from A to B, the slope becomes less negative and equals zero at point B. When you move from B toward C the slope becomes increasingly positive. At some point I about midway between C and D the slope stops increasing and starts to decrease (i.e., it starts to become less positive). This point where the slope makes the transition from increasing magnitude to decreasing magnitude is called the *point of inflection*. From the point of inflection through D and up to E the slope of the function becomes smaller and smaller until it again becomes zero at point E. After point E the slope again becomes negative, and increasingly so.

The curve in Figure 6.1 has two "bends," one at B and one at E. The bends are points where the slope changes direction, from negative to positive at B and from positive to negative at E. The arc around the first bend described by ABC is said to be *upwardly concave*, and the arc around the second bend DEF is *downwardly concave*.

As we have seen, the slope of the function in Figure 6.1 is constantly changing throughout the range of X. The usual interpretation of a slope is the change in Y per *unit* increase in X. In this function, however, if X changes by one unit, the slope of the tangent (the slope of the function at a particular point) changes. Thus, we have to use a leap of the imagination to interpret the slope of the function at a particular point. We have to imagine that if X increased by an infinitesimal amount, the slope of the tangent times this incredibly small increase in X would give the change in Y. Thus, the slope of the tangent at a point gives the *instantaneous rate of change* in Y at that point.

Since there are an infinite number of different slopes for a nonlinear relationship such as that shown in Figure 6.1, we cannot evaluate and interpret them all. Although we might evaluate the slope at a few key points in the function, we will usually be more interested in describing and interpreting the pattern of the nonlinear effects of X on Y. In Figure 6.1, X has a negative effect for low values of X (around A), little or no effect for somewhat higher values around B, a positive effect throughout much of the range of X described by the intermediate values of X associated with the arc CD, little or no effect around E, and a negative effect for high values of X around point F. It is also important to note that although the slope is the same at points E and B, Y is greater at E than at B; Y is also greater at F than at A even though the slopes at those points are about equal. As this nonlinear function suggests, nonlinear effects may be much more complex and intriguing than linear effects.

A General Test for Nonlinearity

Describing an Empirical Nonlinear Relationship

In a linear relationship between empirical measures of X and Y, the values of Y never fall exactly along a straight line. The same is true for a nonlinear relationship. Therefore, the first issues we must explore are what we mean by

FIGURE 6.2 Nonlinear Relationship Between X and the Conditional Means of Y

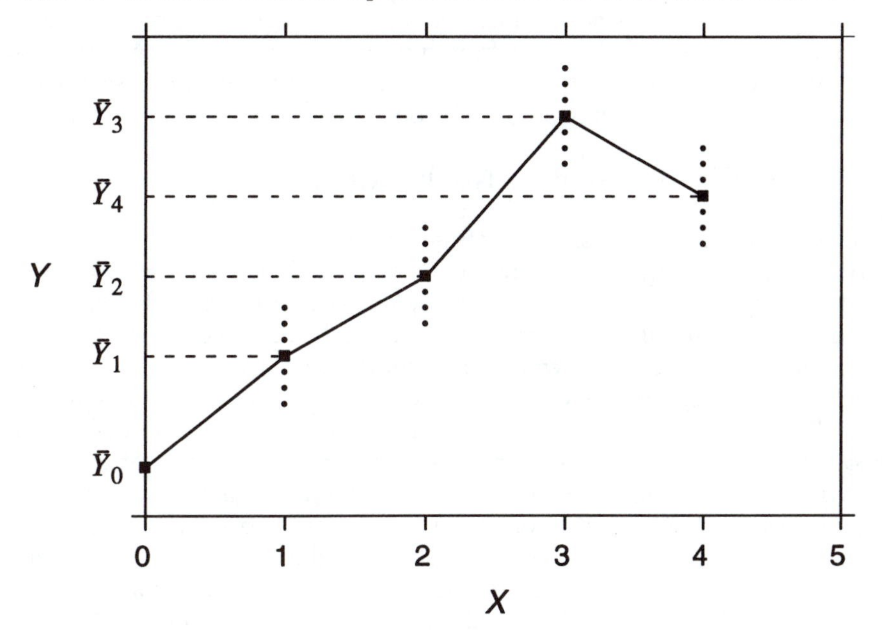

a nonlinear relationship for our empirical observations and how we test for or detect it. A later issue will be how to determine a relatively simple mathematical function that describes the nonlinearity reasonably well.

Figure 6.2 shows a potential nonlinear relationship between observed sample values of X and Y. The plot shows that for each value of X there is a spread in the values of Y. The distribution of Y at each value of X has a mean, which again is called the *conditional mean* of Y. Figure 6.2 shows a line connecting each of the adjacent conditional means of Y. This line, however, is not straight throughout the range of X. When the conditional means of Y do not fall on a single straight line, we say there is an empirical nonlinear relationship. The line connecting the conditional means in Figure 6.2 exactly describes the nonlinear relationship in the sample. The slope of the nonlinear relationship is positive from $X = 0$ to $X = 3$, although the magnitude of the slope changes somewhat over this range (thus, the XY relationship is nonlinear in the range 0 to 3). The slope then becomes negative between $X = 3$ and $X = 4$. In this case, it is quite clear from the pattern of the conditional means of Y that there is a nonlinear relationship in this sample.

It is not clear, however, whether the relationship is nonlinear in the population from which the sample was selected. Due to sampling error, we would never get a perfectly linear relationship in the sample when the population relationship is linear. Thus, we have to determine whether the departure from linearity in the sample is sufficiently great to assure us that there is little likelihood that the sample was drawn from a population in which the relationship was linear. We need to conduct a statistical test to evaluate the possibility that the nonlinear sample relationship occurred by chance.

The Nominalization Test for Nonlinearity

Since a nonlinear relationship is defined by the conditional means of Y, we need to specify a model such that for each value of X, the predicted value of Y equals the conditional mean of Y for that value of X. Such a model will exactly describe the nonlinear relationship present in the sample data. In the discussion of dummy variables, it was shown that if $g - 1$ dummy variables are created for the g categories of the nominal variable, the predicted value of Y for each category of the nominal variable will equal the mean of Y for that category. We can use the same procedure for an interval or ordinal variable as for a nominal variable. That is, we create a dummy variable for each value of the interval/ordinal X except one. If, however, the interval X is a continuous variable for which all fractional values between the minimum and maximum values can occur, then this method clearly cannot be used. Also, if X is a discrete variable (it takes on only certain values) with a large number of values, this procedure will be impractical. Alternate procedures for handling such variables will be noted later. Many social science variables, however, have relatively small numbers of discrete values. We will cover cases of this type first.

In the first example, we will use a discrete interval-level variable with a manageable number of values. The data are from the 1974 General Social Survey, a probability sample of the adult population of the United States. The dependent variable Y is the respondents' reports of the number of children they consider to be ideal for a family to have (CHLDIDEL). The independent variable X is the actual number of children the respondent has (CHILDS). The range of values for X is 0 to 8, where 8 represents 8 or more children. Therefore, X has $g = 9$ discrete values.

In order to nominalize X, $g - 1 = 8$ dummy variables will be created. These dummy variables are labeled X_i in the coding scheme outlined below. The first column shows the $g = 9$ discrete values of X = CHILDS. The other eight columns show how each value of X will be coded on each of the X_i dummy variables. For example, a person with six children (i.e., $X = 6$) will be coded as 1 on dummy variable X_6, and persons with fewer than six children or more than six children will have a code of 0 on X_6. A dummy variable has not been created for those with $X = 0$ children; such persons are coded as 0 on each of the eight dummy variables. Thus, persons without any children are the reference group in this analysis.

Dummy Variables for X = CHILDS

X	X_1	X_2	X_3	X_4	X_5	X_6	X_7	X_8
0	0	0	0	0	0	0	0	0
1	1	0	0	0	0	0	0	0
2	0	1	0	0	0	0	0	0
3	0	0	1	0	0	0	0	0
4	0	0	0	1	0	0	0	0
5	0	0	0	0	1	0	0	0
6	0	0	0	0	0	1	0	0
7	0	0	0	0	0	0	1	0
8	0	0	0	0	0	0	0	1

The following SPSS commands may be used to create the dummy variables and to run the regression analyses for the nominalization test of nonlinearity.

```
IF CHILDS EQ 1 X1 = 1
IF CHILDS NE 1 X1 = 0
IF CHILDS EQ 2 X2 = 1
IF CHILDS NE 2 X2 = 0
IF CHILDS EQ 3 X3 = 1
IF CHILDS NE 3 X3 = 0
IF CHILDS EQ 4 X4 = 1
IF CHILDS NE 4 X4 = 0
IF CHILDS EQ 5 X5 = 1
IF CHILDS NE 5 X5 = 0
IF CHILDS EQ 6 X6 = 1
IF CHILDS NE 6 X6 = 0
IF CHILDS EQ 7 X7 = 1
IF CHILDS NE 7 X7 = 0
IF CHILDS EQ 8 X8 = 1
IF CHILDS NE 8 X8 = 0
REGRESSION VARS = CHLDIDEL CHILDS X1 TO X8/
  DEP = CHLDIDEL/ENTER CHILDS/
  DEP = CHLDIDEL/ENTER X1 TO X8
```

In order to conduct a test for nonlinearity, the linear relationship between Y and the interval/ordinal X is estimated first. A second regression is then run for Y and the set of $g - 1$ dummy variables. In the second regression, only the $g - 1$ dummies are used; the original X must not be included in the equation. If X was included in the equation with the dummy variables, perfect multicollinearity would exist. The analysis for our example pertains to married respondents only ($n = 888$). The results for the linear regression and the dummy-variable regression are shown in Table 6.1.

TABLE 6.1 **Regressions of Ideal Number of Children on Number of Children**

Variable	*b*	*F*
1. Linear		
CHILDS	.187	115.94
Constant	2.190	
R^2	.11572	
2. Dummy variables		
X1	−.004	.00
X2	−.172	2.55
X3	.277	6.05
X4	.437	11.02
X5	.875	37.57
X6	1.054	29.86
X7	.460	2.89
X8	1.642	51.21
Constant	2.463	
R^2	.15862	20.71

The linear equation shows that the ideal number of children increases by .187 for each additional child that the respondent has. The F test for the slope (which is equivalent to the F test for R^2 in the linear case) is significant at $p < .01$ (the critical $F_{1,886} = 6.66$ for $\alpha = .01$ and the critical $F_{1,886} = 3.85$ for $\alpha = .05$). Thus, X has a significant positive linear effect on Y.

The important question here, however, is whether there is a nonlinear function that will fit the data even better than the significant linear function. The dummy-variable equation provides the first answer to that question. The slope for each dummy variable indicates the difference between the mean of the persons coded 1 on that dummy and the mean of the reference group, respondents with no children. The intercept a is the mean for those with zero children. The mean ideal number of children for respondents with no children is $a = 2.463$. The intercept plus the slope for a particular dummy variable gives the predicted value of Y for persons coded 1 on that dummy variable. For example, the predicted value of Y for persons with 6 children is

$$\begin{aligned}\hat{Y} &= 2.463 - .004X_1 - .172X_2 + .277X_3 + .437X_4 + .875X_5 \\ &\quad + 1.054X_6 + .460X_7 + 1.642X_8 \\ &= 2.463 - .004(0) - .172(0) + .277(0) + .437(0) + .875(0) + 1.054(1) \\ &\quad + .460(0) + 1.642(0) \\ &= 2.463 + 1.054 = 3.517\end{aligned}$$

FIGURE 6.3 Predicted Values of Y (CHLDIDEL) from the Dummy-Variable Regression

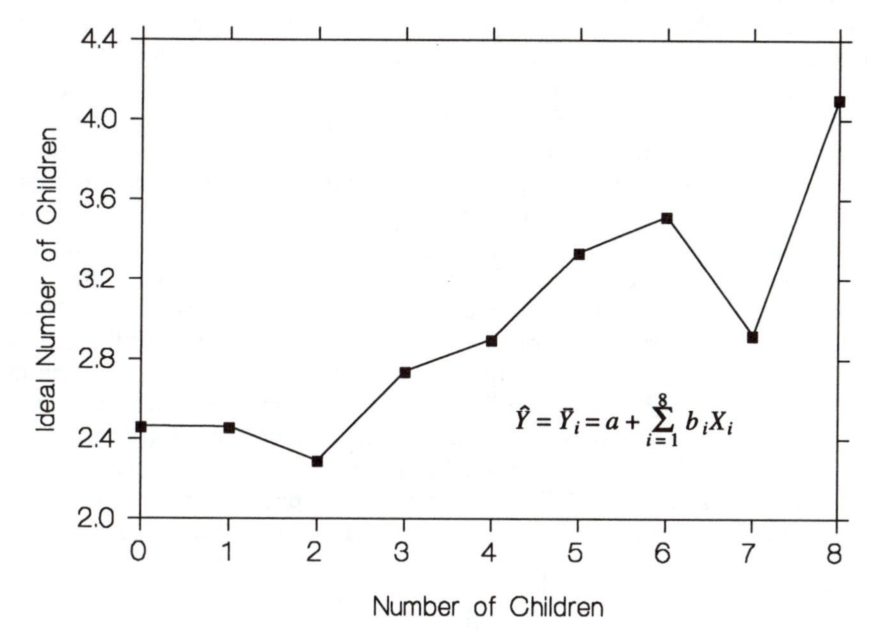

This predicted value equals the mean for that group; in other words, it is the conditional mean for respondents with a given number of children (i.e., 6). Figure 6.3 shows a plot of the predicted values, or conditional means, for each value of X.

Clearly, there is a nonlinear pattern for the conditional means. The F value for each dummy-variable slope in the regression output tests whether the mean of the persons at each level of X is significantly different from the mean of the reference group. Those with one and two children are not significantly different in their ideal number of children from those with no children. Respondents with three, four, five, six, and eight or more children prefer significantly more children than those with no children. Inexplicably, those with seven children do not differ significantly from those with no children. The big dip in the plot in Figure 6.3 for $X = 7$ shows this apparently nonlinear pattern.

The plot of the means in Figure 6.3 seems to show that there is little or no change in Y from $X = 0$ to $X = 2$, but there is a more-or-less linear increase in Y from $X = 2$ to $X = 8$, except for the big dip at $X = 7$. The tests of the slopes for the dummy variables do not tell us, however, whether there is a departure from linearity. If the relationship were linear, we would probably still find that those with three or more children differed significantly from those with no children. Furthermore, those with values of X close to $X = 0$ (such as $X = 1$ and $X = 2$) might not differ significantly from those with $X = 0$ even if there were a true linear relationship; the sample size n might not be large enough to separate significantly those who are close together on X, even though the overall linear relationship might be significant.

The tests of the dummy slopes, however, are not the reason that we are conducting the dummy-variable analysis. The dummy regression is used mainly to see how much the explained variance increases when we use a model that captures all of the nonlinearities that are present in the sample data. The variance explained by the dummy-variable equation is the maximum amount that can be explained by X; since it predicts the conditional mean of Y for each value of X, we can do no better than that. The regression output shows that R^2 increased from .11572 for the linear equation to .15862 for the dummy-variable equation. Since the dummy equation uses every piece of information about Y that can be gleaned from X, the explained variance will always go up. The increase, which is caused by departures from linearity in the sample, however, may have occurred by chance. Therefore, a significance test for the increase in R^2 is needed. The following F test can be used to determine if the use of $g - 1$ dummy variables captures a significant departure from linearity:

$$F = \frac{(R^2_{Y\cdot 12...(g-1)} - r^2_{YX})/g - 2}{(1 - R^2_{Y\cdot 12...(g-1)})/n - g} \tag{6.1}$$

This formula is based on Equation 4.14, the general formula for F. In the numerator there is the increase in explained variance derived from using the dummy-variable equation, divided by the degrees of freedom. The numerator degrees of freedom equals the difference between the number of X's in the dummy-variable equation and the number of X's in the linear equation (one); therefore, $df_1 = (g - 1) - 1 = g - 2$. The denominator equals the proportion of variance not explained by the dummy-variable equation divided by its degrees of freedom; since the number of independent variables equals $g - 1$, $df_2 = n - (g - 1) - 1 = n - g$. For our example data, the solution for F is

$$F = \frac{(R^2_{Y\cdot 12345678} - r^2_{YX})/9 - 2}{(1 - R^2_{Y\cdot 12345678})/888 - 9} = \frac{(.15862 - .11572)/7}{(1 - .15862)/879} = 6.40$$

The critical $F_{7,879} = 2.66$ for $\alpha = .01$. Thus, the level of significance of F is $p < .01$. We therefore reject the null hypothesis that Y is a linear function of X in favor of the alternative hypothesis that Y is some nonlinear function of X.

The nominalization test is a global test for any type of nonlinear relationship. It does not identify the best mathematical specification of the nonlinear function. Estimating the specific mathematical form of nonlinearity will be taken up next. It is often valuable, however, to conduct the global test for nonlinearity before attempting to define a more economical and specific mathematical form for that relationship. If the number of values of X is too large to make the nominalization test practical, there is an alternative procedure (Berry and Feldman 1985, pp. 54–56) that involves collapsing the large number of values into a more manageable number of categories.

Estimation of Power Polynomial Functions

Characteristics of Power Polynomials

If Y is a nonlinear function of X, the form of the relationship might be any one of an almost infinite number of types of mathematical functions. In order to select the nonlinear function that is most valid for the variables under consideration, the investigator would need to examine the scatterplot of the relationship between X and Y. Theory might also be developed sufficiently to provide guidelines for the appropriate functional form. Finally, knowledge of the characteristics of various types of mathematical functions is needed in order to make an intelligent selection.

After selecting what is believed to be the most appropriate function or functions, the investigator would then want to conduct a statistical analysis to determine how well the function fits the data or which function best fits the data. Least-squares regression can be used to estimate the parameters of some nonlinear functions, after appropriate transformations of the variables have been made. The parameters of other nonlinear functions, however, cannot be estimated with ordinary least squares. Power polynomials are a family of mathematical functions whose parameters can readily be estimated with ordinary multiple regression. These functions are very flexible in their ability to describe or approximate many types of nonlinear relationships. If you do not have any theoretical knowledge about the most appropriate mathematical function, power polynomials will often provide a satisfactory description of any nonlinear patterns present in the data.

The general equation for the family of power polynomials is

$$Y = a + b_1X + b_2X^2 + b_3X^3 + \cdots + b_kX^k \qquad \textbf{(6.2)}$$

Equation 6.2 is called a polynomial of *order* k, where k is the highest power of X included in the equation. When $k = 1$, we have a first-order polynomial or the familiar linear function

$$Y = a + b_1X \quad \text{[linear]}$$

When $k = 2$, a second-order polynomial, the function is called a *quadratic* equation:

$$Y = a + b_1X + b_2X^2 \quad \text{[quadratic]}$$

The third-order polynomial is a *cubic* equation,

$$Y = a + b_1X + b_2X^2 + b_3X^3 \quad \text{[cubic]}$$

All polynomials of order 2 or greater are nonlinear functions; squaring X, raising X to the power 3, or raising X to any higher power are all nonlinear transformations of X. For example, when a set of scores (e.g., 1, 2, 3, 4) is squared ($1^2 = 1, 2^2 = 4, 3^2 = 9, 4^2 = 16$), the differences between the squared values are not proportional to the differences between the original scores (e.g., $4/2 = 2$, but $4^2/2^2 = 4$).

Power polynomials are very useful because they can be used to describe nonlinear relationships with any number of bends in the curve. *A polynomial of order k will have a maximum of $k - 1$ bends* ($k - 1$ points at which the slope of the curve changes direction); for example, a cubic equation can have two bends. The signs and magnitudes of the b_i coefficients will determine where the curve is concave downward and where it is concave upward. For example, the two quadratic equations below (which are evaluated for $X = 0$ to $X = 3$) each have one bend, but the bends have opposite concavity.

$Y = 3 + 2X - X^2$	$Y = 2 - 2X + X^2$
$Y = 3 + 2(0) - (0)^2 = 3 + 0 + 0 = 3$	$Y = 2 - 2(0) + (0)^2 = 2 - 0 - 0 = 2$
$Y = 3 + 2(1) - (1)^2 = 3 + 2 - 1 = 4$	$Y = 2 - 2(1) + (1)^2 = 2 - 2 + 1 = 1$
$Y = 3 + 2(2) - (2)^2 = 3 + 4 - 4 = 3$	$Y = 2 - 2(2) + (2)^2 = 2 - 4 + 4 = 2$
$Y = 3 + 2(3) - (3)^2 = 3 + 6 - 9 = 0$	$Y = 2 - 2(3) + (3)^2 = 2 - 6 + 9 = 5$

Notice that the two equations have opposite-signed coefficients for X and opposite-signed coefficients for X^2, but the magnitude of the coefficients is the same across equations. The differences in signs will cause the two curves to bend in opposite directions, as shown in Figure 6.4. (The difference between the intercepts does not affect the shapes of the curves.) A close look at the computations above shows why the signs and magnitudes of the coefficients cause the curves to bend as they do. The positive slope for X in the first equation causes Y initially to rise between $X = 0$ and $X = 1$ because $+2(1)$ is greater in absolute value than $-(1)^2$. When $X = 2$, however, $-(2)^2$ equals $+2(2)$, and thus Y decreases. At $X = 3$, $-(3)^2$ is greater in absolute value than $+2(3)$, causing an even greater decrease in Y. In sum, the positive coefficient for X and the negative coefficient for X^2 cause the curve to rise initially and then to fall. Just the opposite pattern occurs for the second equation; the negative coefficient for X and the positive coefficient for X^2 cause this function to fall initially and then to rise.

A third-order polynomial, or cubic equation, can have two bends. One bend will be concave upward, and one will be concave downward. The signs of the coefficients will determine whether the upward concavity or the downward

FIGURE 6.4 Two Quadratic Equations with Opposite Concavity

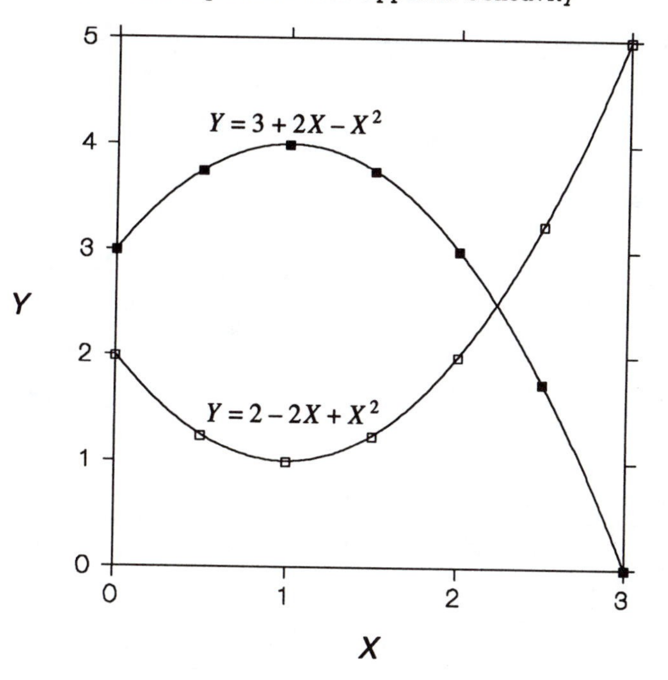

concavity occurs first (for lower values of X). The equation in the following table provides an example.

Y	=	4	+	$2X$	−	X^2	+	$.1X^3$										
Y	=	4	+	2(0)	−	$(0)^2$	+	$.1(0)^3$	=	4	+	0	−	0	+	.0	=	4.0
Y	=	4	+	2(1)	−	$(1)^2$	+	$.1(1)^3$	=	4	+	2	−	1	+	.1	=	5.1
Y	=	4	+	2(2)	−	$(2)^2$	+	$.1(2)^3$	=	4	+	4	−	4	+	.8	=	4.8
Y	=	4	+	2(3)	−	$(3)^2$	+	$.1(3)^3$	=	4	+	6	−	9	+	2.7	=	3.7
Y	=	4	+	2(4)	−	$(4)^2$	+	$.1(4)^3$	=	4	+	8	−	16	+	6.4	=	2.4
Y	=	4	+	2(5)	−	$(5)^2$	+	$.1(5)^3$	=	4	+	10	−	25	+	12.5	=	1.5
Y	=	4	+	2(6)	−	$(6)^2$	+	$.1(6)^3$	=	4	+	12	−	36	+	21.6	=	1.6
Y	=	4	+	2(7)	−	$(7)^2$	+	$.1(7)^3$	=	4	+	14	−	49	+	34.3	=	3.3

In this equation the sign for X is positive, the sign for X^2 is negative, and the sign for X^3 is positive. You should examine carefully the computations to see how the signs for these terms govern the rise and fall of Y. For small values of X, the X term itself has the greatest effect on Y, causing it to rise. At intermediate values of X, the squared term has the greatest effect on Y, causing it to fall.

FIGURE 6.5 Third-Order (Cubic) Polynomial

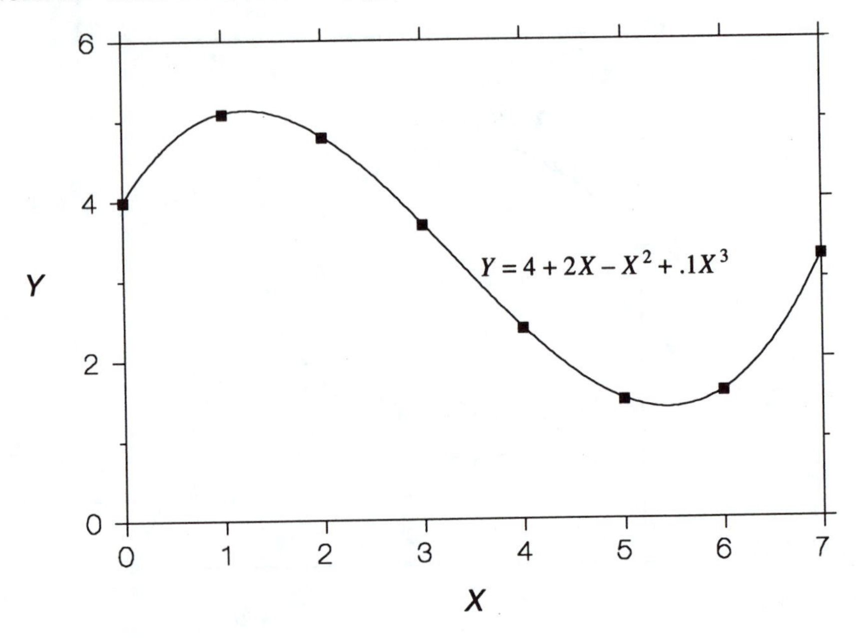

Finally, at high values of X, the cubic term in combination with X itself has the greatest effect, causing Y to rise again. The plot of the function is shown in Figure 6.5. The curve is initially concave downward and then concave upward.

We have said that a kth-order polynomial will have $k - 1$ bends. These bends, however, will not always occur over the range of positive values of X, as occurred in the above examples. The quadratic equation in the following table provides an example.

Y	=	2	+	$2X$	+	X^2								
Y	=	2	+	$2(-3)$	+	$(-3)^2$	=	2	−	6	+	9	=	5
Y	=	2	+	$2(-2)$	+	$(-2)^2$	=	2	−	4	+	4	=	2
Y	=	2	+	$2(-1)$	+	$(-1)^2$	=	2	−	2	+	1	=	1
Y	=	2	+	$2(0)$	+	$(0)^2$	=	2	+	0	+	0	=	2
Y	=	2	+	$2(+1)$	+	$(+1)^2$	=	2	+	2	+	1	=	5
Y	=	2	+	$2(+2)$	+	$(+2)^2$	=	2	+	4	+	4	=	10
Y	=	2	+	$2(+3)$	+	$(+3)^2$	=	2	+	6	+	9	=	17

The coefficients in this equation have the same values as those in the first two quadratics we examined, except X and X^2 both have the same signs (positive). Because of this, the curve will not have a bend for positive values of X.

FIGURE 6.6 Quadratic with No Bend for Positive Values of X

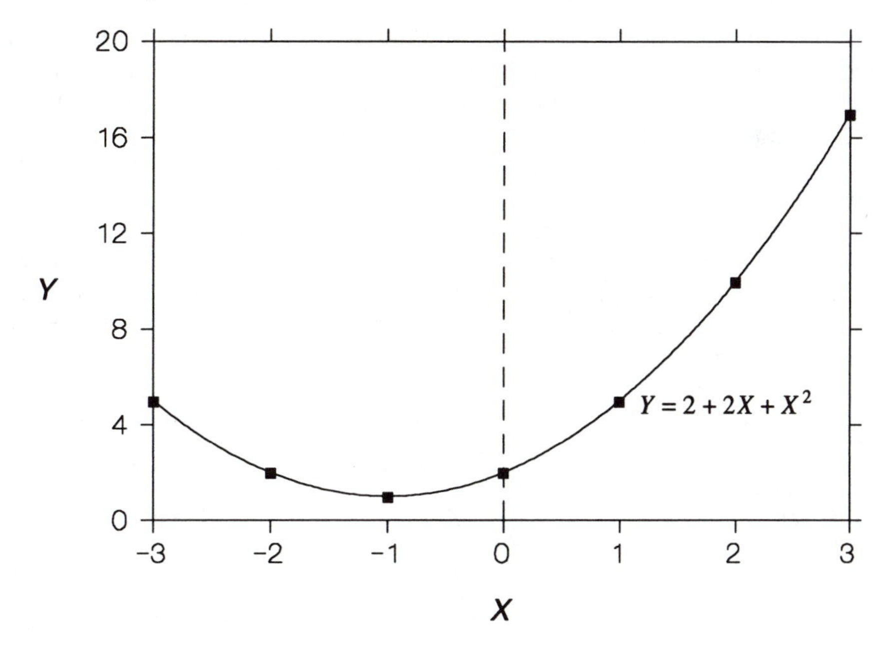

We can see, however, that there is a bend in the negative range of X (see Figure 6.6). This characteristic of polynomials has important implications for data analysis. If our observed values of X were all positive and there were no bends in the relationship between X and Y over the observed positive values of X, the quadratic polynomial would reflect this pattern in the data and would not have a bend for positive values of X (as in Figure 6.6). Notice, however, that the XY relationship is not linear in Figure 6.6 for the positive values of X. This shows that a quadratic could be used to describe a nonlinear data relationship that doesn't change direction through the range of observed values of X. The quadratic would still have one bend, but it would not occur in the range of the observed X.

Hierarchical Polynomial Regression

We have examined some of the characteristics of power polynomial functions and seen their flexibility for describing various types of nonlinear relationships. Now we address the issue of how to use regression both to estimate the parameters of second-order or higher-order polynomials and to determine which order of polynomial fits the data best.

The general form of a multiple regression equation (shown below) shows that it consists of linear relationships between each independent variable and Y:

$$\hat{Y} = a + b_1X_1 + b_2X_2 + \cdots + b_kX_k$$

There is a single slope for each of the different independent variables. This indicates that no matter where a unit change in X_i occurs over the entire range of values of X_i, the associated change in Y will equal b_i. Thus, the multiple regression equation is constrained to estimate only linear effects for each of the variables in the equation. Yet the definition of a nonlinear effect is that the slope or effect of X_i is different for different values of X_i.

The solution to this problem was briefly noted earlier. We can use regression to estimate the coefficients of a nonlinear relationship if we can make appropriate transformations of an X that has a nonlinear effect. The example below shows how we can transform a nonlinear relationship of the form $Y = a + bX^2$ into a linear relationship by creating a new variable $V = X^2$.

$Y = 1 + .5X^2$	$V = X^2$	$Y = 1 + .5V$
$Y = 1 + .5(0)^2 = 1.0$	$V = (0)^2 = 0$	$Y = 1 + .5(0) = 1.0$
$Y = 1 + .5(1)^2 = 1.5$	$V = (1)^2 = 1$	$Y = 1 + .5(1) = 1.5$
$Y = 1 + .5(2)^2 = 3.0$	$V = (2)^2 = 4$	$Y = 1 + .5(4) = 3.0$
$Y = 1 + .5(3)^2 = 5.5$	$V = (3)^2 = 9$	$Y = 1 + .5(9) = 5.5$
$Y = 1 + .5(4)^2 = 9.0$	$V = (4)^2 = 16$	$Y = 1 + .5(16) = 9.0$

The above nonlinear function is not a quadratic because it does not contain X to the first power. It does, however, show a nonlinear relationship between X and Y. The values of Y are computed over the range of $X = 0$ to $X = 4$. A new variable V is computed that equals X^2. The right-hand column then shows that Y and V have a linear relationship for which the intercept and coefficient for V are the same as in the original nonlinear function for X and Y. The plot of the nonlinear (Y, X) function and the linear (Y, V) function are shown in Figure 6.7.

If we wanted to estimate the coefficient b for a nonlinear equation of the form $Y = a + bX^2$ for some set of data, we could compute $V = X^2$ and use an ordinary regression program to get the least-squares solution for $\hat{Y} = a + bV$. This logic implies that if we want to estimate the coefficients for a third-order polynomial $Y = a + b_1X + b_2X^2 + b_3X^3$, we would compute the following new variables:

$$V = X^2 \quad \text{and} \quad W = X^3$$

We would then use the two new variables, along with X itself, in the following regression equation:

$$\hat{Y} = a + b_1X + b_2V + b_3W$$

The partial coefficients for this linear regression equation would equal the coef-

FIGURE 6.7 *Y* as a Nonlinear Function of *X* and a Linear Function of *V*

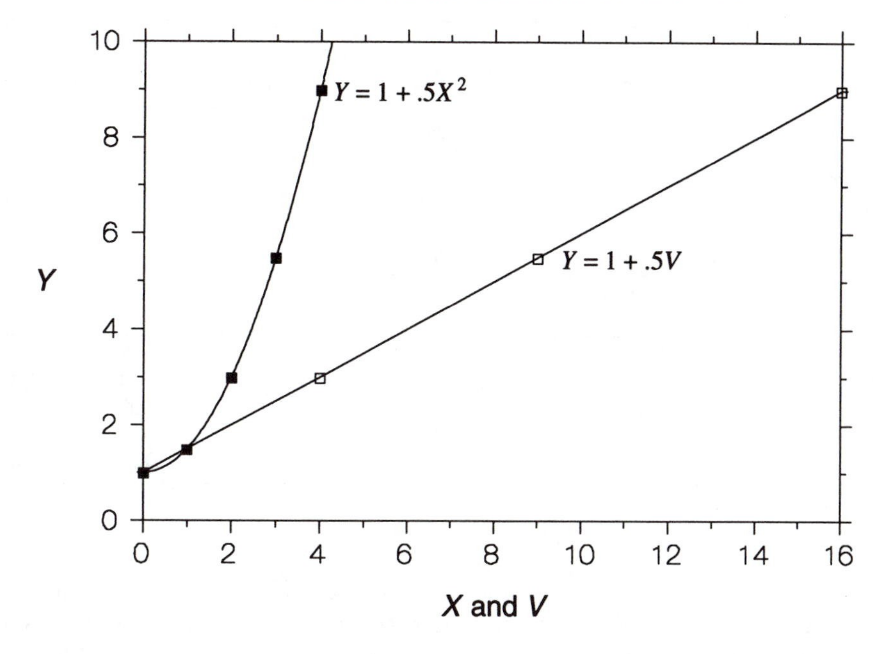

ficients for the cubic polynomial. We would be "tricking" the regression program into believing that we had three different independent variables, each of which is linearly related to Y. In fact, however, the regression equation contains only one conceptual variable, but this single conceptual variable is represented by three sets of scores: the original set of raw X scores and two nonlinear transformations of these raw scores, V and W. There is an analogy between what we are doing here, i.e., using three variables to represent a single conceptual variable, and what was done previously in the nominalization test, i.e., using $g - 1$ dummy variables to represent the single interval-level variable. Instead of using dummy-coded transformations of X, we are now using power transformations of X.

To find the polynomial that provides the best fit to the data, we will normally use a hierarchical testing procedure. The hierarchical tests start with the simplest polynomial, the linear function, and then proceed step by step to add variables representing each successive higher power of X, testing at each step whether the power of X added at that step significantly improves the fit (i.e., significantly increases the explained variance). That is, we will test whether the quadratic regression provides an improved fit over the linear regression, whether the cubic regression fits better than the quadratic equation, and so on.

TABLE 6.2 Regression of Ideal Number of Children on Powers of Children

	Correlations			
	CHLDIDEL	*CHILDS*	*CHILDSQ*	*CHLDCUBE*
CHLDIDEL	1.000	.340	.358	.335
CHILDS	.340	1.000	.935	.839
CHLDSQ	.358	.935	1.000	.973
CHILDCUBE	.335	.839	.973	1.000

	Regression Equations					
	Step 1		*Step 2*		*Step 3*	
Variable	*b*	*F*	*b*	*F*	*b*	*F*
CHILDS	.187	115.94	.022	.21	−.237	5.74
CHILDSQ	—	—	.025	13.08	.124	13.61
CHLDCUBE	—	—	—	—	−.009	9.03
Constant	2.190		2.360		2.480	
R^2	.11572	115.94	.12860	65.31	.13741	46.94

The relationship between number of children (CHILDS) and ideal number of children (CHLDIDEL) that was examined earlier will provide the first illustration of the procedure. The nominalization test lead us to conclude that there is a nonlinear relationship between the two variables. Now we want to find out which order of the power polynomial functions, if any, best describes this nonlinear relationship. The following SPSS commands may be used to make the appropriate variable transformations and to enter the variables into the appropriate regression equations.

```
COMPUTE CHILDSQ = CHILDS**2
COMPUTE CHLDCUBE = CHILDS**3
REGRESSION DESCRIPTIVES/
  VARS = CHLDIDEL CHILDS CHILDSQ CHLDCUBE/
  STAT = DEFAULTS TOL/
  DEP = CHLDIDEL/
  ENTER CHILDS/ENTER CHILDSQ/ENTER CHLDCUBE
```

CHILDSQ equals the square of the independent variable, and CHLDCUBE equals the independent variable raised to the third power. The variables are entered into the regression equation in three steps, with the linear equation estimated first, the quadratic equation estimated second, and the cubic equation estimated third. The results of the regression analysis are given in Table 6.2.

The correlation matrix shows that the independent variables are very highly correlated. This is always the case in polynomial regression, because the independent variables are all powers of a single conceptual variable. Thus, multicollinearity will always be high for power polynomial regression analyses.[2]

As we saw earlier, CHILDS has a significant positive effect on ideal number of children in the linear equation. In Step 2, the quadratic equation, we are interested first in the test for the slope of the squared variable, which is the variable added to the equation in this step. The F test for this variable tells us whether the quadratic significantly increased the explained variance over that explained by the linear equation. The formula for F as a test of a partial slope in an equation with two independent variables is determined from Equation 4.14 to be

$$F = \frac{(R^2_{Y\cdot12} - r^2_{Y1})/1}{(1 - R^2_{Y\cdot12})/n - 3}$$

This test evaluates whether adding a particular independent variable to the equation significantly increases the explained variance over that explained by the other variables in the equation. Evaluating this equation for the squared term in Step 2, we get

$$F = \frac{(.1286 - .11572)/1}{(1 - .1286)/885} = 13.08$$

This result equals the F value given for the slope of CHILDSQ. The critical $F_{1,885} = 6.66$ for $\alpha = .01$; thus, the level of significance is $p < .01$. Since the squared aspect of the independent variable has been found to increase the explained variance significantly, we know the quadratic fits better than the linear function. The t test for the squared variable would have done just as well as F, but the F test makes it absolutely clear that we are testing the improved fit provided by the quadratic equation.

The regression coefficients give us the following quadratic equation:

$$Y = 2.36 + .022X + .025X^2 + e$$

The coefficients for both powers of X are positive. Thus, this function will not

---

2. High multicollinearity means that each power variable will have very little unique variance (that is, variance that is independent of the other power variables). Thus, the tolerance, or $1 - R_i^2$, will be very low. With third- or higher-order polynomials, the tolerance is very likely to be less than .01. The SPSS default level for tolerance is .0001, which means that the program will enter a variable into the equation if at least 1/10,000 of its variance is independent of the other variables. According to Cohen and Cohen (1983:238), some programs may encounter problems with computational accuracy when the multicollinearity is very high. They recommend transforming the X scores to deviation scores $(X - \overline{X})$ before creating the power variables. When the deviation scores are powered, the correlations between even- and odd-powered variables will be reduced appreciably because negative deviation scores will become positive when raised to an even power but will remain negative when raised to an odd power. This may reduce problems associated with extremely high multicollinearity.

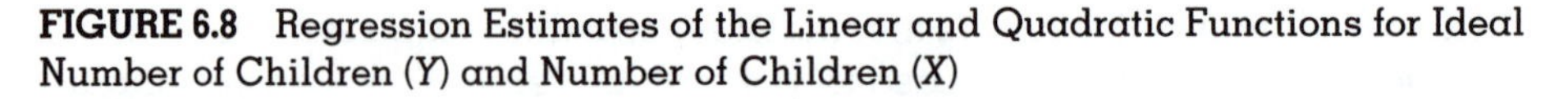

FIGURE 6.8 Regression Estimates of the Linear and Quadratic Functions for Ideal Number of Children (Y) and Number of Children (X)

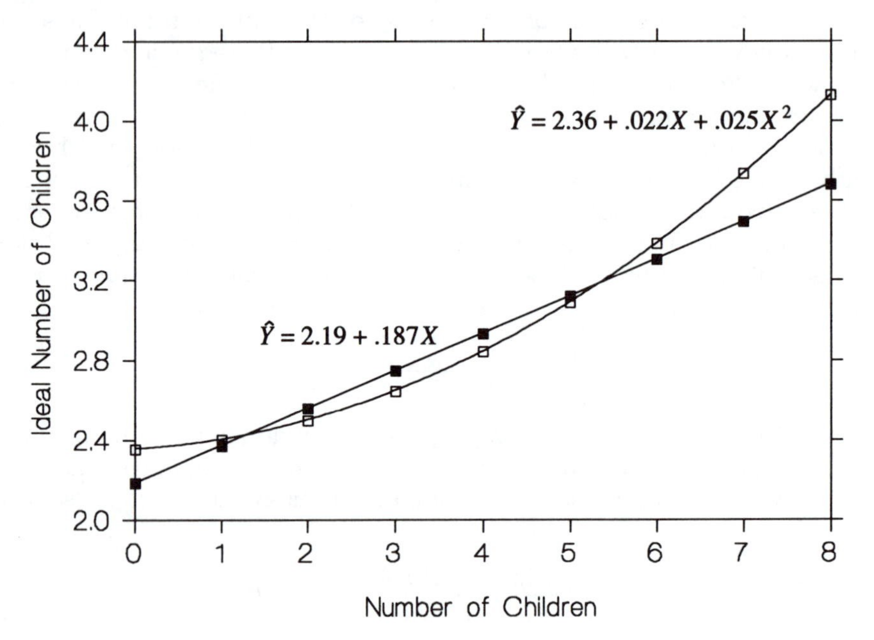

have a bend (a change in the direction of the slope), for positive values of X. Since all of our observed values of X are positive, there will be no bend in the observed range of X. The plots of the quadratic and linear functions are shown in Figure 6.8. The improved fit provided by the quadratic occurs mainly in the lower range of X. The quadratic better describes the fact that there is little or no change in Y over the range of $X = 0$ to $X = 2$ than does the linear function (see the plot of the conditional means in Figure 6.3). The quadratic, however, does not describe the inexplicable dip in Y at $X = 7$. To summarize the picture presented by the quadratic, those with two or fewer children differ very little with respect to their preferred number of children. As the number of children increases to three and higher, however, the ideal number of children increases at an *increasingly rapid rate*.

Next, we examine the F test for the increase in explained variance created by adding the third power of children (CHLDCUBE) to the equation in Step 3. By using the R^2's for Steps 2 and 3, plus the appropriate degrees of freedom, we get

$$F = \frac{(R^2_{Y\cdot 123} - R^2_{Y12})/1}{(1 - R^2_{Y\cdot 123})/n - 4} = \frac{(.13741 - .1286)/1}{(1 - .13741)/884} = 9.026$$

FIGURE 6.9 Regression Estimate of the Cubic Equation for Ideal Number of Children (Y) and Number of Children (X)

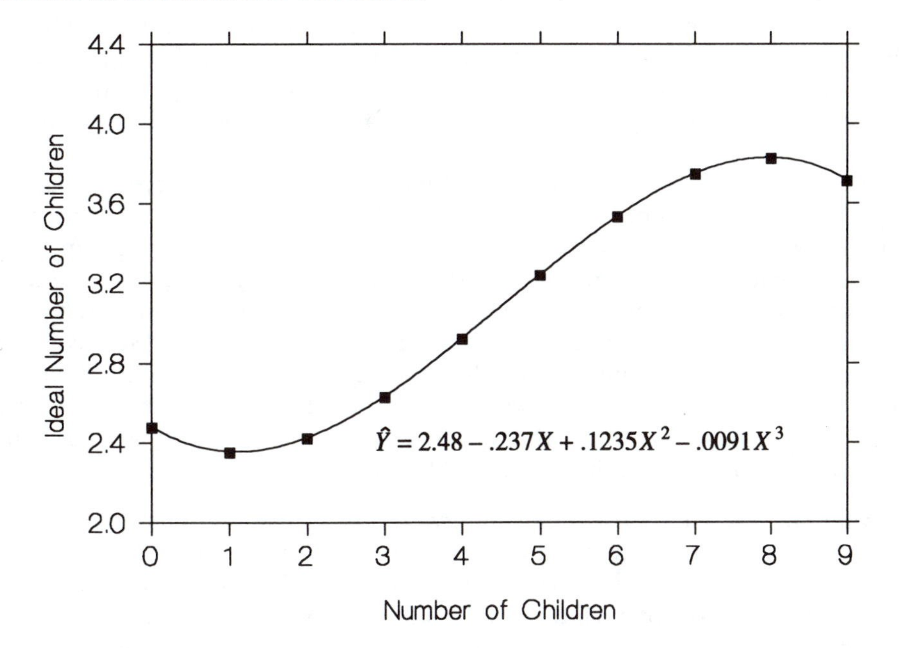

This F is also significant at $p < .01$, and thus we conclude that the cubic equation fits better than the quadratic form. Note again that this F value is identical to the F value provided by the regression output for the cubic variable. The cubic equation is thus estimated to be

$$Y = 2.48 - .237X + .124X^2 - .009X^3 + e$$

The fact that the signs of the coefficients for the odd powers of X are negative and the sign for the even power is positive indicates that the curve may bend twice over the positive range of X, with Y initially decreasing, then rising, and finally falling again at higher values of X. Figure 6.9 shows the plot of the cubic equation.

Notice that the curve does not have two bends in the observed range of X. The second bend does not occur until X changes from 8 to 9, and 8 is the maximum value that the measured variable takes (8 equals 8 or more children). This curve fits the data differently from the quadratic in several respects. First, there is a slight bend (concave upward) at $X = 1$, which roughly fits the dip in the conditional means shown in Figure 6.3. The curve still reflects the fact, however, that at small values of X there is little or no change in Y. The other difference between the cubic and quadratic occurs in the upper range of values. Whereas the quadratic shows that Y increases at an increasing rate (the curve gets steeper) for higher values of X, the cubic shows that the rate of increase in Y slows down at higher values of X (the curve gets flatter), although the slope of the cubic equation remains positive throughout this upper range. The dip in the conditional mean of Y at $X = 7$ undoubtedly influences this declining rate of increase, although the cubic never bends to fit the dip at $X = 7$.

Although we might proceed to test the fourth-order polynomial, it seems theoretically wise to stop at this point. The cubic function makes sense because it describes a pattern of a decreasing rate of increase in the ideal number of children as the actual number of children gets relatively large; we would not expect the ideal number of children to move toward infinity for those with many children, as the quadratic predicts. It also does not seem theoretically sound to try to fit the dip in ideal number of children for those with seven children, unless we were able to come up with some astute theoretical reason for such a dip. Thus, for this example, proceeding to test the fourth-order polynomial would appear to be a case of trying to over-fit the data.

Summary. At each step of hierarchical testing of the polynomial model, examine the F test (or t test) for the slope of the power of X added in that step. Since the power of X added at each hierarchical step is always the highest power of X in the equation, the crucial test for the coefficients in any polynomial equation is always the test for the coefficient of the highest power of X. If the coefficient for the highest power is not significant, drop it from the equation and use the next-lower-ordered polynomial. One caveat, however, is in order. It may sometimes occur that the kth power of X is not significant, but if the next-higher-ordered polynomial is estimated, the $k + 1$ power of X may be significant. For example, if there is truly a cubic function with two bends in the data, the quadratic function might not fit any better than the linear, but if we proceed to the cubic equation, the third power of X may be significant. Thus, it is often a good idea to let your testing be guided by the patterns that may be apparent in a scatterplot of the data rather than blindly following the hierarchical procedure.

A Quadratic Example

An additional example may be helpful. The data for this example also come from the 1974 General Social Survey, from which 460 employed males have been chosen for analysis. The dependent variable is Y = PRESTIGE. PRESTIGE is the occupational prestige scores of the respondents' reported occupations, which range from approximately 0 to 100. The independent variable is X = EDUCATION, measured in years of school completed, with a range of 0 to 20. It is expected that men with more education will get higher-prestige jobs. But should we really expect a linear effect of education on occupational prestige?

FIGURE 6.10 Linear and Quadratic Equations for Prestige (Y) and Education (X)

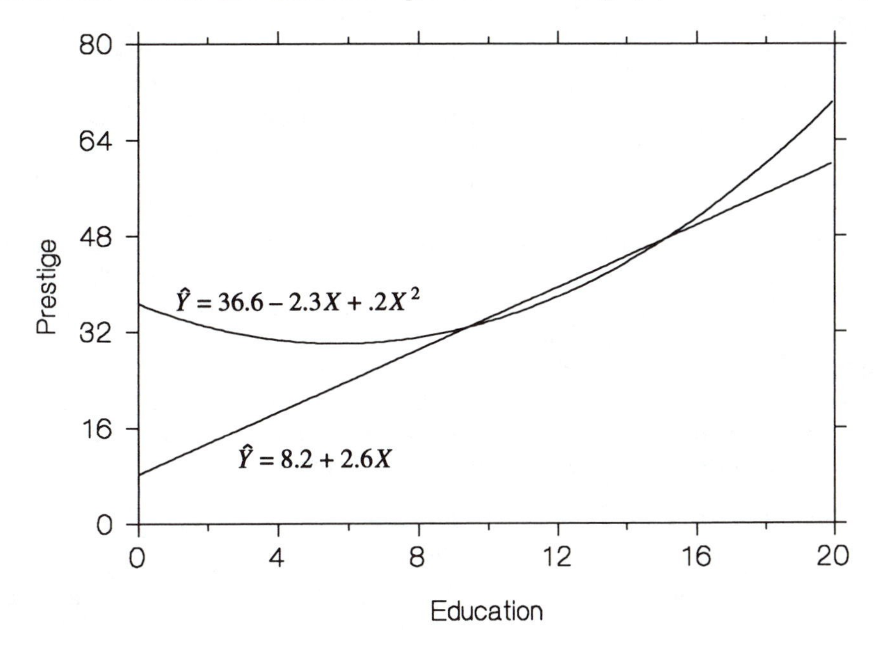

A linear effect would mean that completing junior high school (about eight years of education) would cause the same increase in occupational prestige as completing college (about sixteen years of schooling).

Figure 6.10 shows a plot of the linear and quadratic equations for prestige and education. Both the linear and quadratic equations are significant (not shown); that is, education was found to have a significant positive effect on prestige when the bivariate slope was tested and the squared value of education was found to be significant in the quadratic equation. The linear plot, of course, implies that an increase in education from 0 to 1 years of schooling has the same effect on prestige (2.6 points) as an increase in education from 15 to 16 years (i.e., completing the first grade has the same effect as completing the last year of college). The quadratic plot shows, however, that there is little or no change in prestige over the range of 0 to about 10 years of schooling. And because Y rises at an increasing rate as education increases above 8 or 10 years of schooling, the increase in prestige that results from completing high school (about 12 years) is not as great as the increase caused by completing college (about 16 years).

As described above, the quadratic curve makes better theoretical sense than the more frequently used linear function. Notice, however, that there is one bend in the range of the observed education scores. It shows a slight decrease in prestige as education increases from 0 to 6 years. This seems unlikely and probably should not be taken literally. When the cubic equation is fitted to the data (it is not shown because it does not significantly improve the fit), the only difference between it and the quadratic is that the cubic function is almost perfectly flat (horizontal) from 0 to 6 years of schooling. We would not use the cubic, however, because it is not significant and it increased the explained variance by less than .01. This shows a slight flaw in the use of polynomials to fit all types of nonlinearities; in order for the quadratic equation to give a good description of the increasing rate of growth in prestige in the middle to upper ranges of education, it had to bend slightly in the lower ranges and thus somewhat distort the pattern for 0 to 6 years of education. The point here is that we should not take every bend and undulation in the polynomial function seriously, because power polynomials will not always exactly fit or describe the empirical nonlinearities that are present in the data. Although we should bear this in mind, we will often find that power polynomials are capable of providing very good approximations to nonlinear relationships.

Equations with Two Power Polynomials

In Chapters 3 and 4, we examined the effects of years of labor force experience and years of education on the annual earnings of males who were employed

FIGURE 6.11 Earnings Regressed on Experience and Education

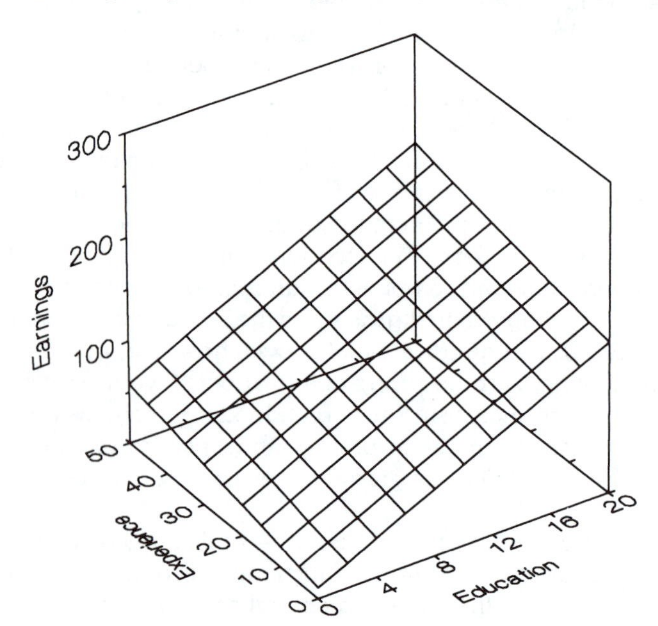

TABLE 6.3 Earnings as a Quadratic Function of Experience and Education

**** MULTIPLE REGRESSION ****

Multiple R	.478	Analysis of Variance			
R Square	.229		DF	Sum of Squares	Mean Square
Adjusted R Square	.223	Regression	4	.336811E + 06	84202.646
Standard Error	45.802	Residual	542	.113702E + 07	2097.826
		F = 40.138		Signif F = .000	

------------------Variables in the Equation---------------------------------

Variable	B	SE B	Beta	Tol	T	Sig T
EDUC	−1.470	2.793	−.095	.044	−.526	.599
EXPER	5.113	.803	1.075	.050	6.368	.000
EDUCSQ	.356	.116	.550	.044	3.073	.002
EXPERSQ	−.087	.016	−.908	.050	−5.397	.000
(Constant)	15.455	19.259			.803	.423

as wage and salaried workers. The multiple regression equation showed that both experience and education had significant positive effects on earnings when earnings was expressed as a linear function of each of these variables. The multiple regression plane (Figure 3.7) is reproduced here as Figure 6.11.

An examination of the partial scatterplots shown in Figure 3.6 indicates the possibility of nonlinear relationships between both earnings and experience and earnings and education. The following multiple regression equation, which specifies earnings as a quadratic function of both experience and education, may be used to estimate these nonlinear relationships:

$$\hat{Y} = a + b_1X_1 + b_2X_2 + b_3X_1^2 + b_4X_2^2$$

The results for this equation are shown in Table 6.3. The squared variables for experience and education are both significant at $p < .05$. Thus, we can conclude that there is a nonlinear relationship between earnings and each of these variables. The signs of the slopes of education and education squared are negative and positive, respectively, whereas the signs for experience and experience squared are positive and negative, respectively. Therefore, we would infer that the patterns of the nonlinear relationships are not the same for education and experience. The plot for the quadratic multiple regression equation is shown in Figure 6.12.

The shape of this surface resembles what is called a *saddle* function. Increases in education are accompanied by very slight decreases in earnings for very low levels of education (between 0 and 4 years of education). After four years of education, earnings increase as education increases, and increasingly

FIGURE 6.12 Earnings as a Quadratic Function of Both Experience and Education

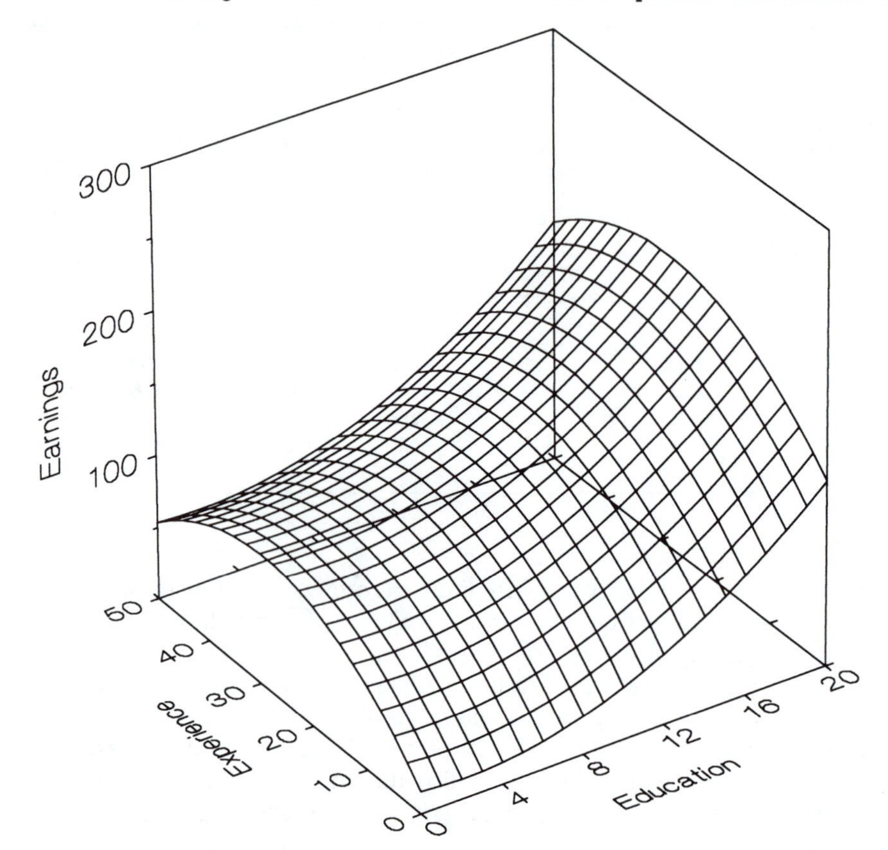

so. Thus, the effect of twelve years of education is estimated to be greater than eight years, the effect of sixteen years is greater than twelve years, and the effect of twenty years is greater than sixteen years. A quite different pattern occurs for experience, however. Earnings increase at first as experience increases, but after about thirty years of experience the regression surface shows earnings decreasing with further increases in experience.

These nonlinear patterns can be seen more clearly when the relationship between each independent variable and the dependent variable is shown in a two-dimensional plot (Figure 6.13). These plots are not simple bivariate quadratic relationships. They are plots showing the quadratic relationship between earnings and one independent variable for selected values of the other independent variable. For example, if an experience value of 30 is entered into the

FIGURE 6.13 (a & b) Earnings as Quadratic Functions of Experience and Education

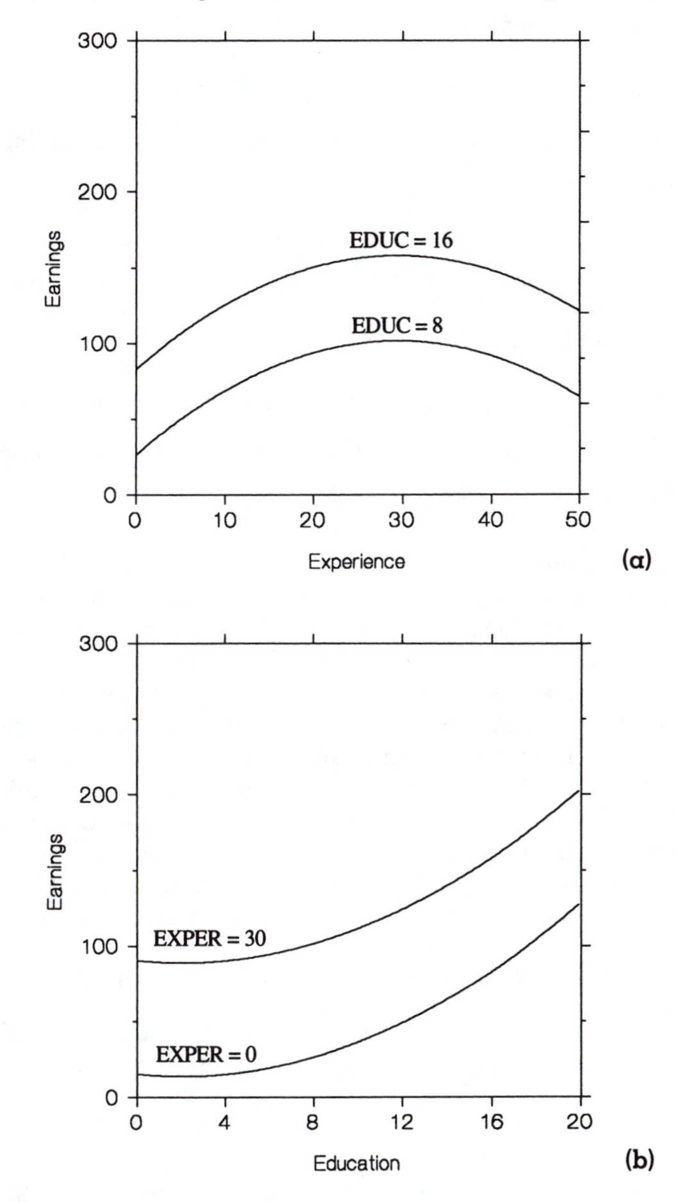

regression equation, we get the following quadratic equation between earnings and education for men with 30 years of experience:

$$\begin{aligned}\widehat{EARN} &= 15.455 + 5.113EXPER - .087EXPER^2 - 1.470EDUC + .356EDUC^2 \\ &= 15.455 + 5.113(30) - .087(30)^2 - 1.470EDUC + .356EDUC^2 \\ &= 90.545 - 1.470EDUC + .356EDUC^2\end{aligned}$$

Then we can plot the quadratic relationship between earnings and education for 30 years of experience, as has been done in Figure 6.13b. If we substitute 8 years of education into the multiple regression equation, we get the following equation, which is plotted in Figure 6.13a.

$$\begin{aligned}\widehat{EARN} &= 15.455 + 5.113EXPER - .087EXPER^2 - 1.470EDUC + .356EDUC^2 \\ &= 15.455 + 5.113EXPER - .087EXPER^2 - 1.470(8) + .356(8)^2 \\ &= 26.479 + 5.113EXPER - .087EXPER^2\end{aligned}$$

The figures show clearly the pattern of the quadratic relationships. The nonlinear relationship between earnings and experience as estimated by the multiple regression equation is identical at each level of education (Figure 6.13a), although the curve is higher for 16 years of education than for 8 years of education because men with college degrees earn more than those with only a junior-high-school education at each level of experience. The pattern of the nonlinear relationship between earnings and education is the same for number of years of experience. Notice that in Figure 6.13b the two curves appear to be closer together at high levels of education than at low levels of education. This is only an optical illusion. The *vertical* distance between the two curves (the earnings difference) is exactly the same at each number of years of education. The vertical difference between the two curves at each level of education is

$$\begin{aligned}\widehat{EARN}_{30} - \widehat{EARN}_{0} &= [15.455 + 5.113(30) - .087(30)^2 - 1.470ED + .356ED^2] \\ &\quad - [15.455 + 5.113(0) - .087(0)^2 - 1.470ED + .356ED^2] \\ &= [15.455 + 5.113(30) - .087(30)^2] - 15.455 \\ &= 75.09\end{aligned}$$

Thus, men with thirty years of experience are predicted to earn $7,509 more (earnings are coded in hundreds of dollars) than men with zero years of experience at each level of education. In sum, the nonlinear effect of experience is constant across levels of education, and the nonlinear effect of education is constant across years of experience.

Transformations for Linearity

If Y is a nonlinear function of X but there are no bends in the curve (i.e., the slope does not change direction), Y is said to be a *monotonic* function of X. Y might be either a positive monotonic function of X (e.g., Figure 6.13b) or a negative monotonic function of X. In these cases, it may be possible to make a transformation of either X or Y that will bring the relationship between the two variables into linearity. More specifically, it may be possible to find an X' that is a nonlinear transformation of X such that Y is a linear function X' or to find a Y' that is a nonlinear transformation of Y such that Y' is a linear function of X. In either of these cases, it would not be necessary to use power polynomials to represent the nonlinear relationship between Y and X; instead, either X' or Y' alone could be used in the regression equation.

Figure 6.14 shows an example of transforming X to achieve linearity. In this case, Y is a positive monotonic function of X in which the slope of the curve increases as X increases. Transforming X by taking $X' = X^2$ produces a linear relationship between Y and X'. This transformation of X is effective because squaring X "stretches" larger values of X disproportionately in comparison to smaller values of X (except for $0 < X < 1$, where smaller X values are shrunken disproportionately), thereby "pulling" the nonlinear relationship into a straight

FIGURE 6.14 Transformation to Linearity by Taking $X' = X^2$

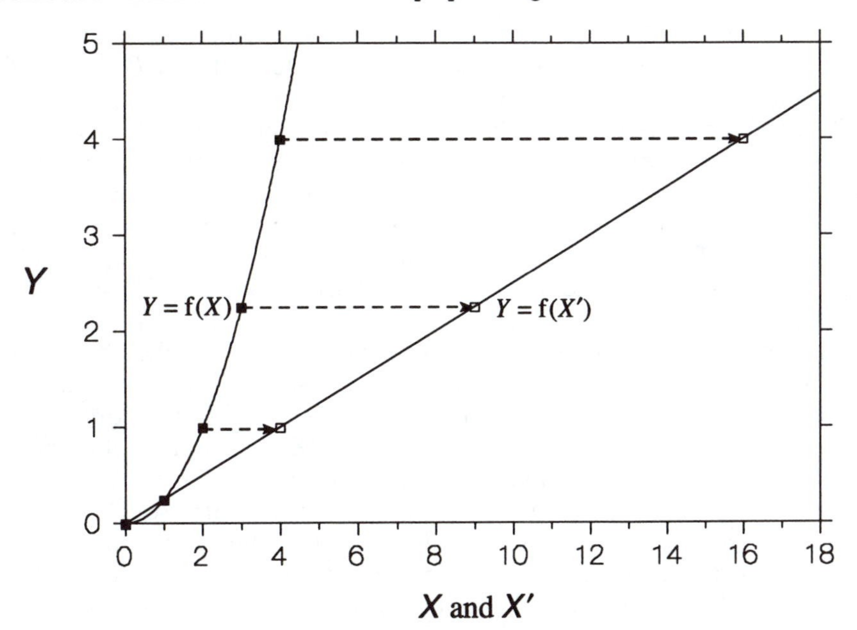

FIGURE 6.15 Transforming to Linearity by Taking $Y' = \sqrt{Y}$

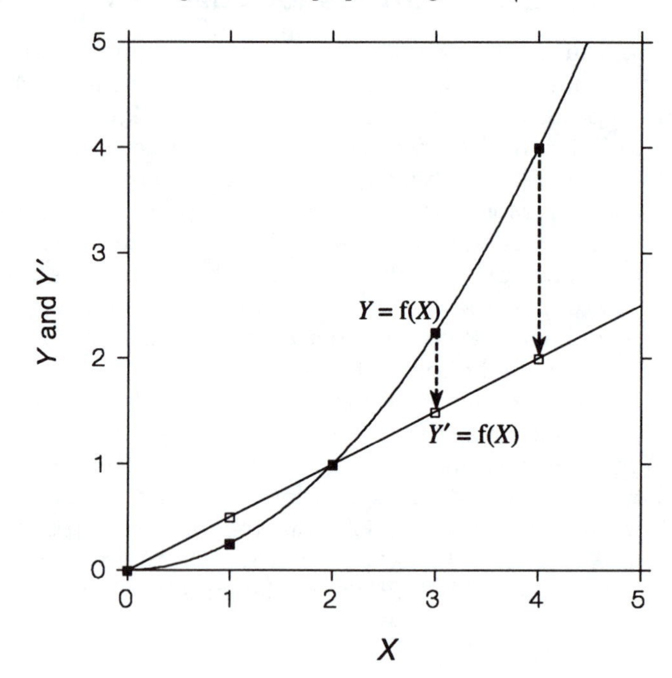

line. The transformation $X' = X^2$ produces a perfectly linear relationship because the nonlinear relationship was constructed to be $Y = (1/4)X^2$, and thus $Y = (1/4)X'$.

The same nonlinear relationship may also be linearized by transforming Y (Figure 6.15). If we transform Y such that $Y' = \sqrt{Y}$, the relationship between Y' and X becomes linear. This result occurs because taking the square root of Y "shrinks" larger values of Y disproportionately in comparison to smaller values of Y (except for $0 < Y < 1$, where smaller Y are disproportionately stretched), again pulling the curve into a straight line. Since $Y = (1/4)X^2$, $Y' = (1/2)X$.

The preceding example suggests a general strategy for linearizing monotonic nonlinear relationships that are positive throughout. If the effect of X on Y is greater for larger values of X (i.e., the slope of Y on X increases as X increases), then we want a transformation of X that will stretch larger values of X proportionately more than smaller values of X. Or, we may also linearize the relationship by transforming Y so that large values of Y will be shrunken proportionately more than smaller values of Y. This is what we did in the above example. If, on the other hand, the nonlinear relationship is such that the effect of X decreases as X increases (i.e., the slope of Y on X is smaller for larger values of X), then we need a different kind of transformation. In this case we need to shrink larger values of X proportionately more than smaller values, or we may transform Y by stretching larger values of Y more than smaller values of Y. The trick is to find the most appropriate mathematical transformation for making a particular nonlinear relationship approximately linear. Squaring X (or Y) or taking the square root of X (or Y) will not always provide the best transformation.

Power Transformations

Table 6.4 shows some illustrative *power transformations* of X that are often useful in linearizing relationships between X and Y, including the X^2 and $\sqrt{X}$ transformations we have just examined. The power transformations are given as $X' = X^p$, where p can be either positive or negative. In Table 6.4, $p = -2$, -1, $-.5$, $.5$, 2, and 3. These transformations are expressed in terms of X, but they could be applied to either X or Y, depending upon the needs of the investigator (see below). When X has a power equal to unity ($p = 1$) it is equal to X itself ($X^1 = X$). Powers greater than unity increase X, and powers less than unity decrease X (Table 6.4). Fractional powers of X are roots of X, such as the square root of X, where $X^{.5} = X^{1/2} = \sqrt{X}$. Negative powers ($p < 0$) involve reciprocals of powers of X, e.g., $X^{-1} = 1/X$, $X^{-2} = 1/X^2$, and $X^{-.5} = 1/X^{.5} = 1/\sqrt{X}$. Notice that when X is greater than 1 and p is negative, X' will always be a fraction and the order of the values of X' will be reversed from the order of X. Therefore, Table 6.4 gives $X' = -X^p$ when $p < 0$ to preserve the order of the values of X.

Table 6.4 does not contain the 0th power of X because $X^0 = 1$ for all values of X. Instead, Table 6.4 includes $\ln X$, the "natural" logarithm of X, which may also be expressed as $\log_e X$, where $e \approx 2.7183$.[3] Using the natural log of X in

TABLE 6.4 Illustrative Power Transformations of X

$-X^{-2}$	$-X^{-1}$	$-X^{-.5}$	$\ln X$	$X^{.5}$	X	X^2	X^3
−4.000	−2.000	−1.414	−0.693	0.707	**0.500**	0.250	0.125
−1.000	−1.000	−1.000	0.000	1.000	**1.000**	1.000	1.000
−0.250	−0.500	−.707	0.693	1.414	**2.000**	4.000	8.000
−0.111	−0.333	−.577	1.099	1.732	**3.000**	9.000	27.000
−0.062	−0.250	−.500	1.386	2.000	**4.000**	16.000	64.000
−0.040	−0.200	−.477	1.609	2.236	**5.000**	25.000	125.000
−0.028	−0.167	−.408	1.792	2.449	**6.000**	36.000	216.000

3. If b raised to the ath power equals X ($b^a = X$), then a is called the logarithm to the base b of X. More formally, we would write $\log_b X = a$. For example, if $X = 100$, the logarithm of X to the base 10 equals 2 ($\log_{10} 100 = 2$) because $10^2 = 100$. The natural logarithm of 100 ($\log_e 100 = \ln 100$), however, is 4.6052, because $e^{4.6052} = (2.7183)^{4.605} = 100$.

FIGURE 6.16 The Ladder of Power Transformations: $X' = (X^p - 1)/p$ for $p \neq 0$, and $X' = \ln X$ for $p = 0$ (Tukey 1977, p. 90; Judd and McClelland 1989, p. 519; Fox 1991, p. 46)

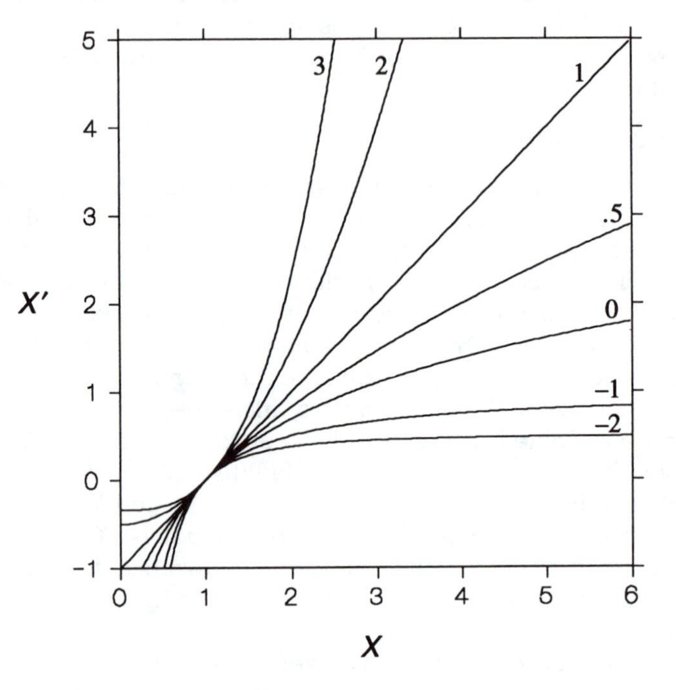

place of the 0th power is reasonable because the closer p gets to 0, the more X^p resembles the natural logarithm function (Fox 1991, p. 47).

The relationship between X' and X for each value of p in Table 6.4 (except $p = -.5$) is shown graphically in Figure 6.16. As Table 6.4 and Figure 6.16 show, each successive power above $p = 1$ leads to a greater proportionate stretching of large values of X than small values of X. Consequently, the intervals between large values of X are stretched more than equivalent intervals between small values of X. Thus, the transformations are stronger as p moves further above unity. Using values of p that are further and further above unity is called *moving up the ladder of power transformations* (Tukey 1977). X^3 is a rather strong transformation that might not have much practical utility. Values of p between 1 and 2 or between 2 and 3 may also be utilized.

Each successive power below $p = 1$ leads to greater proportionate shrinking of large values of X relative to small values of X. The consequence of *moving down the ladder of power transformations* is that the intervals between large values of X are shrunken more than equivalent intervals between small values of X. This is the opposite of the effect of moving up the ladder. But one factor is the same whether we move up or down the power ladder; the large values of X, and the intervals between the large values of X, are affected the most, either by greater stretching or by greater shrinkage.

All of the values of X in Table 6.4 are positive. The presence of negative values of X can present problems in using the power transformations. First, roots and logs of negative numbers are undefined. Second, negative numbers raised to even powers become positive, and the original order of the values of X is lost. Third, negative numbers raised to odd powers remain negative, but the transformation is stronger for large negative numbers (low values of X) than for small negative numbers (higher values of X). Consequently, before the transformation can be carried out, a sufficiently large positive number should be added to X to make all values positive.

We now return to the question of which power transformation should be used in a particular situation. Thus far, only positive nonlinear functions have been examined (Figures 6.14 and 6.15). Figure 6.17 shows four generic types of monotonic nonlinear functions. The lower right-hand corner shows a positive relationship with the slope of Y on X increasing as X increases. This is the type of curve we examined in Figures 6.14 and 6.15. The arrow for this curve points in the direction that the curve "bulges," which is down and to the right. The upper right-hand corner shows a negative function that becomes more negative as X increases. It bulges up and to the right. The upper left-hand corner shows a positive function that becomes less positive as X increases; it bulges up and to the left. And the lower left-hand corner shows a negative function that becomes less negative as X increases. This function bulges down and to the left.

The "bulging rule" (Tukey 1977) indicates that the direction in which the curve bulges determines whether we move up or down the power transformation ladder. Whether the curve bulges to the right or to the left, for example, tells us whether to move up or down the ladder, respectively, in order to transform X. The curve in Figure 6.14 bulges to the right, so we move up the power ladder to transform X (we "power-up" X). In this example, making $X' = X^2$ brought the relationship into linearity. The upper right-hand curve in Figure 6.17 also bulges to the right, so powering-up X will also be appropriate for this negative relationship. The upper and lower left-hand curves, however, both bulge to the left, so in both these cases we would move down the power ladder in order to transform X, using a square root, reciprocal, or log transformation, for example. The bulging rule works because a curve that bulges to the right is bulging in the direction of higher X values. Thus, we power-up X in order to transform X into higher values. The leftward-bulging curve, on the other hand, is pointing in the direction of lower values of X; thus, we power-down X in order to get smaller values.

We may also attempt to create a linear relationship by transforming Y. If the curve bulges down, as in the lower left-hand and right-hand corners of

FIGURE 6.17 The Bulging Rule for Power Transformations to Linearity [adapted from Tukey (1977, p. 198)]

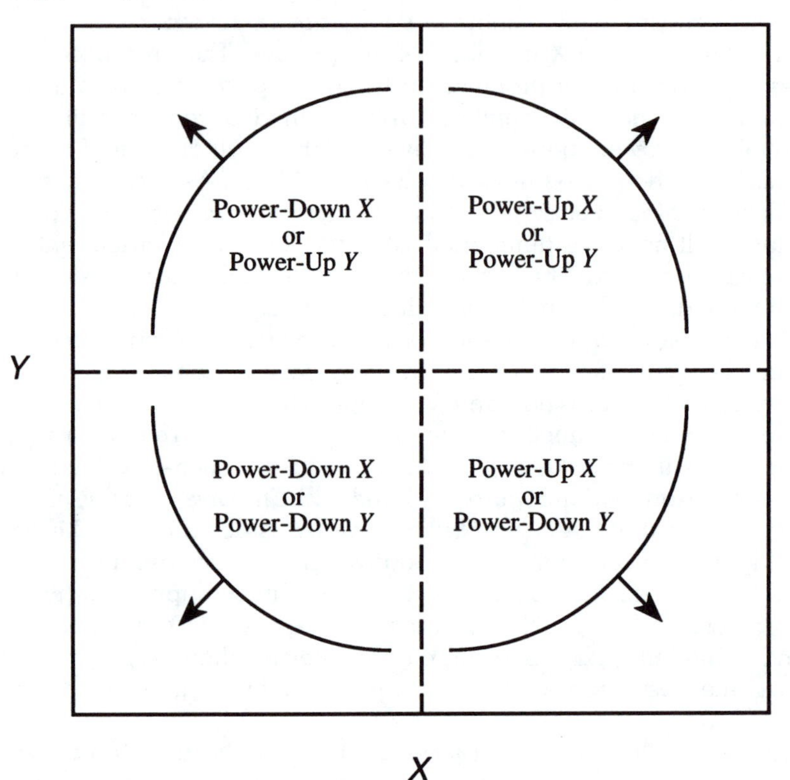

Figure 6.17 and in Figure 6.15, we would power-down Y. Thus, in the example in Figure 6.15 the square root transformation was used. If the curve bulges up, as in the upper left-hand and right-hand corners of Figure 6.17, we would power-up Y by squaring or cubing Y, for example.

The bulging rule tells us which way to move on the power ladder, but it does not tell us which of the transformations in the specified direction will be the most appropriate. If, for example, the bulge points in the direction of moving down the ladder, should a mild transformation such as the square root be used or is a much stronger transformation such as $1/X$ needed? Generally speaking, the stronger the nonlinear relationship is, the stronger the transformation will need to be. Judging this is partly a matter of experience. It is also possible simply to experiment with different-strength transformations and use the one that fits the data best. The best transformation may simply be the one that gives the highest R^2.

Another issue unanswered by the bulging rule is whether to transform X or Y. In a bivariate analysis it may make little or no difference. But in the more typical case in which there are multiple independent variables, it is generally best to transform the X or X's that have monotonic nonlinear relationships with Y. Transforming Y may change its relationship with other X's. Transforming Y might also change the error distribution of Y in such a way that it violates one or more of the least-squares assumptions.

The Example of Siblings and Children

In Chapters 2 and 4 the number of children of ever-married women 35 years of age or older was expressed as a linear function of the women's number of siblings. We now examine whether a nonlinear monotonic function might provide a better expression for the relationship between number of children and number of siblings. In exploring transformations we will use siblings + 1 because several of the transformations are not defined for $X = 0$ (e.g., $\ln X$ and $1/X$). In order to use the bulging rule to determine what kind of transformations might be best, we first examine a scatterplot for the two variables (Figure 6.18). In this plot, the values of both X and Y have been jittered in a similar fashion to Figure 2.9. In this instance, we have experimented with a smaller amount of jittering (i.e., the random numbers added to X and Y are smaller). As a consequence, the cases that are equal on X and equal on Y remain in a tight cluster

FIGURE 6.18 Jittered Scatterplot of Number of Children (Y) and Number of Siblings (X) with a Lowess Smooth ($f = .25$)

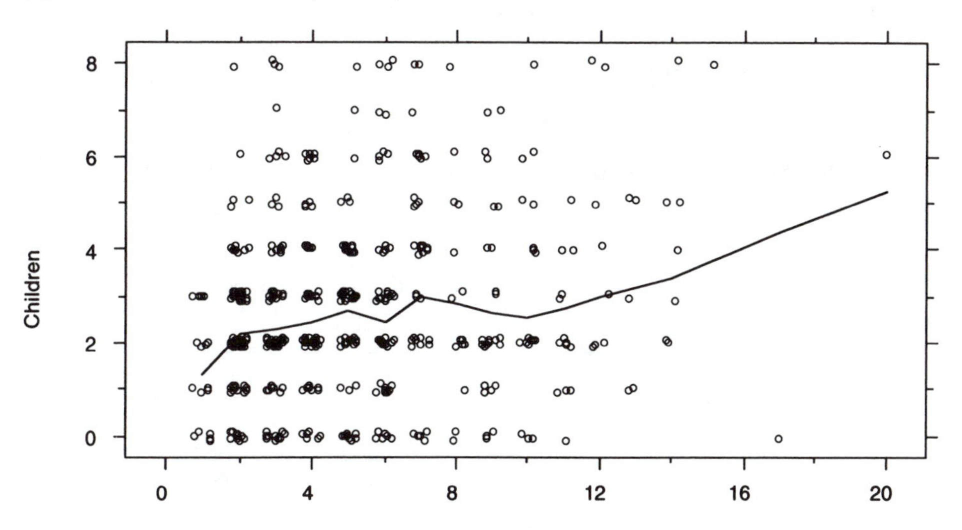

around their true (X, Y) coordinates. This introduces less error into the plot, and it should also make it clear to the viewer that the data are jittered. At the same time, however, the data are spread apart enough that the viewer can readily discern the areas of high and low density in the plot.

Because the data are scattered widely around the plot (indicating a smaller proportion of explained variance) it is very difficult to determine by visual examination alone the shape of any nonlinear relationship that may be present. Therefore, a *lowess* $\hat{Y}$ line has been added to the plot as an aid in examining possible nonlinearity in the relationship. Lowess stands for *lo*cally *w*eighted *s*catterplot *s*moother (Cleveland 1985; Fox 1991). The lowess procedure fits a regression line for each observed X_i that is based on some fraction (f) of the cases (specified by the analyst) with values of X closest to X_i. The greater the value of f selected ($0 < f \leq 1$), the smoother will be the line (if $f = 1$, the lowess curve will be perfectly linear). N regressions are fit to the data, one for each X_i, and each X_i's regression is used to predict $\hat{Y}$ for that case. In Figure 6.18, $f = .25$, which means that each locally fitted regression was based on 25 percent of the cases, a relatively low value that allows the lowess to be rather responsive to irregular patterns in the data (notice how the lowess dips at $X = 6$).

The lowess line rises rather rapidly in the lower range of X, shows little or no pattern of change in the middle range of X, and rises again sharply in the upper range of X. Since there are very few cases in the upper range of X, the most important pattern appears to be the rather rapid increase in X across the lower values of X. Given this, how should we describe the direction of the "bulge" in the data? The bulge is not as clear as the idealized functions shown in Figure 6.17. Nevertheless, the nonlinear lowess bulges up and to the left over the range of X where most of the cases are located. Therefore, the rule is to either power-down X or power-up Y.

The regression equations for three transformations of X down the power ladder were examined (Table 6.5). The square root transformation produced a small increase in R^2 compared to the original X values. The natural log ($\ln X$)

TABLE 6.5 Regression Equations for Transformations of Siblings (X) and Children (Y)

		Transformation			
	(X, Y)	$\sqrt{X}$	ln X	$-1/X$	Y^2
Siblings	.1116	.5619	.6288	2.0431	.8641
Constant	2.2582	1.4997	1.8035	3.3027	7.5276
R^2	.0333	.0374	.0409	.0422	.0360

resulted in another small improvement in fit. The reciprocal of X showed a very slight increase in R^2 over ln X. Finally, powering up Y to Y^2 showed a little better fit than X but not as good as ln X or $1/X$.

Although $1/X$ has a slightly better fit than ln X, for all practical purposes there is little or no empirical difference in the predictive accuracy of these two transformations. For reasons of interpretability, however, ln X is a better choice than $1/X$. The slope for ln X means that if $\Delta(\ln X) = 1$, then $\Delta\hat{Y} = .6288$. This is the standard interpretation of a slope, but it is stated in terms of the transformed X.

It is also possible to give a meaningful interpretation of the slope for ln X in terms of X itself. The slope for ln X equals the change in Y when X increases by a factor of 2.7183, where 2.7183 = e, the base of the natural logarithm.[4] Thus, if siblings increases by 2.7183 times, the number of children is predicted to increase by .6288 children. In general, the slope for any $\log_b X$ equals the predicted change in Y when X increases by a factor of b, the base of the logarithm. The interpretation is particularly appealing when the base of the log is 2, because the slope for $\log_2 X$ equals the amount by which Y is expected to change each time X doubles (changes by a factor of 2). Thus, log transformations allow us to interpret the regression slope in terms of the original units of measurement.

Number of children as linear and log functions of number of siblings (plus 1) is shown in Figure 6.19. The most important difference in the two functions is that in the log function the predicted number of children increases more sharply as the number of siblings increases from 0 to 2. Throughout much of the remainder of the range of X (up to about 12 siblings) the two functions make very similar predictions. Thus, the log function captures the diminishing effect of having an additional sibling, a pattern that the linear function is unable to describe.

The Example of Earnings, Education, and Experience

In another previous example, earnings was expressed as quadratic functions of both education and experience (Table 6.3 and Figure 6.12). It has been stressed that power polynomials, such as the quadratic, provide a very general and effective way of estimating nonlinear relationships. We found that for both education and experience the addition of a squared term significantly improved the fit. Now we want to look more closely at the data to see if a single power transformation of each X might fit the data as well as, or perhaps even better than, the quadratic expressions. The plots of the quadratic curves shown in Figure 6.13 suggest that earnings may be a positive monotonic function of education but not a monotonic function of experience.

4. If $\Delta(\ln X) = \ln X_2 - \ln X_1 = 1$, $e^{(\ln X_2 - \ln X_1)} = e^{(\ln X_2)}/e^{(\ln X_1)} = X_2/X_1 = 2.7183$ since $e^{(\ln X_2 - \ln X_1)} = e^1 = 2.7183$.

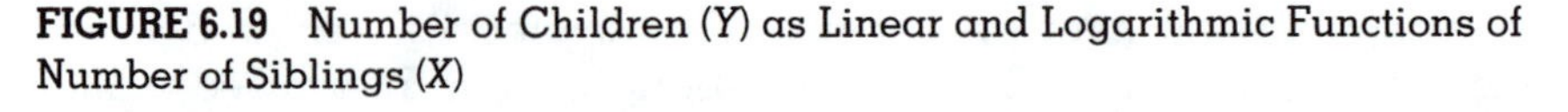

FIGURE 6.19 Number of Children (Y) as Linear and Logarithmic Functions of Number of Siblings (X)

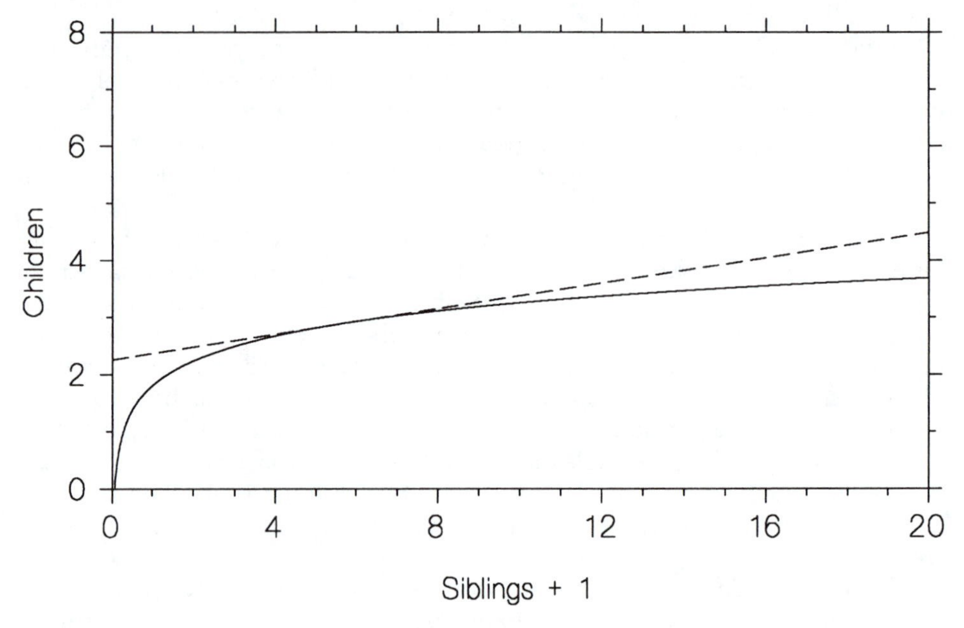

Partial-Residual Plots. When viewing the data for the possibility of monotonic relationships between Y and one X, we need to construct a plot that controls for the effects of other X's on Y. Since it has been determined that there is some form of nonlinear relationship between Y and each X (the significant quadratic expressions), these nonlinear patterns should be controlled. One method of doing this is by constructing *partial-residual plots* (Fox 1991, pp. 56–58). The first step in constructing partial-residual plots is to calculate the residuals from a multiple regression equation containing all of the X's. It may often be sufficient to use an equation containing only the linear terms for each X, such as $\hat{Y} = a + b_1X_1 + b_2X_2$, where X_1 equals education in this example. Sometimes, however, nonlinearity is more effectively displayed if we use the residuals from an equation containing quadratic expressions. Since we already know that the quadratic terms are significant for both education and experience, we will calculate the residuals from $\hat{Y} = a + b_1X_1 + b_2X_2 + b_3X_1^2 + b_4X_2^2$:

$$e = Y - \hat{Y} = Y - a - b_1X_1 - b_2X_2 - b_3X_1^2 - b_4X_2^2$$

The next step is to calculate two partial-residual variables, one for education and one for experience, as follows:

$$e^{(1)} = (Y - \hat{Y}) + b_1X_1 + b_3X_1^2$$

$$e^{(2)} = (Y - \hat{Y}) + b_2X_2 + b_4X_2^2$$

Each of these partial-residuals equals the total residual plus the effect of one X. That is, one of the components in the original equation is added back into the residual. In this case, $e^{(1)}$ is the total residual plus the quadratic effect of education. Thus, $e^{(1)}$ includes variance contributed by education but excludes the effects of the other X's, i.e., experience. It is called a partial-residual variable because it excludes the variance in Y that is related to the partial effects of experience, but it includes variance that might have a nonlinear relationship to education.

The partial-residual plot for education is constructed from $e^{(1)}$ and X_1 (Figure 6.20). To aid visual interpretation, a lowess curve has been added to the plot. The lowess curve is clearly monotonic, showing a positive relationship between the partial-residuals of income and education, with the positive slope increas-

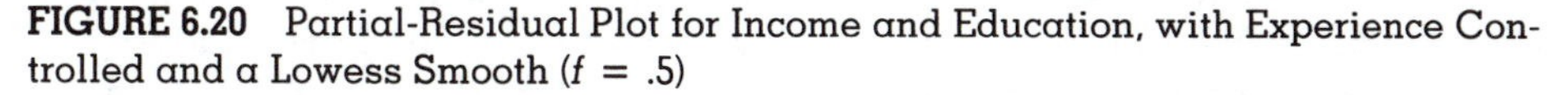

FIGURE 6.20 Partial-Residual Plot for Income and Education, with Experience Controlled and a Lowess Smooth (f = .5)

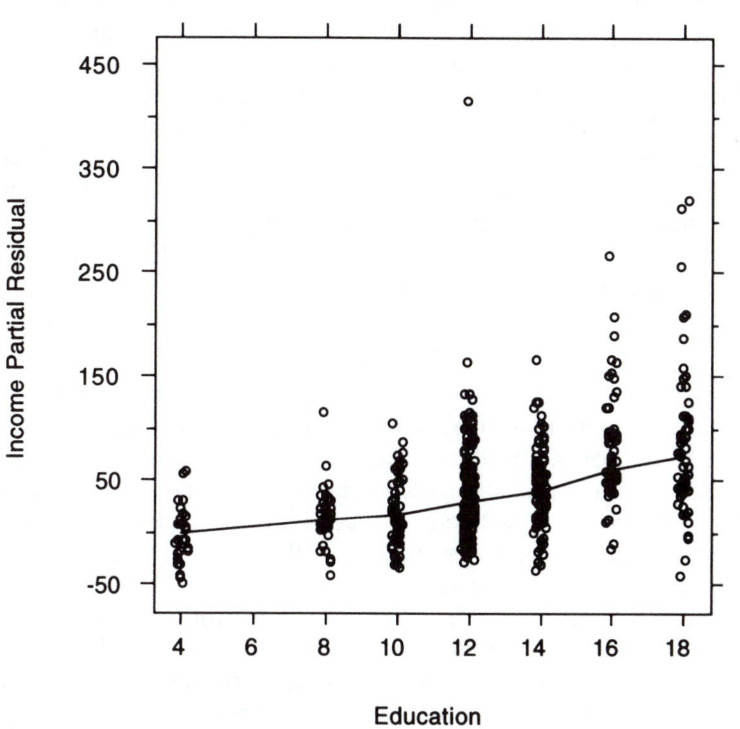

TABLE 6.6 **Regression Results for Selected Transformations of Education and Income**

	X_1	X_1^2	$X_1^{2.5}$	X_1^3	$\sqrt{Y}$	ln Y	$-1/X$
Constant	−29.186	6.918	16.149	23.103	4.299	3.535	−.021
Exper (X_2)	5.004	5.105	5.120	5.120	.219	.040	.0004
Exper²	−.084	−.087	−.088	−.089	−.004	−.001	−.0000
Educ (X_1)	6.899	.297	.065	.015	.297	.054	.005
R^2	.215	.228	.228	.226	.225	.215	.140

ing as education increases. This curve clearly bulges down and to the right. Thus, the bulging rule indicates that we should power-up education or power-down income.

Table 6.6 shows the regression results for several transformations of education and income. Going up the power ladder to transform education shows that the best fit (greatest R^2) occurs for squaring education and raising education to the 2.5th power. Both of these transformations have a better fit than either of the transformations of income that move down the power ladder ($\sqrt{Y}$, ln Y, $-1/X$). Therefore, X^2 appears to be the best transformation because it is simpler than the equally well fitting $X^{2.5}$. Furthermore, using only the square of education produces only a slightly worse fit than the quadratic equation (R^2 = .229 in Table 6.3), and that fit is not significantly worse, since X_1 was not significant with X_1^2 in the equation. Thus, X_1^2 alone adequately captures the monotonic relationship between income and education.

The partial-residual plot for income and experience confirms that the relationship between experience and income is not monotonic (Figure 6.21). Therefore, transformations of experience were not explored. The quadratic terms for experience should be retained in the equation.

Summary

In a nonlinear relationship, the effect or slope of an independent variable is not constant but instead varies as a function of its own values. Unlike a linear relationship with only one form, nonlinear relationships may take on many different functional forms. Before exploring some of the more common nonlinear forms, we learned how to conduct a general significance test for nonlinearity that does not require a specification of the form of nonlinearity. This test—called *nominalization*—requires representing a quantitative variable with a set of dichotomous nominal variables (e.g., dummy variables) and using an F test to determine whether they significantly improve the fit over the linear function. When a sufficiently large number of dummy variables is used, the nominalization test is capable of detecting most forms of nonlinearity.

FIGURE 6.21 Partial-Residual Plot for Income and Experience, with Education Controlled and a Lowess Smooth (f = .5)

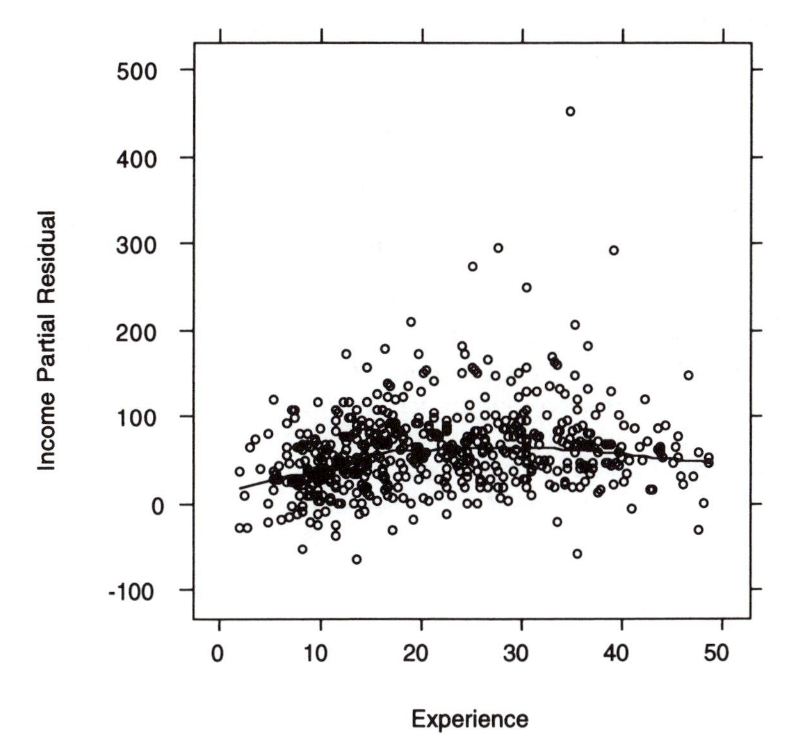

Next, we examined a family of related nonlinear functions called power polynomials. These functions are very flexible in that they can describe a wide range of different shapes of nonlinearity and they are more parsimonious (i.e., they use fewer degrees of freedom) than the nominalization test. Only in very rare cases would a higher-order polynomial than a cubic equation be needed, and typically a quadratic equation is sufficient. A hierarchical procedure was described for determining the order of the equation with the best fit.

Both nominalization and power polynomials involve using more than one variable in the regression equation to represent the nonlinear effect of an independent variable. In certain types of nonlinear relationships (i.e., monotonic nonlinear relationships), however, you may be able to use a power transformation of the values of X (or Y) in such a way that the values of the transformed variable are linearly related to Y (or X). In this case, you do not need more than one variable in the regression equation to represent the effect of an independent variable. The bulging rule was introduced for determining whether to "power up" (squaring, cubing, etc.) or "power down" (square roots, logarithms, reciprocals, etc.) either X or Y in order to find the nonlinear transformation that provides the best fit. Partial-residual plots were introduced as a method of visually determining the direction of the bulge in the nonlinear relationship when more than one independent variable is included in the regression analysis.

References

Berry, William D., and Stanley Feldman. 1985. *Multiple Regression in Practice*. Beverly Hills: Sage.

Cleveland, William R. 1985. *The Elements of Graphing Data*. Monterey, CA: Wadsworth.

Cohen, Jacob, and Patricia Cohen. 1983. *Applied Multiple Regression/Correlation Analysis for the Behavioral Sciences*. 2nd edition. Hillsdale, New Jersey: Lawrence Erlbaum Associates.

Fox, John. 1991. *Regression Diagnostics*. Newbury Park: Sage.

Judd, Charles M., and Gary H. McClelland. 1989. *Data Analysis: A Model-Comparison Approach*. San Diego: Harcourt Brace Jovanovich.

Tukey, John W. 1977. *Exploratory Data Analysis*. Reading, MA: Addison-Wesley.

7 Nonadditive Relationships

All of the multiple regression equations that we have examined thus far have been *additive* equations. In an additive equation, the effect of one independent variable is the same for all values of the other independent variables. We have seen how the effect of a variable can change as the value of that variable changes; that is, the effect of a variable is a function of the values of that variable. This is called a *nonlinear* relationship. In the nonlinear relationships that have been examined, however, the nonlinear relationship between the dependent variable and one independent variable is exactly the same for all values of the other independent variables (Figures 6.12 and 6.13). Thus, these equations were additive nonlinear equations.

In this chapter, we introduce a frequently used class of *nonadditive* equations that contain products of independent variables. In these equations the magnitude of the effect of one independent variable is a function of the values of one or more other independent variables. Nonadditive effects such as these are frequently called *statistical interactions* or interaction effects. We will begin with simple interactions involving a quantitative variable and a dichotomous nominal variable. It is relatively easy to develop the proper conceptualization of nonadditive effects for such pairs of variables. We will then extend the logic of interaction effects to the case of two quantitative variables. Finally, we will examine more complex interactions involving three independent variables that may be both nominal and quantitative in nature. Before taking up nonadditive equations, however, it will be helpful to take a closer look at the nature of additive equations.

Additive Equations

A linear additive equation with k independent variables is

$$Y = a + \beta_1 X_1 + \beta_2 X_2 + \cdots + \beta_k X_k + \varepsilon \quad \textbf{(7.1)}$$

This equation is linear because the effect of a unit change in each X_i equals β_i, no matter what X_i may be equal to when the unit change occurs. The equation is *additive* because the contribution to Y of each X_i, i.e., $\beta_i X_i$, are added together. Another way of stating the meaning of *additive* is that the effect of each X is independent of the values that the other X's may have in any particular case; for each value of X_j, the effect of X_i will equal β_i. Thus, Y will change by β_i when X_i increases by one unit, regardless of whether $X_j = 1$ or $X_j = 100$.

We will examine more concretely the implication of additivity via an example. The data come from the 1974 General Social Survey. The dependent variable equals the number of correct answers to a ten-item vocabulary test given during the interview (Y = WORDSUM); thus, scores range from 0 to 10. The independent variables are the number of years of schooling completed (X_1 = EDUC) and race (X_2 = RACE), where race is a dummy variable coded 1 for whites and 0 for blacks. The number of cases is 1,468. The results of the regression are given below.

Variable	*b*	*F*
EDUC	.381	570.89
RACE	1.076	45.56
Constant	.454	
R^2	.316	

The results indicate that the vocabulary scores increases by .38 correct words for each additional year of schooling, holding race constant, and that whites average 1.07 more correct words than blacks, holding education constant. Therefore, the linear additive equation is estimated to be

$$\hat{Y} = .454 + .381X_1 + 1.076X_2$$

If the effects of education and race are truly additive, we can determine what the bivariate regression of Y on education would be for each race by setting race equal to 0 and solving and then by setting race equal to 1 and solving as follows:

$$\hat{Y}_B = .454 + .381X_1 + 1.076(0) = .454 + .381X_1 \quad \text{[blacks]}$$

$$\hat{Y}_W = .454 + .381X_1 + 1.076(1) = 1.530 + .381X_1 \quad \text{[whites]}$$

Substituting the scores for blacks and whites into the equation gives *implied*

FIGURE 7.1 Implied Equations for WORDSUM (Y) and EDUC (X)

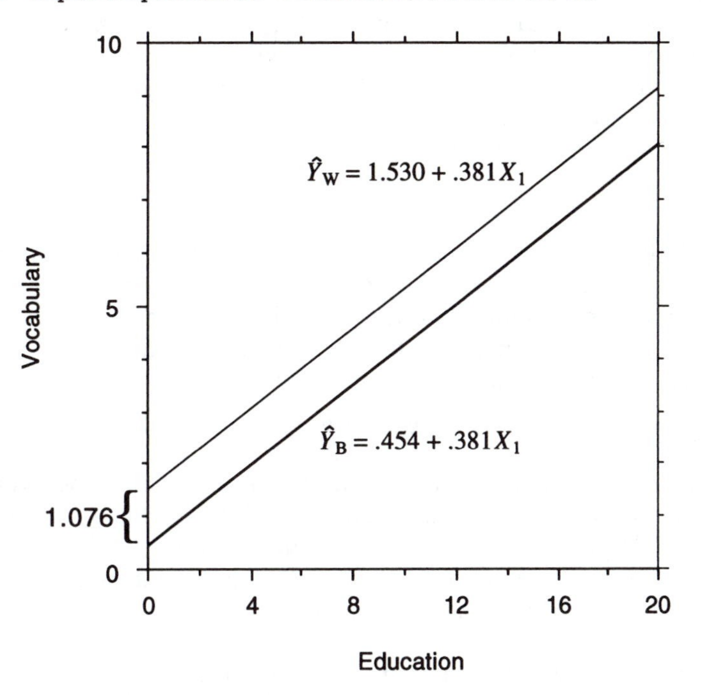

bivariate regression equations for each race that differ only with respect to the intercept; the racial difference in the implied intercepts is equal to the regression slope for the dummy race variable. The two equations that are implied by the assumption that race and education have additive effects are diagrammed in Figure 7.1.

The slopes of the implied equations for blacks and whites are parallel because the additive model specifies that the effect of education is the same for each score on race. These results predict that at each level of education, whites will score 1.076 points higher than blacks on the vocabulary test. This amount is marked on the Y-axis as the difference in intercepts between the two implied equations; again, the difference in intercepts equals the slope for the race dummy variable.

Nonadditivity with a Quantitative and a Nominal X

It must be emphasized that the two equations in Figure 7.1 are not regression equations; they are the equations for each race implied by the additivity specification in the original regression equation, with education and race as the two independent variables. If the additivity assumption is correct, however, we should get the two equations shown in Figure 7.1 if we compute the bivariate regression equations separately for blacks and whites. Instead of running two separate regression equations—one for each race—we can use the following nonadditive equation to derive the same results that we would get from two separate regression equations:

$$Y = \alpha + \beta_1 X_1 + \beta_2 X_2 + \beta_3 X_1 X_2 + \varepsilon \qquad \textbf{(7.2)}$$

Equation 7.2 differs from an ordinary additive equation; in addition to the terms for X_1 and X_2, it contains the product $X_1 X_2$. It is the inclusion of the multiplicative term that makes this equation nonadditive. The effect of a unit increase in X_1, for example, will now depend on the value of X_2. We can rewrite Equation 7.2 to show more clearly what now determines the size of the effect of X_1:

$$Y = \alpha + (\beta_1 + \beta_3 X_2) X_1 + \beta_2 X_2 + \varepsilon \qquad \textbf{(7.3)}$$

The expression in parentheses ($\beta_1 + \beta_3 X_2$) that is multiplied times X_1 represents the variable-sized effect of X_1. If we multiply X_1 times this expression, we get Equation 7.2. How much Y changes per unit increase in X_1 depends on the value of $\beta_1 + \beta_3 X_2$, which in turn depends on the values of the betas and the score that a particular case has on X_2; the effect of X_1 is now a linear function of X_2.

Nonadditivity, however, applies to the effects of both X's. Thus, we can again rewrite Equation 7.2 to show how the effect of X_2 depends on X_1:

$$Y = \alpha + (\beta_2 + \beta_3 X_1) X_2 + \beta_1 X_1 + \varepsilon \qquad \textbf{(7.4)}$$

Equation 7.4 shows that the effect of X_2 is now equal to $\beta_2 + \beta_3 X_1$; that is, the effect of X_2 is a linear function of X_1. When two variables have nonadditive effects on Y, the effect of each one depends on the value of the other; they are two sides of the same coin.

How do we estimate the coefficient β_3, the coefficient for the product term in Equation 7.2? This is the key coefficient in the equation that determines how much the effect of each variable depends on the other; it is the coefficient that makes the equation nonadditive. If β_3 is not equal to zero, then the independent variables have nonadditive effects on Y. The least-squares multiple regression equation $\hat{Y} = a + b_1 X_1 + b_2 X_2 + b_3 X_3$, however, is an additive equation. Thus, we must transform the variables involved in a nonadditive equation in an appropriate fashion so that the nonadditive coefficients in Equation 7.2 can be estimated by an additive regression equation. The solution is rather straightforward; we compute the following new variable:

$$X_3 = X_1 X_2$$

The new variable X_3, which equals the product of the original two variables,

TABLE 7.1 Nonadditive Effects of Race and Education on Vocabulary

```
COMPUTE RACEDUC = RACE* EDUC
REGRESSION
  VARS = WORDSUM RACE EDUC RACEDUC/
  DEP = WORDSUM/ENTER
```

Variable	*b*	*F*
EDUC	.213	23.03
RACE	−.989	3.44
RACEDUC	.192	16.43
Constant	2.226	
R^2	.323	

can then be included in the regression equation to get estimates of the coefficients in nonadditive Equation 7.2.

Returning to our example with education and race as independent variables, the SPSS commands in Table 7.1 can be used to get the appropriate regression estimates for Equation 7.2. The F test for the product term RACEDUC indicates whether adding the nonadditive term to the previous additive equation for RACE and EDUC significantly improves the fit to the data. This F value is computed according to the usual formula as follows:

$$F = \frac{(R^2_{Y\cdot123} - R^2_{Y\cdot12})/1}{(1 - R^2_{Y\cdot123})/n - k - 1} = \frac{(.323 - .316)}{(1 - .323)/1464} = 16.43$$

Although the increase in explained variance caused by adding the product variable is not great, it is highly significant at $p < .01$ (the significance results from a relatively large sample size and the fact that the explained variance $R^2_{Y\cdot123}$ is relatively large). The significance of the product variable leads to the conclusion that X_1 and X_2 have nonadditive effects on Y. This is usually referred to as a statistical interaction. The term *interaction* almost always refers to a situation in which two or more independent variables have nonadditive effects on Y.

It should also be emphasized that the F test for the product variable only represents a test for interaction when the two variables used to compute the product are also included in the equation; the product variable carries an interaction when X_1 and X_2 are partialed out of the product X_1X_2. The resulting residuals for the product variable represent unique variance that is not additively related to either X_1 or X_2.

The regression coefficients in Table 7.1 represent the following nonadditive equation:

$$\hat{Y} = 2.226 + .213X_1 - .989X_2 + .192X_1X_2$$

If we rearrange this equation into the form of Equation 7.3, we have

$$\hat{Y} = 2.226 + (.213 + .192X_2)X_1 - .989X_2$$

The expression in parentheses $(.213 + .192X_2)$ shows how the effect of X_1 is a linear function of X_2; in this case, the effect of education is expressed as a positive linear function of race. As race increases by one unit (from Black = 0 to White = 1), the *effect* of education on the vocabulary score increases by .192. If we enter the score for blacks (0) into the equation, we can derive the regression equation that shows the effect of education on vocabulary for blacks:

$$\hat{Y}_B = 2.226 + [.213 + .192(0)]X_1 - .989(0) = 2.226 + .213X_1 \quad \text{[blacks]}$$

Entering the score for whites (1) gives the following equation for whites:

$$\hat{Y}_W = 2.226 + [.213 + .192(1)]X_1 - .989(1) = 1.236 + .405X_1 \quad \text{[whites]}$$

The slope for whites is almost twice as large as the slope for blacks; vocabulary scores increase about twice as much for each additional year of schooling that whites complete as they do for blacks. The difference between the white and black education slopes equals .192, which is the value of the regression coefficient for the product variable X_3. Thus, the slope for the product term represents the difference in the effects of one independent variable for people differing by one unit on the other independent variable. The bivariate regression equations for blacks and whites are shown in Figure 7.2.

These equations are exactly equal to the bivariate equations that we would get if we made separate regression runs for blacks and whites. Notice that the racial gap in vocabulary $\hat{Y}_W - \hat{Y}_B$ increases in favor of whites as education increases beyond high school graduation; blacks fall further behind whites the further they go in school, although the vocabulary of both races increases with more education (i.e., the slope for education is positive for both blacks and whites). Also, notice that below about five years of education the equations predict that blacks will have higher vocabulary scores than whites. This prediction is probably not very valid, since it doesn't make theoretical sense. It probably results from the fact that there aren't many cases at this extremely low end of the education scale. Thus, in order to fit the difference in slopes between blacks and whites in the middle to upper range of education, the regression solution apparently distorts the slopes in the very low range of education. The difference between the black and white intercepts is shown in Figure 7.2 as equaling .989, in favor of blacks; this value equals the slope of the dummy race variable in the nonadditive regression equation.

The slope for the product term in the nonadditive regression equation is the most important aspect of the output. However, how do we interpret the slopes for X_1 and X_2? The slope for each one of these variables alone represents its effect when the other variable is equal to zero. Thus, the slope for RACE (−.989)

FIGURE 7.2 Regression Equations for Whites and Blacks

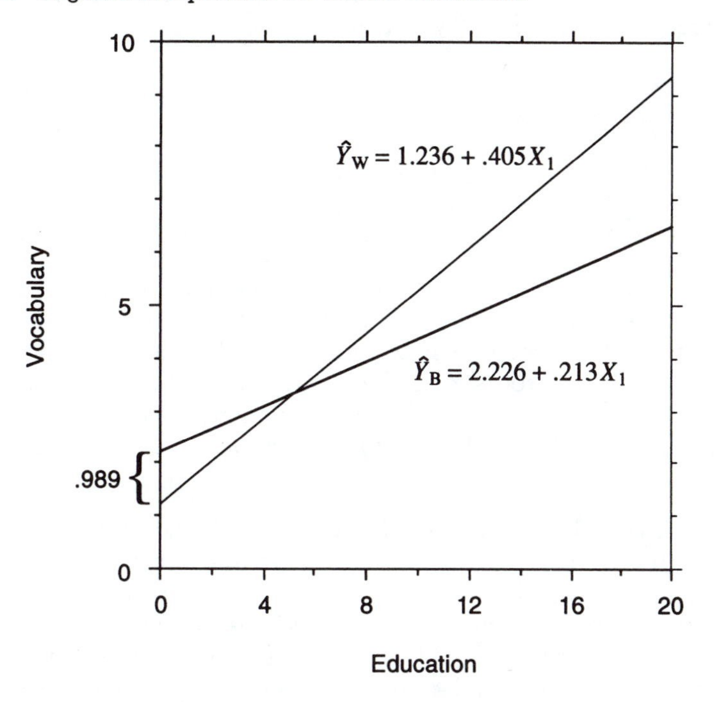

is the predicted effect of race when EDUC = 0; we noted that this was the difference in intercepts in Figure 7.2. The slope for EDUC (.213) is the predicted effect of education when RACE = 0; thus, it is the slope of the education variable for black respondents, as shown in Figure 7.2. As we saw for the difference in intercepts in Figure 7.2, it may often be the case that the estimated effect of one variable when the other is zero may not be meaningful. Sometimes a value of zero is outside the range of observed scores on a variable (e.g., Age = 0), and sometimes it is an extreme score for which there are very few observed cases (e.g., Education = 0). Therefore, it usually is not advisable to interpret the slopes for X_1 and X_2 in a nonadditive equation that contains X_1X_2. There are ways, however, to transform the X's so that their slopes will be more meaningful. This will be covered later. Nevertheless, the emphasis in most nonadditive equations is on the slope for the product variable(s).

We have noted that a nonadditive, or interaction, effect is a two-sided coin. Thus far, we have focused on how the effect of education differs by race. The flip side is that the effect of race differs by education level. If we return to the nonadditive equation that has been estimated, we may enter various values of education and solve to see the effect of race for each of these values. Although the education variable can take on many different values (unlike the dichotomous race variable), 12 years and 16 years will be used to illustrate the procedure.

$$\hat{Y} = 2.226 + .213X_1 - .989X_2 + .192X_1X_2$$

$$\hat{Y} = 2.226 + .213(12) - .989X_2 + .192(12)X_2 = 4.767 + 1.320X_2 \quad [\text{EDUC} = 12]$$

$$\hat{Y} = 2.226 + .213(16) - .989X_2 + .192(16)X_2 = 5.630 + 2.090X_2 \quad [\text{EDUC} = 16]$$

The solution for twelve years of education indicates that the effect of race equals a 1.3-word difference in vocabulary in favor of whites for high school graduates; the solution for sixteen years of education indicates that the white advantage increases to 2.09 words for college graduates. Thus, we can look at the interaction from two directions: higher education increases the effect of race (to the advantage of whites) and race (white) increases the positive effect of education.

A Nominal *X* with Three or More Categories. If one of the variables (e.g., X_2) is a nominal variable with $g > 2$ categories, then it may be represented by $g - 1$ dummy variables ($D_1, D_2, \ldots, D_{g-1}$). Product variables would then be created between each dummy variable and the quantitative variable ($X_1D_1, X_1D_2, \ldots, X_1D_{g-1}$) and they would be included in an equation containing each of the dummy variables and the quantitative variable. Products would not be formed between dummy variables, however. If $g = 3$, for example, we would use

$$\hat{Y} = a + b_1X_1 + b_2D_1 + b_3D_2 + b_4D_1X_1 + b_5D_2X_1$$

In this equation, b_4 may be interpreted as the difference between Group 1 and Group 3 (the reference group) in the effect of X_1 and b_5 is the difference between Groups 2 and 3 in the effect of X_1. The bivariate regression equation for Y and X_1 can be solved for a particular group by substituting the dummy variable codes for that group into the equation.

Two Quantitative *X*'s

The interpretation of an interaction effect between two quantitative variables is a straightforward extension of the interpretation of nonadditivity for one quantitative and one dichotomous variable. Assume that Y and X_1 still represent vocabulary scores and years of education, respectively. However, imagine that X_2 stands for age rather than race. The equation below is a hypothetical regression equation for these variables:

$$\hat{Y} = 2 + .05X_1 + .02X_2 + .002X_1X_2$$

FIGURE 7.3 Positive Interaction Between Two Quantitative Variables

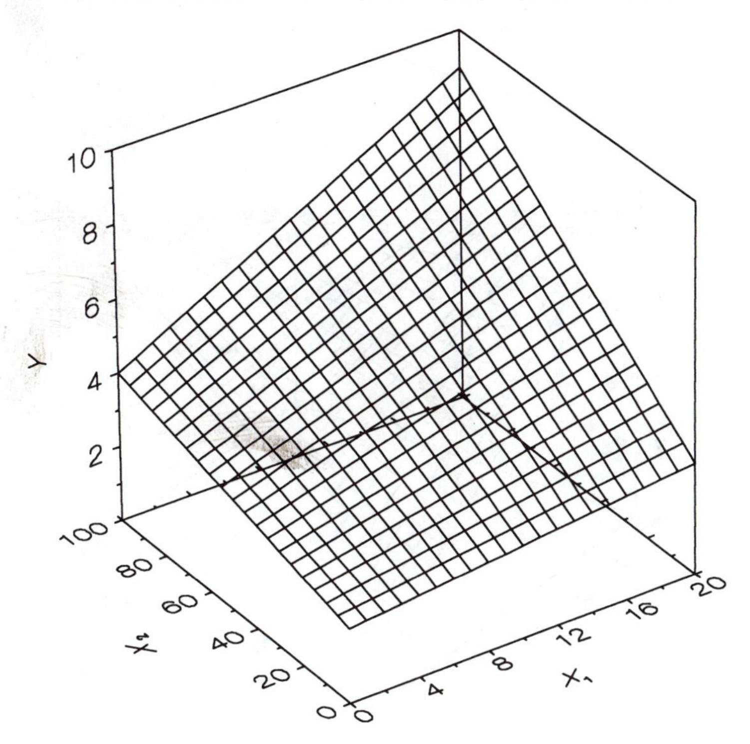

The positive slope for X_1 indicates the effect of education on vocabulary when X_2 equals zero. Likewise, the positive slope for X_2 indicates the effect of age on vocabulary when X_1 equals zero. The positive slope for the X_1X_2 product means that there is a positive nonadditive/interaction effect of X_1 and X_2. More specifically, the positive interaction effect means that for each unit increase in X_2 the *effect* of X_1 increases by .002. It also means that if X_1 increases by one unit, the effect of X_2 increases by .002. Since both X_1 and X_2 have positive slopes when the other variable equals zero, the positive interaction means that each will have increasingly strong positive effects on Y as the other increases. This can readily be seen in Figure 7.3.

The bottom left edge of the regression surface shows the slope of Y on X_2 when X_1 equals 0; we know from the equation that this slope equals .02. The lines that are parallel to the lower left edge show increasingly steep slopes as X_1 increases. It is clear that the slope of the line forming the upper right-hand edge of the surface (at $X_1 = 20$) is a much steeper positive slope than the slope of the lower left-hand edge. With respect to the effect of X_1 on Y, the lower right-hand edge shows the slope of Y on X_1 when $X_2 = 0$ (a fictitious situation if this

FIGURE 7.4 **Negative Interaction Between Two Quantitative Variables**

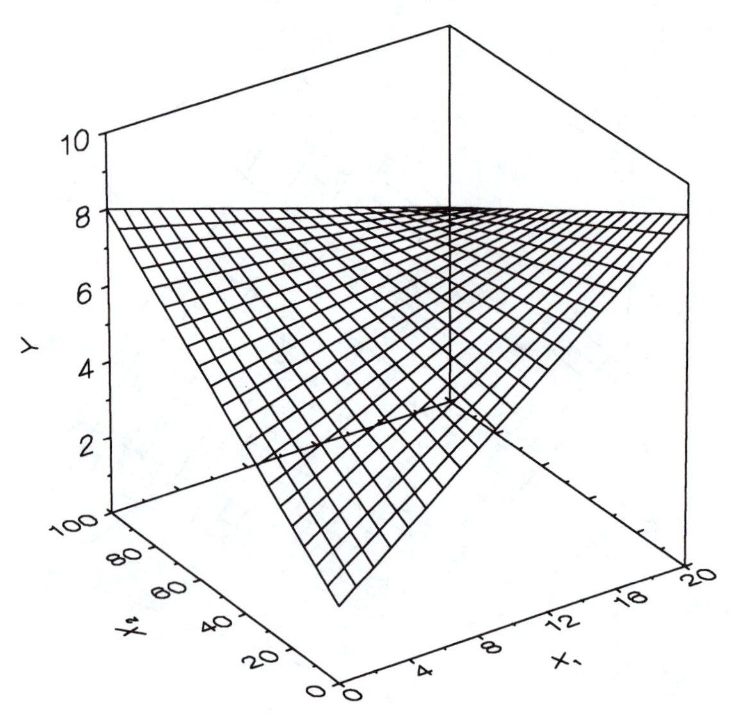

is zero years of age). It can also be seen that the lines that are parallel to the lower right-hand edge become steeper and steeper as X_2 increases, culminating with the steepest slope at the upper left-hand edge ($X_2 = 100$).

The next equation illustrates the surface of a regression equation with a negative interaction effect:

$$\hat{Y} = 2 + .36X_1 + .06X_2 - .005X_1X_2$$

The plot of this surface (Figure 7.4) looks much different from the surface for the positive interaction in Figure 7.3. The near left-hand edge and the near right-hand edge both have steeper positive slopes than those in Figure 7.3. As X_1 increases, the lines representing the slopes of Y on X_2 become increasingly flatter and then become negative. The negative slope for Y on X_2 when $X_1 = 20$ can be readily discerned at the rear right-hand edge of the surface. The same pattern occurs for the slope of Y on X_1 as X_2 increases.

The surfaces shown in Figures 7.3 and 7.4 do not exhaust all the forms that positive and negative interaction surfaces can take. Other shapes can be produced if, for example, b_1, b_2, or both are negative. One thing that all positive interaction effects have in common, however, is that the slope of each variable becomes greater as the other variable increases. This may take the form of Figure 7.3, where positive slopes become more positive, but it can also take the form of negative slopes either becoming less negative or becoming positive. All negative interaction effects have the characteristic that the slope of one variable becomes smaller as the other variable increases. This can mean that a positive slope becomes less positive or even negative (e.g., Figure 7.4), or it can mean that a negative slope becomes more negative.

More Complex Interactions

When there are three or more independent variables, the number and complexity of potential nonadditive effects increases dramatically. The following equation shows all possible nonadditive effects for the case of three independent variables:

$$Y = \alpha + \beta_1 X_1 + \beta_2 X_2 + \beta_3 X_3 + \beta_4 X_1 X_2 + \beta_5 X_2 X_3 + \beta_6 X_1 X_3 + \beta_7 X_1 X_2 X_3 + \varepsilon \quad \textbf{(7.5)}$$

The addition of X_3 to the analysis creates three possible pairs of products. The β_i coefficients for the products of two variables are called *two-way interactions*. There is also a product of all three variables; the β_i for this product is called a *three-way* interaction. The terms in Equation 7.5 differ in terms of their *order*; the single-variable terms have the lowest order, the two-variable products have the second-lowest order, and the three-variable product has the highest order. In order to use additive regression analysis to estimate the coefficients of Equation 7.5, four new variables will be needed:

$$X_4 = X_1 X_2$$

$$X_5 = X_2 X_3$$

$$X_6 = X_1 X_3$$

$$X_7 = X_1 X_2 X_3$$

To provide an example of the analysis of these more complex interactions, we will add age (X_3 = AGE) to the independent variables that we have considered above. Table 7.2 shows the SPSS commands and output that are needed for a number of tests that will be considered.

Equation 1 in Table 7.2 contains only education and age as independent variables; the reason for estimating this equation will be shown later. Equation 2 is an additive equation containing all three independent variables. Each has a significant positive effect on vocabulary. Equation 3 contains the product variables for race times education and race times age. The race times education interaction is again significant. The race times age interaction also is sig-

TABLE 7.2 Selected Additive and Nonadditive Equations for WORDSUM

```
COMPUTE RACEDUC = RACE*EDUC
COMPUTE RACEAGE = RACE*AGE
COMPUTE AGEDUC = AGE*EDUC
COMPUTE RACEDAGE = RACE*EDUC*AGE
REGRESSION DESCRIPTIVES/
  VARS = WORDSUM RACE AGE EDUC RACEDUC
          RACEAGE AGEDUC RACEDAGE/
  DEP=WORDSUM/ENTER EDUC AGE/
  ENTER RACE/ENTER RACED RACEAGE/
  ENTER AGEDUC/ENTER RACEDAGE
```

	Equations				
Variable	(1) *b*	(2) *b*	(3) *b*	(4) *b*	(5) *b*
EDUC (X_1)	.409*	.393*	.203*	.298*	.278*
RACE (X_2)	—	.948*	−2.179*	−2.336*	−2.576
AGE (X_3)	.020*	.019*	.000	.019	.015
RACEDUC (X_4)			.213*	.215*	.237
RACEAGE (X_5)			.019*	.023*	.027
AGEDUC (X_6)				−.002*	−.0015
RACEDAGE (X_7)					−.0004
Constant	.286	−.292	2.500	1.489	1.694
R^2	.3133	.3317	.3397	.3417	.3418
n	1436				

*$p \leq .05$

nificant; the positive coefficient for RACEAGE indicates that the effect of age on vocabulary is stronger (more positive) for whites than for blacks.

We now examine a test to see if there is a significant difference between races for the regression equation consisting of vocabulary predicted by education and age. It is often of interest whether separate regression equations have to be estimated for two or more groups (such as black–white or male–female) or whether the groups can be pooled and a single equation estimated. The appropriate equations in Table 7.2 for computing this F test are Equations 1 and 3. The F test is as follows:

$$F = \frac{(R^2_{Y\cdot 12345} - R^2_{Y\cdot 13})/3}{(1 - R^2_{Y\cdot 12345})/n - k - 1} = \frac{(.3397 - .3133)/3}{(1 - .3397)/1430} = 19.02$$

The test is to determine whether adding the dummy variable for race and the product of race times education and race times age significantly increases the explained variance. The above F test, with 3 and 1,430 degrees of freedom, is significant at $p < .01$. Therefore, we reject the null hypothesis that the equation is the same for blacks and whites. Substituting 0 for blacks and then 1 for whites into Equation 3 in Table 7.2 allows us to solve for the separate racial equations as follows:

$$\hat{Y} = 2.5 + .203\text{EDUC} - 2.179\text{RACE} + .0003\text{AGE} + .213\text{RACEDUC} + .019\text{RACEAGE}$$

$$\hat{Y} = 2.5 + .203\text{EDUC} - 2.179(0) + .0003\text{AGE} + .213(0)\text{EDUC} + .019(0)\text{AGE}$$
$$= 2.500 + .203\text{EDUC} + .0003\text{AGE} \quad \text{[blacks]}$$

$$\hat{Y} = 2.5 + .203\text{EDUC} - 2.179(1) + .0003\text{AGE} + .213(1)\text{EDUC} + .019(1)\text{AGE}$$
$$= .321 + .416\text{EDUC} + .0193\text{AGE} \quad \text{[whites]}$$

The slopes for EDUC and AGE in Equation 3 represent the effects of those variables when RACE = 0; the intercept in Equation 3 is the predicted score when all variables equal 0 (including race), and thus it is the intercept for blacks. The intercept for whites equals the intercept of Equation 3 plus the slope for race; the slopes for education and age of whites equal the slopes for education and age in Equation 3 plus the slopes for education times race and age times race, respectively.

The form of the F test conducted above for racial differences in the regression equation is equivalent to what is widely known as the *Chow test* (Chow 1960).[1] The Chow test is a global test for group differences in a regression equation, including differences in both the intercept and slopes. Since the Chow test is significant, we should not use an *additive* model with blacks and whites pooled together, such as Equations 1 or 2. We can, however, use the total sample of blacks and whites combined if we include appropriate product terms and a dummy variable for race, as in Equation 3.

The education times age interaction is included in Equation 4. This two-way interaction is also significant. This interaction involves two interval variables, in contrast to the interactions between an interval variable and a dummy variable that have been tested thus far. The negative slope indicates that the effect of education on vocabulary decreases as age increases or the positive effect of age on vocabulary decreases as education increases. We can also use Equation 4 to conduct a test of the null hypothesis that all two-way interactions are zero, i.e., to test whether an additive model (Equation 2) can be rejected in favor of a model with the three two-way interactions (Equation 4). We would test the increase in explained variance caused by adding the three two-way interactions by using the following F test:

$$F = \frac{(R^2_{Y\cdot123456} - R^2_{Y\cdot123})/3}{(1 - R^2_{Y\cdot123456})/n - k - 1} = \frac{(.3417 - .3317)/3}{(1 - .3417)/1429} = 7.29$$

1. The Chow test assumes that the variance of ε is the same in both groups. If not, see Kamenta (1986, p. 421) for how to correct this heteroskedasticity with weighted least squares.

Since this F is significant at $p < .01$, we would reject the additive model in favor of a model that contains all of the specified two-way interactions. When we are adding a number of interaction variables to the model, this type of F test can be used to gain some protection against finding one or more of the interactions to be significant by chance. We would look at the tests for the individual product variables only after finding the F test for all product variables to be significant.

Equation 5 contains a product variable for age times education times race, which provides a test of the single three-way interaction that is possible when we have three independent variables. A three-way interaction can be interpreted as showing how a two-way interaction between X_1 and X_2, for example, depends on the level of X_3, that is, how the racial difference in the effect of education (a two-way interaction) differs by age. The slope for the race times education times age term in Equation 5 is negative. This means that the positive race times education interaction decreases as age increases; the greater effect of education for whites as compared to blacks is not as pronounced among older persons. The three-way interaction, of course, can be looked at from three sides of the coin: it can also be interpreted to say that the race times age interaction decreases as education increases or that the negative age times education interaction is even more negative for blacks than for whites. In this case, however, we do not need to interpret the three-way interaction because it is not significant. It should be emphasized that the product of three variables provides a test of a three-way interaction only when all of these variables and all pairs of products for these variables are also included in the equation (i.e., all lower-order terms), as in Equation 5.

Since the three-way interaction is not significant, it should be dropped from the equation; thus, Equation 4 is the best equation for these three variables. Notice that the two-way interactions are not significant when the race times education times age product is in the equation (Equation 5). This is because there is usually a relatively high amount of multicollinearity when product variables are included in a regression equation, much the same as for power polynomial equations. Since the three-way interaction is not significant, the three-variable product is highly redundant with the two-variable products, and thus none of them are significant. When the nonsignificant three-variable product is removed (Equation 4), all two-way interactions are significant.

Interactions with Deviation Scores

It was noted that the slopes for X_1 and X_2 may not be very meaningful in an equation containing the product variable X_1X_2. This is because the slope for X_1, for example, is the predicted effect of X_1 when $X_2 = 0$. The interpretation of these slopes may be more meaningful if we convert raw scores for interval or ordinal variables to deviation scores, e.g., $X_1 - \overline{X}_1$. After the variable has been transformed to deviation scores, a zero means that the person is at the mean, which is a *meaningful* value. There is no advantage to transforming a dummy

TABLE 7.3 Nonadditivity with Deviation Scores

```
COMPUTE EDUC = EDUC - 11.868
COMPUTE AGE = AGE - 44.263
COMPUTE RACEDUC = RACE*EDUC
COMPUTE RACEAGE = RACE*AGE
COMPUTE AGEDUC = AGE*EDUC
REGRESSION VARS = WORDSUM EDUC
  RACE AGE RACEDUC RACEAGE AGEDUC/
  DEP = WORDSUM/ENTER
```

Variable	*b*
EDUC	.214*
RACE	1.221*
AGE	−.003
RACEDUC	.215*
RACEAGE	.023*
AGEDUC	−.002*
Constant	4.878
R^2	.3417

* $p \leq .05$

variable to deviation scores, however, since the mean will equal p, a value that neither group will have. Thus, we will transform education and age to deviation scores and leave the race variable coded as a dummy variable. The product variables will equal the product of deviation scores or a deviation score times a dummy variable score. The equation with all two-way interactions will be

$$Y = \alpha + \beta_1(X_1 - \overline{X}_1) + \beta_2 X_2 + \beta_3(X_3 - \overline{X}_3) + \beta_4(X_1 - \overline{X}_1)X_2 + \beta_5 X_2(X_3 - \overline{X}_3) + \beta_6(X_1 - \overline{X}_1)(X_3 - \overline{X}_3) + \varepsilon$$

In this equation, β_1 represents the effect of X_1 (education) when $X_2 = 0$ (blacks) and $X_3 - \overline{X}_3 = 0$ (X_3 is at its mean); thus, β_1 is the effect of education for blacks with a mean age. This can be seen by setting $X_2 = 0$ and $X_3 - \overline{X}_3 = 0$ in the above equation:

$$\begin{aligned} Y &= \alpha + \beta_1(X_1 - \overline{X}_1) + \beta_2(0) + \beta_3(0) + \beta_4(X_1 - \overline{X}_1)(0) + \beta_5(0)(0) \\ &\quad + \beta_6(X_1 - \overline{X}_1)(0) + \varepsilon \\ &= \alpha + \beta_1(X_1 - \overline{X}_1) + \varepsilon \end{aligned}$$

The SPSS commands in Table 7.3 can be used to estimate the equation with three two-way interactions and with age and education converted to deviation scores. Notice that the slopes for the product variables in Table 7.3 have the same values as the slopes for the product variables in Equation 4 of Table 7.2. The slopes of the highest-order interaction terms will never be changed by transforming some or all variables to deviation scores. The slopes for education, race, and age, however, have values in Table 7.3 that are different from those in Table 7.2. The slope for race was -2.336 in Table 7.2 , which predicted that blacks who are zero years of age and have zero years of education will have higher vocabulary scores than whites of the same age and education! In Table 7.3, however, the slope for race is 1.221, which means that for persons with an average education (11.9 years) and an average age (44.3 years), whites will have a higher score on the vocabulary test than blacks. This effect is significant. The slope for education equals .214. This means that among black persons (RACE = 0) with a mean age (Deviation AGE = 0), education will have a significant positive effect on vocabulary. The slope for age is not significant, which indicates that age does not affect vocabulary for blacks with a mean level of education. A final important point about the equation shown in Table 7.3 is that the explained variance is the same as in Equation 4 of Table 7.2; a linear transformation of the scores never affects R^2.

We have seen that transforming *X*'s to deviation scores can improve the meaningfulness of the slopes for these variables in a nonadditive equation, without affecting the slopes or tests of significance for the highest-order products in the equation. You don't have to transform the *X*'s to deviation scores, however, because other values might produce just as meaningful interpretations. If you subtract 16 from EDUC, for example, then 0 on the transformed variable will represent a person with 16 years of education. The slopes for race, for example, would then indicate the effect of this variable for persons with 16 years of education and AGE = 0.

Summary

In this chapter we have learned how to estimate and interpret models in which the effects of two or more independent variables are nonadditive. Nonadditivity means that the effect of one variable is a function of the values of one or more other variables. Thus, the effect or slope of a variable is not a constant but instead is a variable that changes as the values of other variables change. Nonadditive effects, however, are reciprocal. That is, if the slope of one variable is a function of a second variable, then the slope of the second variable must in turn be a function of the first variable.

Products of variables may be used to estimate nonadditive effects. The slope of the product of two variables (i.e., a two-way *interaction* effect) represents the change in the effect of one variable as the other variable increases by one unit. The slope of the product of three variables (i.e., a three-way interaction effect) represents the change in a two-way interaction as a third variable increases by one unit. For the slope of a product variable to represent an interaction effect, however, it is necessary that the equation also include all of the variables that are included in the product plus all possible lower-order products between these variables. The procedures and interpretations are basically the same for quantitative, nominal, and combinations of quantitative and nominal variables.

We also examined how centering the *X*'s at their means before forming their products provides a more meaningful interpretation of the regression slopes for the individual variables and lower-order products. Centering, however, does not affect the slope and significance test of the highest-order product.

You should now appreciate the analogy between nonlinearity and nonadditivity. In nonlinearity the effect of a variable changes as its own values change, whereas in nonadditivity the effect of a variable changes as values of another variable change. In both cases, products may be used to estimate these variable effects. In nonlinearity, we use products of a variable with itself (i.e., squares, cubes, etc.), and in nonadditivity, we use products of a variable with other variables. It is possible to estimate even more complex models in which nonlinear effects are linear or even nonlinear functions of other variables, i.e., nonadditive nonlinear models (Aiken and West 1991).

This chapter concludes the coverage of material that is customarily presented as regression analysis proper. These regression statistics and techniques might be used for purposes of description, prediction, exploration, or explanation. The perspective of this book is that multiple regression is well suited and commonly used for nonexperimental causal analysis, i.e., for explanation. The partial slopes in a regression equation are often treated as estimates of the effects of independent variables on a dependent variable. In the final two chapters we examine more explicitly and extensively several additional ways in which multiple regression may be used for causal analyses.

References

Aiken, Leona S., and Stephen G. West. 1991. *Multiple Regression: Testing and Interpreting Interactions*. Newbury Park: Sage.

Chow, G. C. 1960. "Tests of Equality Between Sets of Coefficients in Two Linear Regressions." *Econometrica* 28:591–606.

Kamenta, Jan. 1986. *Elements of Econometrics*. New York: Macmillan.

8 Causal Analysis I

In Chapters 2 through 4, we covered the basic statistical concepts of regression analysis, including both descriptive and inferential statistics. These concepts constitute the core of linear regression analysis. In Chapters 5 through 7, we saw how to deal with three common problems in linear regression—nominal independent variables, nonlinear relationships, and nonadditive relationships. No new statistical concepts were introduced in these chapters. Instead, the focus was on how to use transformations of variables—such as dummy variables, powers of variables, and products between variables—in order to incorporate nominal variables, nonlinearity, and nonadditivity into regression equations.

In Chapters 8 and 9 we move beyond the material that is customarily presented as regression analysis proper. The perspective of this book is that multiple regression is well suited and commonly used for nonexperimental causal analysis—that is, for explanation. When the variables that are included in a multiple regression equation are carefully selected (see the discussion in Chapter 1) the partial slopes may be treated as estimates of the effects of two or more independent variables on a single dependent variable. The final two chapters extend this perspective to systems of equations or *multi-equation* causal models. The multi-equation models with which we will be involved consist of two or more dependent variables (*endogenous variables*), at least one of which must be an independent variable with respect to one or more of the other endogenous variables (see Figure 1.2). The analysis will be restricted to linear and additive equations.

In this chapter we will see how to use multi-equation models to calculate a new kind of causal effect, an *indirect effect*. In single-equation models only *direct effects* exist. The distinction between direct and indirect effects also allows us to calculate a *total effect*, which is the sum of the direct effect and all indirect effects of one variable on another. Ordinary least-squares regression can be used to estimate the parameters (effects) in *recursive* multi-equation models but not in *nonrecursive* models (i.e., models with feedback loops). Thus, learning to distinguish between recursive and nonrecursive models and understanding why OLS is invalid for estimating the parameters of nonrecursive models are also important objectives of this chapter. Before taking up the analysis of multi-equation models, however, it will be helpful to look more closely at some characteristics of single-equation models.

Single-Equation Models

Multiple regression analysis may be used to estimate the effects of several different independent variables on a single dependent variable. A single multiple regression equation represents the simplest type of causal model. It is often helpful to use causal diagrams, or path diagrams, to show the structure of causal models, especially when we are concerned with more complex models that will be introduced shortly. The path diagram for a causal model that would be estimated with a regression equation containing three independent variables is shown in Figure 8.1.

Some of the rules and principles that are used in path diagrams were presented in the section on causal diagrams in Chapter 1. You should review that section. In Figure 8.1, the single-headed arrows represent causal paths, or effects, between variables. The double-headed curved lines represent covariances or correlations between variables. The dependent variable Y is an *en-*

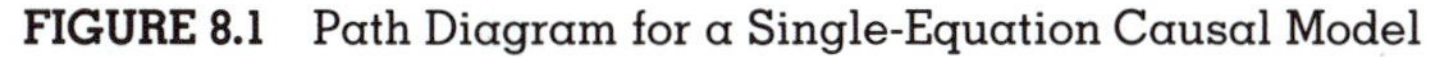

FIGURE 8.1 Path Diagram for a Single-Equation Causal Model

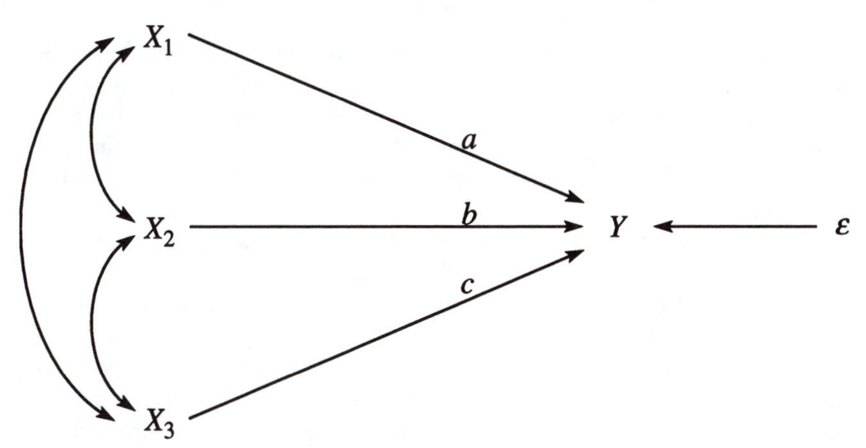

dogenous variable; the X's and ε *are exogenous* variables. The lower case letters a, b, and c on the causal paths represent the effect of each X on Y (the change in Y per unit increase in X). These letters are used instead of the usual β's because the subscript notation for the β's becomes somewhat unwieldy when we are dealing with more complex models. The causal effects of the X's are estimated with the following regression equation:

$$\hat{Y} = a + b_1X_1 + b_2X_2 + b_3X_3 \quad \textbf{(8.1)}$$

The intercept of the regression equation usually is not shown in the path diagram. The estimates of a, b, and c in the diagram are b_1, b_2, and b_3, respectively. Since the b's are unstandardized slopes, this indicates that the effects shown in the diagram are unstandardized effects. In the causal modeling literature, unstandardized effects are often called **structural coefficients**. Because all of the causal effects are estimated by Equation 8.1, the model shown in Figure 8.1 is called a *single-equation model*.

Notice that in Figure 8.1 there is no symbol on the path from the error term to Y. Since the error term represents the effects of unmeasured variables, we have to make an assumption about the scale of this hypothetical construct before we can specify the value of its effect on Y. The convention is to assume that ε has the same scale as Y; e.g., if Y is measured in dollars, then the scale for ε is also taken to be dollars. The consequence of this assumption is that the effect of a unit change in ε is equal to unity; e.g., if ε increases by one dollar, then Y increases by one dollar. Since the effect of ε is equal to unity, it can be omitted from the path diagram.

Causal models may also be specified in terms of standardized variables and coefficients (Figure 8.2). In this diagram, A, B, and C represent the causal effects of the standardized variables (the z_i). The standardized effects A, B, and C would be estimated with the standardized regression equation

$$\hat{z}_Y = B_1z_1 + B_2z_2 + B_3z_3 \quad \textbf{(8.2)}$$

The standardized effects for a causal model are often called **path coefficients** because the original principles of path analysis developed by biologist Sewell Wright (1921) were based entirely on standardized coefficients. Path coefficients are often symbolized by p_i. We will not use this notation, however. Instead, we will use our normal symbols B_i to represent the estimates of the standardized effects in causal models.

There is another difference between the diagrams for standardized and unstandardized models. In a standardized model we treat all of the variables as being standardized to have a unit variance and standard deviation (i.e., z scores). Thus, in Figure 8.2 it is also understood that the error term z_ε has a unit variance and standard deviation. The consequence of this specification is that the effect of ε on Y will not be equal to unity, as in the unstandardized model; an effect of unity for ε would mean that an increase in ε of one standard devia-

FIGURE 8.2 Path Diagram for a Standardized Causal Model

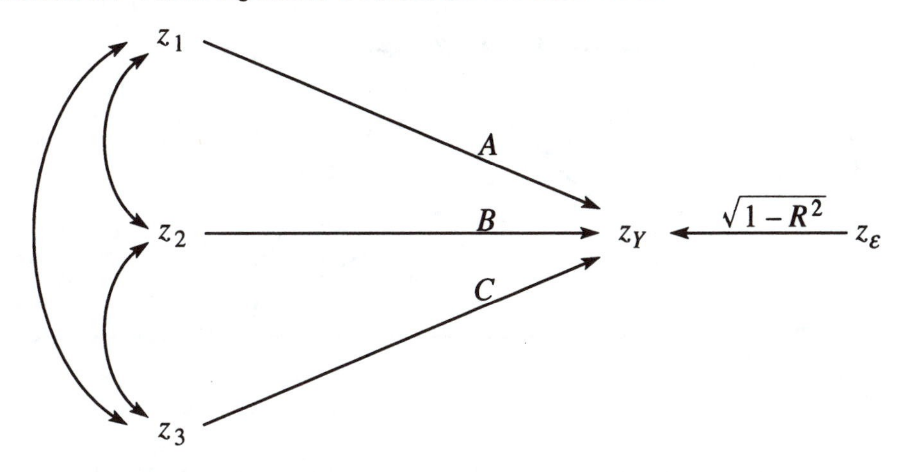

tion would cause a one-standard-deviation increase in Y, a perfect correlation. Instead of a unity effect, the effect of the standardized ε equals $\sqrt{1 - R^2}$, i.e., the square root of the residual proportion of variance; this effect equals the multiple correlation between the unmeasured variables and Y. Thus, if the path diagram represents a standardized model, the value of $\sqrt{1 - R^2}$ would be entered on the path from z_ε to z_Y, as shown in Figure 8.2.

A Two-Equation Causal Model

A single-equation causal model does not include any causal effects between the independent variables; they may be correlated, but the model does not specify that these correlations are due to causal effects. If theory or known temporal sequence indicates that one or more of the X's may be dependent on one or more of the other X's, then the causal model in Figure 8.1 can be elaborated to include additional causal specifications. Let us assume that there is reason to believe that X_3 is affected by X_1 and X_2. Figure 8.3 shows this elaborated model.

Figure 8.3 contains two endogenous variables (X_3 and X_4) and four exogenous variables (X_1, X_2, ε_3, and ε_4). Because there are two endogenous variables in the model, we will no longer label one of them Y; thus, the original variable labeled Y in Figure 8.1 is now called X_4.[1] Endogenous variable X_4 is a dependent variable only; it is not specified as having an effect on any of the other variables in the model. Endogenous variable X_3 is a dependent variable rel-

1. In some notational systems, the endogenous variables are symbolized by Y's and the observed exogenous variables are represented by X's (e.g., Jöreskog and Sörbom 1989). We do not use this system because it would unduly complicate the notation in this chapter and Chapter 9.

FIGURE 8.3 Causal Model with Two Endogenous Variables

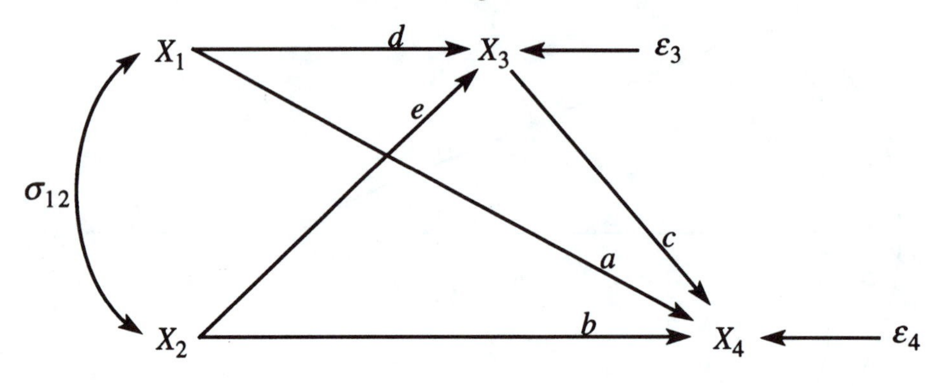

ative to X_1 and X_2, but it is an independent variable relative to X_4. Thus, this causal model reveals a new type of variable (i.e., X_3). X_3 is an *intervening* variable; it intervenes between X_1 and X_4 and also between X_2 and X_4.

How do we estimate the causal effects for the model shown in Figure 8.3? They can be estimated by running a regression for each endogenous variable (i.e., a variable that has arrows pointing at it). The variables that have arrows pointing at a particular endogenous variable will be the independent variables in the regression equation for that endogenous variable. In Figure 8.3, X_3 and X_4 will be dependent variables in two separate regression equations. The regression equations are

$$\hat{X}_4 = a_{4\cdot123} + b_{41\cdot23}X_1 + b_{42\cdot13}X_2 + b_{43\cdot12}X_3 \quad \textbf{(8.3)}$$

$$\hat{X}_3 = a_{3\cdot12} + b_{31\cdot2}X_1 + b_{32\cdot1}X_2 \quad \textbf{(8.4)}$$

Equation 8.3 provides the estimates of a, b, and c in Figure 8.3. This is the same as Equation 8.2, except that we are now calling the dependent variable X_4 instead of Y. This means that a, b, and c will have the same values in Figure 8.3 that they had in Figure 8.1. Equation 8.4 provides the estimates for d and e in Figure 8.3. These are the new parameters that were not estimated in the single-equation model. The model in Figure 8.3 is called a two-equation model because the diagram specifies two separate causal equations embedded in a single model. Finally, notice that the error terms ε_3 and ε_4 in Figure 8.3 are not correlated with X_1 and X_2 and also are not correlated with one another (there are no double-headed curved lines connecting the error terms with any variables in the model). This satisfies the regression assumption that unmeasured variables are not correlated with the independent variables in a regression equation (the absence of a correlation between ε_3 and ε_4 indicates that ε_4 is not correlated with X_3). Thus, if this specification is correct, Equations 8.3 and 8.4 will provide unbiased estimates of the causal parameters in Figure 8.3.

Direct, Indirect, Total, and Spurious Effects (DITS)

What do we gain from introducing the two-equation model relative to what we would learn from the single-equation model? First, of course, we have learned about the effects of X_1 and X_2 on X_3. Just as importantly, however, we can now investigate **indirect effects**. Indirect effects are formed by compound causal paths or chains of paths in which one or more intervening variables mediate the effect of one variable on another. There are two indirect effects specified in Figure 8.3. One indirect effect is shown by the following compound path:

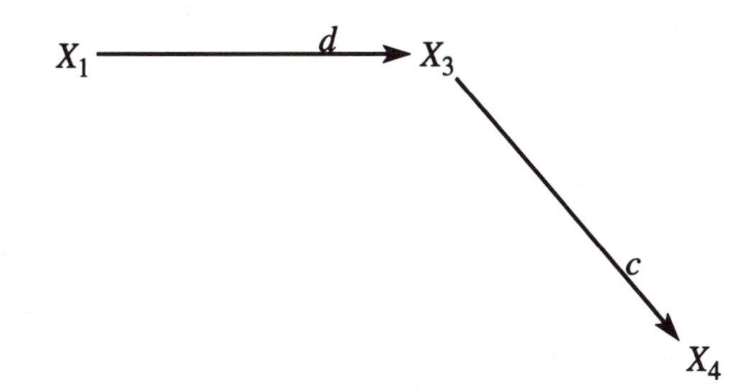

In this indirect effect, a change (or difference) in X_1 causes a change in X_3, and this change in X_3 then causes a change in X_4. *The notation for this indirect effect is* $\boldsymbol{I}_{431}$. The left-hand variable in the subscript is the final dependent variable in the compound path (X_4), the second is the intervening variable (X_3), and the third is the independent variable or source of the indirect effect (X_1). Thus, the order of the variables in the subscript moves backward along the path to the origin of the effect. The other indirect effect is represented as follows:

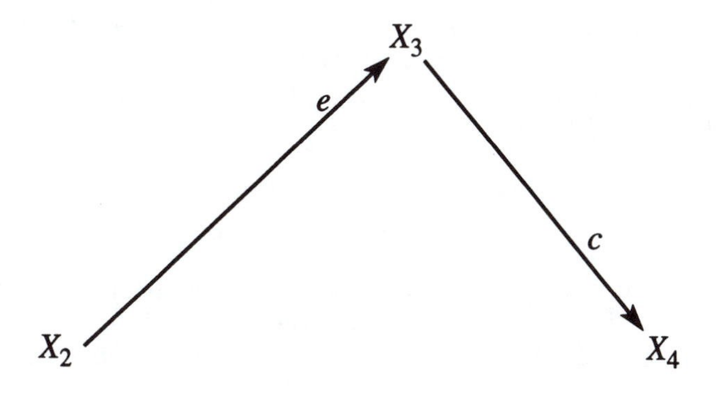

This indirect effect shows that a change in X_2 causes a change in X_3 that in turn causes a change in X_4. The notation is I_{432}.

Just as a direct effect (b or β) is defined as the change in the dependent variable per unit increase in the independent variable, *the value of an indirect effect is the change in the dependent variable at the end of a chain produced by a one-unit increase in the independent variable at the origin of the chain.* Thus, I_{431} equals ΔX_4 that is caused by ΔX_3 that is caused by $\Delta X_1 = 1$. To calculate an indirect effect, we first determine the change in the intervening variable produced by a one-unit increase in the source variable, and then we calculate the change in the final variable produced by this change in the intervening variable. Remembering that the change in a dependent variable equals the slope (effect) of the independent variable times the change in the independent variable, I_{431} and I_{432} are determined as follows:

I_{431}	I_{432}
$\Delta X_1 = 1$	$\Delta X_2 = 1$
$\Delta X_3 = d \cdot \Delta X_1 = d \cdot 1 = d$	$\Delta X_3 = e \cdot \Delta X_2 = e \cdot 1 = e$
$\Delta X_4 = c \cdot \Delta X_3 = c \cdot d$	$\Delta X_4 = c \cdot \Delta X_3 = c \cdot e$
$I_{431} = d \cdot c$	$I_{432} = e \cdot c$

The value of an indirect effect equals the product of the effects along the compound path or chain that links the source variable with the last variable in the chain. When there is one intervening variable, the indirect effect will equal the product of two coefficients, as shown above. If a compound path contains two intervening variables, the indirect effect will equal the product of three coefficients, and so on. The sign of the indirect effect will be positive if there are no negative coefficients or if there are an even number of negative coefficients. The sign of the indirect effect will be negative if there are an odd number of negative coefficients along the chain.

In addition to the indirect effects of X_1 and X_2 on X_4, the two exogenous variables also have **direct effects** on X_4, which equal a and b, respectively. *Direct effects are not mediated by any intervening variables.* We will use $\boldsymbol{D}_{ij}$ to indicate the direct effect of X_j on X_i. When an independent variable has both direct and indirect effects on a particular dependent variable, we can sum the two types of effects to determine the **total effect**, which will be designated by $\boldsymbol{T}_{ij}$. The total effects of X_1 and X_2 on X_4 are

$$T_{41} = D_{41} + I_{431} = a + dc$$

$$T_{42} = D_{42} + I_{432} = b + ec$$

If the indirect effect has the same sign as the direct effect, the total effect will be larger in absolute value than the direct effect. If the sign of the indirect effect is opposite the sign of the direct effect, the total effect will be either smaller in absolute value than the direct effect or it will be opposite in sign from the direct effect. If the intervening variable X_3 is redundant with the independent variable, the indirect effect will have the same sign as the direct effect; if suppression exists between the intervening variable and the independent variable, however, the indirect effect will be opposite in sign from the direct effect.

Not all of the causal relationships in a causal model will have both direct and indirect components. For the model in Figure 8.3, the following pairs of variables have only direct causal relationships, and thus the total effect equals the direct effect:

$$T_{43} = D_{43} = c$$

$$T_{31} = D_{31} = d$$

$$T_{32} = D_{32} = e$$

For any pair of independent and dependent variables in a causal model, the difference between the bivariate (zero-order) slope and the total effect equals the **spurious slope** or spurious "effect":

$$\text{Spurious Slope} = b_{ij} - T_{ij} = S_{ij}$$

The above equation is written in terms of the unstandardized regression slope; therefore, it pertains to a model containing structural coefficients. If we were working with a standardized model (i.e., path coefficients), we would use the standardized regression slope B_{ij} (or r_{ij}) to compute the spurious association.

The spurious component indicates the amount of the bivariate association (slope) between two variables that is not due to the effect of one variable on the other. Since indirect effects are included in the total effect, they are not counted as spurious association; indirect effects and direct effects are equally valid types of causal effects. Most commonly, the spurious component of association is thought of as having the same sign as the total effect. This would occur when the bivariate slope b_{ij} has the same sign and is larger in absolute value than the total effect. In this case, failure to control for other causes of X_i, which either are correlated with X_j or are causes of X_j, would lead to overestimates of the effects of X_j. However, if these other causes are suppressing the relationship between X_i and X_j, b_{ij} may be opposite in sign from the total effect or may have the same sign but be smaller in absolute value. In this case, the bivariate association either will be underestimating the strength of the causal effect or will be giving a wrong-signed estimate of the effect. Thus, spuriousness may involve overestimates, underestimates, or wrong-signed estimates of causal effects.

Direct (D_{ij}), indirect (I_{ij}), total (T_{ij}), and spurious (S_{ij}) effects will be referred to simply as DITS. The calculation of DITS in complex models can become tedious and error prone. We will see, however, how DITS calculations can be expedited through the use of simplified forms of causal models.

Reduced-Form Models and DITS

Causal models may be simplified by removing all of the paths leading from one endogenous variable to another. This produces a causal diagram that contains only paths from the exogenous variables to the endogenous variables. The result of this reduction is to eliminate all compound paths from the diagram. Such models are called **reduced-form models**. Reduced-form models represent only the effects of the exogenous variables. Figure 8.3 has only a single path connecting the endogenous variables, the path from X_3 to X_4. Removing this path gives Figure 8.4.

When the paths between endogenous variables are removed, the paths from the exogenous variables to the endogenous variables may no longer represent effects that are *independent* of other endogenous variables. In the full diagram of the causal system (Figure 8.3), the path from X_1 to X_4 represents an effect that is independent of X_3 because the model includes a path from X_3 to X_4. In the reduced-form model, however, there is no path from X_3 to X_4 to represent the effect of X_3; thus, the path from X_1 to X_4 does not represent an effect that is independent of X_3. Stated differently, the path from X_1 to X_4 no longer represents the *direct* effect of X_1 on X_4 because the indirect or compound path connecting X_1 to X_4 via X_3 is no longer represented.

In a reduced-form diagram, the causal path between an exogenous variable and an endogenous variable represents the direct effect plus the sum of any indirect effects that may exist between the pair of variables. In other words, *the paths in the reduced-form model represent the total effects of the exogenous variables*. In the case of Figure 8.4, since the path from X_3 to X_4 has been removed, the paths from X_1 and X_2 to X_4 stand for the total effects that were calculated from the parameters in Figure 8.3. These total effects are shown on the paths in Figure 8.4. Notice that the effects of X_1 and X_2 on X_3 in the reduced-form diagram are the same as their original direct effects; there were no inter-

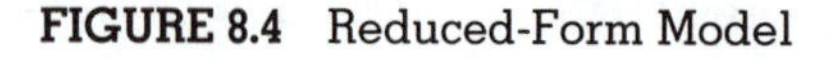
FIGURE 8.4 Reduced-Form Model

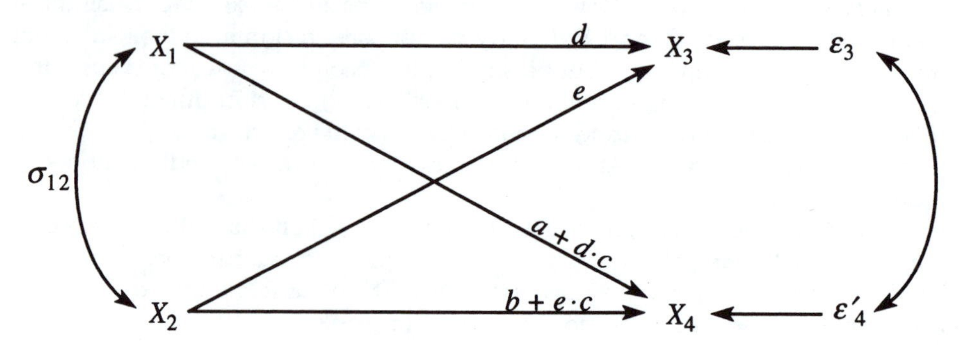

vening variables between the exogenous variables and X_3 to be eliminated in the reduced diagram.

Since the reduced-form diagram contains two endogenous variables, it is still a two-equation model. The effects in Figure 8.4 are estimated with the following regression equations:

$$\hat{X}_3 = a_{3\cdot 12} + b_{31\cdot 2}X_1 + b_{32\cdot 1}X_2 \quad \textbf{(8.5)}$$

$$\hat{X}_4 = a_{4\cdot 12} + b_{41\cdot 2}X_1 + b_{42\cdot 1}X_2 \quad \textbf{(8.6)}$$

There are only two X's in the equation for X_4 (8.6) because the path from X_3 to X_4 has been dropped from the diagram. In Equation 8.6 the slope $b_{41\cdot 2}$ estimates the effect of X_1 on X_4 that is independent of X_2, since X_2 is in the equation and is thus held constant. This slope, however, does not estimate an effect that is independent of X_3 because X_3 is not included in the equation. The regression slopes in Equation 8.6 will equal the effects shown in Figure 8.4 (assuming no sampling error occurs).

$$b_{41\cdot 2} = a + d\cdot c \qquad b_{42\cdot 1} = b + e\cdot c$$

Again, since the equation for X_4 does not contain X_3, the above regression slopes will not be estimates of the direct effects of the exogenous variables. Instead, they will equal the sum of the direct effect plus the indirect effect that passes through X_3, i.e., the total effect.

It should be emphasized that the reduced-form model of Figure 8.4 does not represent an alternative causal model to that specified in Figure 8.3. It is instead a simplification of Figure 8.3 achieved by omitting the path from X_3 to X_4. The absence of a path from X_3 to X_4 does not mean that X_3 does not have an effect on X_4. It was omitted to simplify the causal analysis. A reduced-form model has meaning only when compared to the full model from which it is derived.

Notice in Figure 8.4 that the error term for X_4 is now written as ε_4'. The variance of ε_4' will be larger than the variance of ε_4 in the three-variable equation because X_3 is not included in the equation. Thus, the variance in X_4 that is uniquely explained by X_3 is now contained in the error term ε_4'. Furthermore, since the unique variance in X_3 is the amount that is not related to X_1 and X_2, the error term for X_3 is the source of this unique variance. Therefore, ε_3 is the cause of the unique variance in X_3 that causes some of the variance in X_4 that is now summarized by ε_4'. As a consequence, ε_3 and ε_4' will be correlated. This is shown by the doubled-headed curved line connecting ε_3 and ε_4' in the reduced-form model in Figure 8.4. The error terms in Figure 8.4, however, are not of primary concern to us. We are principally interested in the total effects of the exogenous variables shown in the reduced-form diagram.

The model represented by Figure 8.4 is a valid model for the total effects of the exogenous variables on X_4. When these total effects are estimated with the above regression equation, however, the ability to distinguish between the direct and indirect effects is lost. However, if we estimated only the model shown in Figure 8.4 (that is, if we never estimated Figure 8.3, possibly because we did not have a measure of X_3), the estimates of the total effects of X_1 and X_2 would not be biased by failure to control for the intervening variable X_3. Discussions of regression assumptions (see Chapter 4) always emphasize that the failure to control for a variable that is a cause of Y and that is correlated with the independent variables included in the regression equation will lead to biased estimates of the effects of the variables included in the equation (e.g., $b_{41 \cdot 2}$ and $b_{42 \cdot 1}$ will be biased). This is because the effect of the omitted variable (e.g., X_3) will be part of the error term and thus will be correlated with the included X's. Consequently, part of this effect will be picked up by the b's for the included variables, and this will create biased estimates of their *direct effects*. If the omitted variable is an intervening variable, however, the amount of the bias will be equal to the indirect effects of the included X's, which are dc and ec for X_1 and X_2, respectively. We now see why we must qualify the consequences of violating the assumption that e is uncorrelated with the X's. If the omitted variables are intervening variables, their omission will not lead to biased estimates of the *total effects* of the included variables. If, on the other hand, the omitted variables are either causes of the included variables or are simply correlated with the included variables, the regression slopes will be biased.

There is a practical reason for introducing reduced-form models. Alwin and Hauser (1975) have shown how reduced-form equations can be used to expedite the calculations of indirect effects and total effects for more complex models. We will illustrate the basic principles of the Alwin-Hauser method with the relatively simple model in Figure 8.3. The method involves running a series of regression equations, including the bivariate equations, the reduced-form equation (8.6), and the full equation (8.3). The following SPSS commands may be used.

```
REGRESSION VARS = X1 X2 X3 X4/
  DEP = X4/ ENTER X1/
  DEP = X4/ ENTER X2/ ENTER X1/ ENTER X3
```

The above statements will give us two bivariate equations, one for X_4 and X_1 and one for X_4 and X_2. The commands will also give us the reduced-form equation for X_4 as predicted by X_1 and X_2. Finally, they will give us the full equation for X_4 regressed on X_1, X_2, and X_3.

$$\hat{X}_4 = a_{41} + b_{41}X_1$$

$$\hat{X}_4 = a_{42} + b_{42}X_2$$

$$\hat{X}_4 = a_{4\cdot12} + b_{41\cdot2}X_1 + b_{42\cdot1}X_2$$

$$\hat{X}_4 = a_{4\cdot123} + b_{41\cdot23}X_1 + b_{42\cdot13}X_2 + b_{43\cdot12}X_3$$

The regression coefficients from these equations can be used to get the total effects, indirect effects, and spurious association in terms of structural coefficients, as follows:

$$b_{41\cdot2} = a + d\cdot c = T_{41}$$
$$b_{42\cdot1} = b + e\cdot c = T_{42}$$
$$b_{41\cdot2} - b_{41\cdot23} = (a + d\cdot c) - a = d\cdot c = I_{431}$$
$$b_{42\cdot1} - b_{42\cdot13} = (b + e\cdot c) - b = e\cdot c = I_{432}$$
$$b_{41} - b_{41\cdot2} = b_{41} - T_{41} = S_{41}$$
$$b_{42} - b_{42\cdot1} = b_{42} - T_{42} = S_{42}$$

In this case, we have to make one subtraction for each exogenous variable to get its indirect effect, instead of multiplying the structural coefficients along the compound paths. We do not have to add indirect effects and direct effects together to get the total effect; it can be read off the printout for the reduced-form equation. Although there is only a small savings in computations for this simple model, the savings can be considerable for more complex models. Most importantly, perhaps, familiarity with reduced-form models sharpens our understanding of causal modeling.

A Three-Equation Causal Model

It is possible to elaborate further the causal model of Figure 8.3 by adding a causal path between the two exogenous variables. Let us assume that theory justifies specifying a causal path from X_1 to X_2. This gives Figure 8.5.

Figure 8.5 represents a three-equation model with three endogenous variables and only one exogenous variable, other than the three error terms. The three equations for estimating the parameters of Figure 8.5 are

$$\hat{X}_4 = a_{4\cdot123} + b_{41\cdot23}X_1 + b_{42\cdot13}X_2 + b_{43\cdot12}X_3 \quad \textbf{(8.7)}$$

$$\hat{X}_3 = a_{3\cdot12} + b_{31\cdot2}X_1 + b_{32\cdot1}X_2 \quad \textbf{(8.8)}$$

$$\hat{X}_2 = a_{21} + b_{21}X_1 \quad \textbf{(8.9)}$$

The parameters for X_4 and X_3 in Figure 8.5 are identical to those in Figure 8.3. Thus, Equations 8.7 and 8.8 for X_4 and X_3, respectively, are identical to Equations 8.3 and 8.4. The new parameter f and the new variable ε_2 in Figure 8.5 are estimated by using Equation 8.9.

FIGURE 8.5 Three-Equation Causal Model

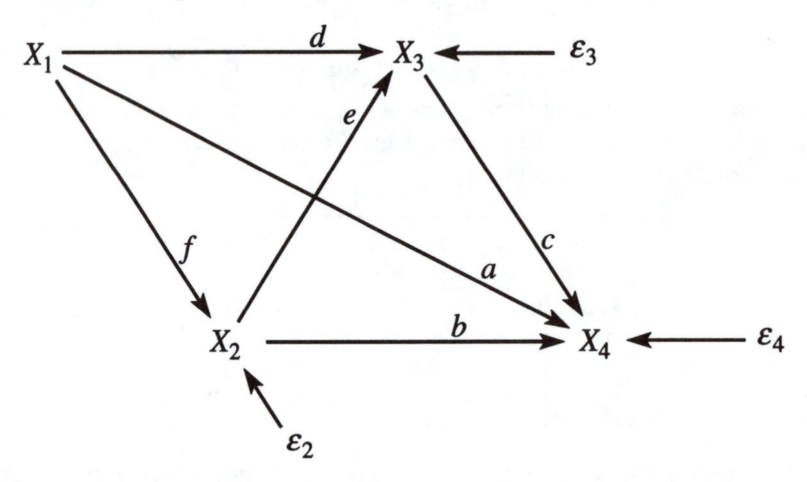

DITS

Although there is only one new structural coefficient in Figure 8.5 (i.e., f), the addition of X_2 as an endogenous variable that intervenes between X_1 and the final two endogenous variables (X_3 and X_4) creates an increase in the complexity of indirect effects. Figure 8.6 shows the four indirect effects that are now present in the model.

Indirect effects I_{431} and I_{432} are the same as in Figure 8.3. The new indirect effects are the ones that pass from X_1 through X_2, I_{421} and I_{4321}. The latter indirect effect involves a chain of three direct effects. The value of this longer indirect effect is determined in the same manner as for the shorter indirect effects.

I_{4321}
$\Delta X_1 = 1$
$\Delta X_2 = f \cdot \Delta X_1 = f \cdot 1 = f$
$\Delta X_3 = e \cdot \Delta X_2 = e \cdot f$
$\Delta X_4 = c \cdot \Delta X_3 = c \cdot (e \cdot f)$
$I_{4321} = f \cdot e \cdot c$

As before, indirect effects equal the product of all structural coefficients (or path coefficients in the case of a standardized model) along the compound path. When there is more than one indirect effect, the total effect of one variable on another will equal the direct effect plus the sum of indirect effects. Since Figure 8.5 does not add any new intervening variables between X_2 and the later endogenous variables (X_3 and X_4), the total effects of X_2 do not differ from those for Figure 8.3. The total effects of X_1, however, are now given by

FIGURE 8.6 Indirect Effects Present in Figure 8.5

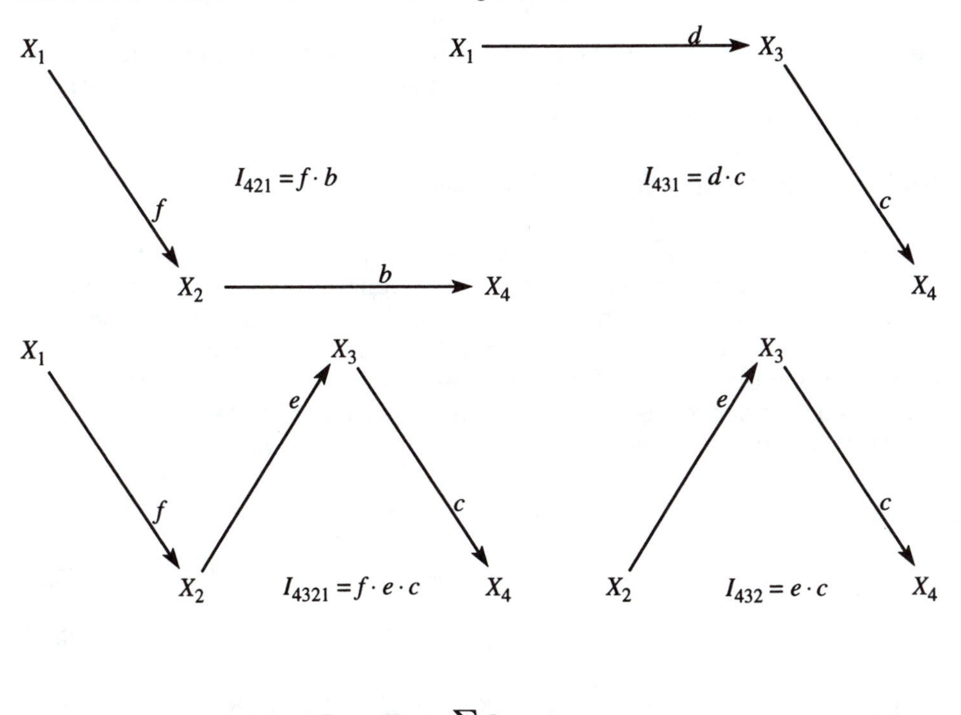

$$T_{ij} = D_{ij} + \sum I_{i...j}$$

$$T_{21} = f$$

$$T_{31} = d + f \cdot e$$

$$T_{41} = a + d \cdot c + f \cdot b + f \cdot e \cdot c$$

The spurious association again equals the bivariate slope minus the total effect ($S_{ij} = b_{ij} - T_{ij}$). The spurious associations between X_2 and X_3, between X_2 and X_4, and between X_3 and X_4 are the same for Figure 8.5 as they were for Figure 8.3. The spurious associations of X_1 with the other variables, however, have changed because the total effects of X_1 have changed.

$$S_{21} = b_{21} - T_{21} = b_{21} - b_{21} = 0$$

$$S_{31} = b_{31} - T_{31} = 0$$

$$S_{41} = b_{41} - T_{41} = 0$$

Figure 8.5 indicates that the total effect of X_1 on X_2 is equal to its direct effect, which equals b_{21} according to Equation 8.9; therefore, all of the association between these variables is nonspurious. Furthermore, since no other variables in the model are correlated with X_1 or antecedent to X_1, all of the association between X_1 and X_3 and between X_1 and X_4 is due to the direct and indirect effects of X_1 on these variables. Therefore, there is no spurious association between X_1 and the other variables in Figure 8.5.

Reduced and Semi-Reduced Models

We can again simplify our model by eliminating all indirect effects from the diagram. The reduced form of Figure 8.5, containing only the paths from the exogenous variable to the endogenous variables, is shown in Figure 8.6. Each endogenous variable in Figure 8.7 is linked by a single causal path to the exogenous variable X_1. As in Figure 8.4, the coefficients represent the total effect of X_1 on each endogenous variable. These effects are estimated with the following bivariate regression equations:

$$\hat{X}_2 = a_{21} + b_{21}X_1 \quad \textbf{(8.10)}$$

$$\hat{X}_3 = a_{31} + b_{31}X_1 \quad \textbf{(8.11)}$$

$$\hat{X}_4 = a_{41} + b_{41}X_1 \quad \textbf{(8.12)}$$

Equation 8.10 is identical to the equation for X_2 in Figure 8.5. Thus, Equation 8.10 is not a reduced-form equation. Equations 8.11 and 8.12 differ from the equations for X_3 and X_4 in the full model because the intervening variables

FIGURE 8.7 Reduced Form of the Three-Equation Model

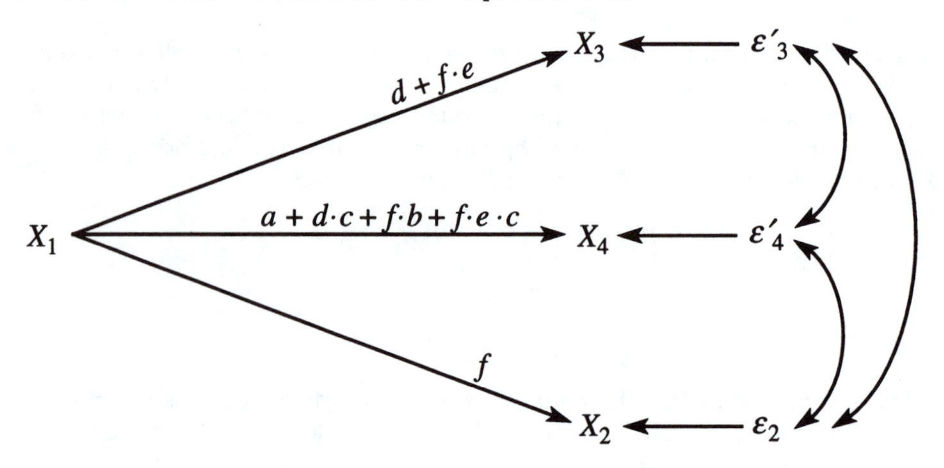

FIGURE 8.8 Semi-Reduced Form of the Three-Equation Model

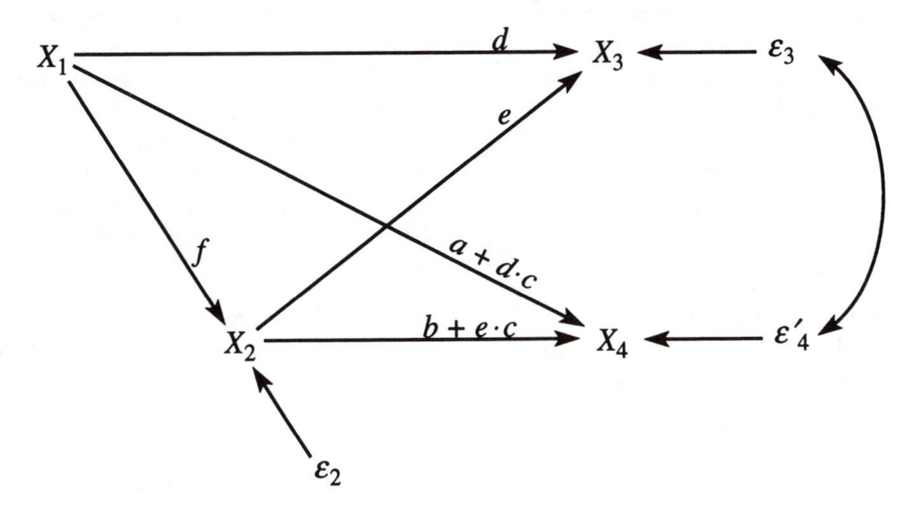

have been removed from these equations. Thus, Equations 8.11 and 8.12 are reduced-form equations. The entire model is called a reduced model because some, but not all, of the equations are reduced. Since the total effects equal the bivariate slopes in this reduced model, we can again see that there is no spurious association between X_1 and the endogenous variables in this model.

The error terms for X_3 and X_4 in Figure 8.7 are different from those in Figure 8.5 because the variance in X_3 uniquely explained by X_2 and the variance in X_4 uniquely explained by X_2 and X_3 become part of the error terms in the reduced-form equations. Also, for reasons analogous to those given for reduced Figure 8.4, the error terms in Figure 8.7 are correlated.

It is also useful to consider the **semi-reduced** model shown in Figure 8.8. The semi-reduced model includes causal paths from the first endogenous variable, X_2, to the later endogenous variables, X_3 and X_4. As in the reduced model, however, the path from X_3 to X_4 is omitted. Thus, the semi-reduced model allows for the indirect effects that pass through the first endogenous variable, but it does not specify indirect effects through later endogenous variables. In semi-reduced models, each endogenous variable (except the first) is affected by all of the exogenous variables plus the first endogenous variable.

The equations for estimating the semi-reduced model are

$$\hat{X}_2 = a_{21} + b_{21}X_1 \tag{8.13}$$

$$\hat{X}_3 = a_{3\cdot12} + b_{31\cdot2}X_1 + b_{32\cdot1}X_2 \tag{8.14}$$

$$\hat{X}_4 = a_{4\cdot12} + b_{41\cdot2}X_1 + b_{42\cdot1}X_2 \tag{8.15}$$

Equations 8.13 and 8.14 are the same as the equations for X_2 and X_3 in the full model (Figure 8.5). These equations, therefore, are neither reduced nor semi-reduced. Only the equation for X_4 (Equation 8.15) is different in the semi-reduced model. Equation 8.15 does not contain X_3, the second endogenous variable, but it does contain X_2, the first endogenous variable. Equation 8.15 is thus a semi-reduced equation. The entire model is called a semi-reduced model, even though not all equations in the model are semi-reduced. Since X_3 is not contained in Equation 8.15, the slope coefficients in the semi-reduced equation equal the direct effects of each variable plus the indirect effects that pass through X_3.

Indirect Effects. By using the Alwin-Hauser method, we can use the differences between the slopes in the reduced, semi-reduced, and full models to compute various indirect effects.

$$\text{Reduced Model } b_{ij} - \text{Semi-Reduced Model } b_{ij}$$

$$= \sum I_{i...j} \text{ Through First Endogenous } X$$

$$b_{41} - b_{41\cdot 2} = (a + d\cdot c + f\cdot b + f\cdot e\cdot c) - (a + d\cdot c)$$

$$= f\cdot b + f\cdot e\cdot c = I_{421} + I_{4321}$$

$$b_{31} - b_{31\cdot 2} = (d + f\cdot e) - d = f\cdot e = I_{321}$$

$$\text{Semi-Reduced } b_{ij} - \text{Full } b_{ij} = \sum I_{i...j} \text{ Through Second Endogenous } X$$

$$b_{41\cdot 2} - b_{41\cdot 23} = (a + d\cdot c) - a = d\cdot c = I_{431}$$

$$b_{42\cdot 1} - b_{42\cdot 13} = (b + e\cdot c) - b = e\cdot c = I_{432}$$

The sum of the indirect effects that pass through the first endogenous variable, in this case X_2, equals the difference between the slope in the reduced-form equation for a particular variable and the slope in the semi-reduced equation for that variable. In our model, X_1 is the only exogenous variable, and thus it is the only variable for which we can calculate the difference between its reduced and semi-reduced coefficients. With respect to the effect of X_1 on X_4, we see that the difference between its reduced and semi-reduced b's equals the sum of two indirect effects, I_{421} and I_{4321}. The first is a two-path chain and the second is a three-path chain; both, however, initially pass through X_2. The difference between the reduced and semi-reduced slopes does not allow us to differentiate between these two indirect effects (I_{421} and I_{4321}); it simply provides a summary of all indirect effects passing through X_2. We would have to multiply the coefficients along the paths to calculate each distinct indirect effect.

We can also use the reduced and semi-reduced models to calculate the indirect effect of X_1 on X_3 (I_{321}) as shown in the formulas above. In this case, there is only one indirect effect that passes through X_2. Finally, the difference between the b's in the semi-reduced and full equations for X_4 can be used to calculate the indirect effects of X_1 and X_2 on X_4 that pass through X_3 (I_{431} and I_{432}), as shown in the above formulas.

Different Models Produce Different DITS

In order to study some of the rules of causal analysis, we have examined the properties of three different types of causal models that can be specified for four variables: the single-equation model shown by Figure 8.1, the two-equation model shown by Figure 8.3, and the three-equation model shown by Figure 8.5. Table 8.1 provides a summary of the various total effects that were derived for each model.

Clearly, the model that we choose can potentially make a big difference in the magnitude of the total effects that we might find for each of the independent variables, with the exception of X_3. If we read across any row in the table except the last one, there is at least one difference between models with respect to the total effect of the independent variable in that row on the dependent variable in that row. We say that there is a potential difference between models, because it is certainly possible that one of the letters representing a difference in effects between models might turn out to be zero, or nearly zero, when we empirically estimate it.

If the model that we choose to estimate can make a big difference in the size of the effects that we estimate, how do we choose between models? It is not valid to try them all and choose the one whose results we like best. Nor is it possible to determine empirically which model fits the data best. The differences between the models is a matter of the number of causal equations that we can theoretically justify. With nonexperimental data there is no way to test whether the causal order that we might specify is valid. All that we can do is use the best theory that is available and the best information about the temporal sequence of the variables that is available to specify certain equations. Once we have specified a model, we can empirically estimate the size of each

TABLE 8.1 Summary of Total Effects

Dependent	*Independent*	*Figure 8.1*	*Figure 8.3*	*Figure 8.5*
X_2	X_1	0	0	f
X_3	X_1	0	d	$d + f \cdot e$
	X_2	0	e	e
X_4	X_1	a	$a + d \cdot c$	$a + d \cdot c + f \cdot b + f \cdot e \cdot c$
	X_2	b	$b + e \cdot c$	$b + e \cdot c$
	X_3	c	c	c

structural or path coefficient. These estimates are valuable for the information they provide about the relative sizes of the different direct and indirect effects specified by the model. But the validity of these estimates is dependent on the validity of the causal order that we have specified.

A Causal Analysis of SES and Self-Esteem

The principles of causal analysis that have been presented will be illustrated with data from the 1986 Akron Area Survey, a telephone survey of residents of Summit County, Ohio. This example uses the same sample of cases ($n = 513$) that were used in the anomia example in Chapter 4. Instead of anomia, however, this example uses *self-esteem* (Rosenberg 1965). The respondents' scores on a self-esteem index (ESTEEM) equal the sum of their coded responses to the four questions shown in Figure 8.9. Note that the scoring has been reversed on the last two questions because the wording of these two items expresses low self-esteem (Figure 8.9). The range of values on ESTEEM is 4 to 16. The other three variables (which were also used in the anomia example) are years of education (1–20), family income (*less than $5,000* = 1; *$5,000–9,999* = 2; *$10,000–14,999* = 3; *$15,000–19,999* = 4; *$20,000–24,999* = 5; *$25,000–34,999* = 6; *$35,000–49,999* = 7; *$50,000* or *more* = 8), and a measure of the respondents' subjective

FIGURE 8.9 Self-Esteem and Subjective Income Questions

Self-Esteem

I feel that I am a person of worth, at least on an equal basis with others. (*strongly agree* = 4; *somewhat agree* = 3; *somewhat disagree* = 2; *strongly disagree* = 1)

I am able to do things as well as most other people. (*strongly agree* = 4; *somewhat agree* = 3; *somewhat disagree* = 2; *strongly disagree* = 1)

I wish I could have more respect for myself. (*strongly agree* = 1; *somewhat agree* = 2; *somewhat disagree* = 3; *strongly disagree* = 4)

I certainly feel useless at times. (*strongly agree* = 1; *somewhat agree* = 2; *somewhat disagree* = 3; *strongly disagree* = 4)

Subjective Income

How seriously do you feel a personal shortage of money these days—a great deal, quite a bit, some, or little or none? (*great deal* = 1; *quite a bit* = 2; *some* = 3; *little or none* = 4)

assessment of the level of their incomes (SHORTINC in Figure 8.9). A high score on SHORTINC indicates a positive assessment of income.

As we did in the previous sections, a two-equation model will first be specified and analyzed according to the principles that have been discussed. Then a three-equation model will be specified to illustrate the additional causal information that can be extracted from such a model. Although other social scientists might argue that causal orderings should be specified that are different from the ones we will analyze, including models with reciprocal causation (see the section on nonrecursive models), we will examine the causal information that can be extracted from the models under the assumption that they are validly specified.

A Two-Equation Model

The two-equation model is shown in Figure 8.10. The following SPSS commands can be used to compute all of the regression equations necessary for estimating the coefficients for the causal model and for computing the various DITS.

```
REGRESSION DESCRIPTIVES = DEFAULTS COV XPROD/
  VARIABLES = ESTEEM SHORTINC INCOME EDUC/
  DEP = SHORTINC/ ENTER INCOME/
  DEP = SHORTINC/ ENTER EDUC/ENTER INCOME/
  DEP = ESTEEM/ ENTER INCOME/
  DEP = ESTEEM/ ENTER SHORTINC/
  DEP = ESTEEM/ ENTER EDUC/ ENTER INCOME/ ENTER SHORTINC
```

The coefficients on the paths in Figure 8.10 come from Equation 5 for ESTEEM and from Equation 3 for SHORTINC in Table 8.2. These are structural

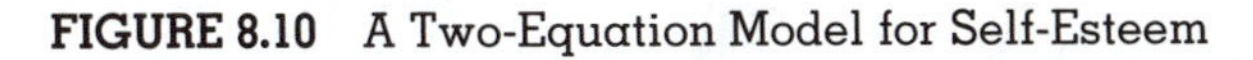
FIGURE 8.10 A Two-Equation Model for Self-Esteem

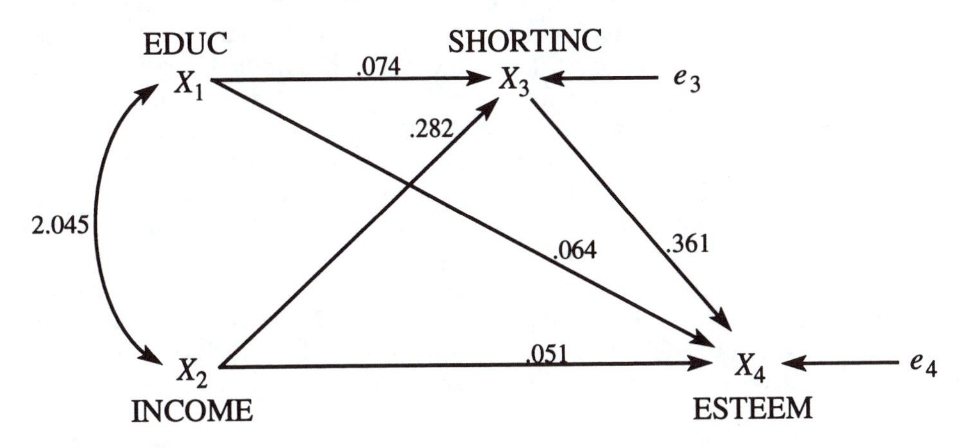

TABLE 8.2 **Unstandardized Regression Equations for INCOME, SHORTINC, and ESTEEM (Standardized Coefficients)**

	ESTEEM Equations				
	1	2	3	4	5
EDUC	.133*	—	—	.091*	.064
	(.165)			(.113)	(.080)
INCOME	—	.196*	—	.152*	.051
		(.184)		(.143)	(.048)
SHORTINC	—	—	.440*	—	.361*
			(.267)		(.219)
Constant	11.590	12.352	11.985	11.358	11.107
R^2	.027	.034	.071	.045	.081
$\hat{s}_\varepsilon$	2.169	2.162	2.119	2.151	2.113

	SHORTINC Equations				*INCOME*
	1	2	3		1
EDUC	.151*	—	.074*		.274*
	(.309)		(.151)		(.364)
INCOME	—	.317*	.282*		—
		(.490)	(.435)		
Constant	1.124	1.501	.696		1.520
R^2	.096	.240	.260		.132
$\hat{s}_\varepsilon$	1.270	1.164	1.150		1.920

*$p \leq .05$

(unstandardized) coefficients. The diagram also shows a covariance of 2.045 between education and income (not given in Table 8.2). SHORTINC is the only variable having a significant ($p < .05$) effect on ESTEEM (Table 8.2). The positive coefficient indicates that the more a person feels he or she has enough income, holding constant education and actual income, the higher will be his or her self-esteem. It is interesting that subjective income is more important than objective income for self-esteem, whereas just the opposite was true for anomia (see Table 4.3). Both education and income, however, have significant positive effects on SHORTINC. We will use all coefficients in Figure 8.10, whether significant or not, for computing direct, indirect, total, and spurious effects (DITS).

Figure 8.11 shows the reduced form of the two-equation self-esteem model. This form shows only the effects of the exogenous variables (education and income) on self-esteem (there is no path from SHORTINC to ESTEEM). The structural coefficients for ESTEEM come from Equation 4 in Table 8.2. They represent the total effects of education and income on self-esteem, and both are signifi-

FIGURE 8.11 Reduced Form of the Two-Equation Model for Self-Esteem

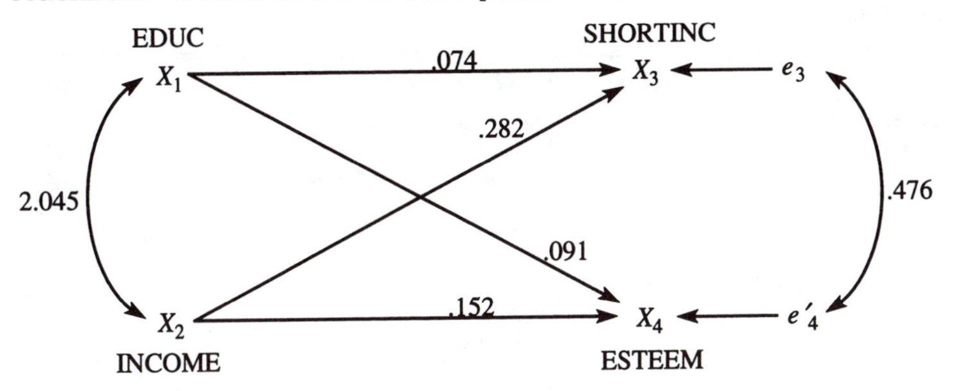

cant. Figure 8.11 also shows a covariance of .476 between the error terms for SHORTINC and ESTEEM. This is the covariance between the regression residuals for SHORTINC and ESTEEM (not shown in Table 8.2).[2]

The direct, indirect, total, and spurious effects (DITS) and their computations are shown in Table 8.3. Two equivalent methods are shown: the *path-diagram* method multiplies the coefficients on compound paths in Figure 8.10 to get indirect effects and sums the direct and indirect effects to get the total effect; the Alwin-Hauser *hierarchical-equations* methods determines the indirect and total effects from the full equation coefficients and reduced-form equation coefficients in Table 8.2.

With respect to the effects of education on self-esteem, the indirect effect is smaller than the direct effect. However, the indirect effect is large enough so that when it is added to the direct effect, the total effect is significant (see reduced equation in Table 8.2). For income, however, its indirect effect on self-esteem is larger than its direct effect. In both these cases a failure to take into account indirect effects would lead the researcher to conclude that educational

2. The following SPSS commands can be used to compute this covariance:

```
REGRESSION DESCRIPTIVES/
  VARIABLES = ESTEEM SHORTINC INCOME EDUC/
  DEP = SHORTINC/ ENTER EDUC INCOME/
  SAVE = RESID (SHORTRES)
REGRESSION DESCRIPTIVES/
  VARIABLES = ESTEEM SHORTINC INCOME EDUC/
  DEP = ESTEEM/ ENTER EDUC INCOME/
  SAVE = RESID (ESTEMRES)
REGRESSION DESCRIPTIVES = DEFAULTS XPROD COV/
  VARIABLES = ESTEEM SHORTINC INCOME EDUC SHORTRES ESTEMRES/
  DEP = ESTEMRES/ ENTER SHORTRES
```

TABLE 8.3 Direct, Indirect, Total, and Spurious Effects (DITS) for the Two-Equation Self-Esteem Model

Dep. Var.	*Indep. Var.*	*DITS Formulas*	*Effects*
ESTEEM	EDUC	Path-Diagram Method	
(X_4)	(X_1)	$D_{41} = b_{41.23} =$	.064
		$I_{431} = b_{31.2}b_{43.12} = (.074)(.361)$	.027
		$T_{41} = D_{41} + I_{431} =$	.091
		$S_{41} = b_{41} - T_{41} = .133 - .091 =$	.041
		Hierarchical-Equations Method	
		$D_{41} = b_{41.23} =$	.064
		$I_{431} = b_{41.2} - b_{41.23} = .091 - .064 =$	.027
		$T_{41} = b_{41.2} =$	.091
		$S_{41} = b_{41} - b_{41.2} = .133 - .091 =$	.041
	INCOME	Path-Diagram Method	
	(X_2)	$D_{42} = b_{42.13} =$	.051
		$I_{432} = b_{32.1}\, b_{43.12} = (.282)(.361)$	.102
		$T_{42} = D_{42} + I_{432} =$	.153
		$S_{42} = b_{42} - T_{42} = .196 - .153 =$	.043
		Hierarchical-Equations Method	
		$D_{42} = b_{42.13} =$	.051
		$I_{432} = b_{42.1} - b_{42.13} = .152 - .051 =$	.101
		$T_{42} = b_{42.1} =$	.152
		$S_{42} = b_{42} - b_{42.1} = .196 - .152 =$	.044
	SHORTINC	Path-Diagram Method	
	(X_3)	$D_{43} = b_{43.12} =$	.361
		$I_{4..3} =$ none	—
		$T_{43} = D_{43} + I_{4..3} =$	.361
		$S_{43} = b_{43} - T_{43} = .440 - .361 =$	.079
SHORTINC	EDUC	Path-Diagram Method	
(X_3)	(X_1)	$D_{31} = b_{31.2} =$	.074
		$I_{3..1} =$ none	—
		$T_{31} = D_{31} + I_{321} =$	.074
		$S_{31} = b_{31} - T_{31} = .151 - .074 =$	.077
	INCOME	Path-Diagram Method	
	(X_2)	$D_{32} = b_{32.1} =$	.282
		$I_{3.2} =$ none	—
		$T_{32} = D_{32} + I_{3..2} =$	.282
		$S_{32} = b_{32} - T_{32} = .317 - .282 =$	.035

attainment and objective income play no causal role in determining self-esteem. Nevertheless, the direct effect of subjective income is greater than the total effects of either education or income itself. For education, income, and subjective income, the spurious component of DITS is less than half as large as each variable's total effect. Also, the total effect of each variable is less than its bivariate slope and of the same sign, an indicator of redundancy among these variables.

A Three-Equation Model

The three-equation model is shown in Figure 8.12. The only difference between it and the two-equation model is that education is now specified as a cause of income. The structural coefficient for this new path is estimated by the regression equation in the last column of Table 8.2.

Figure 8.13 shows the reduced form of the model. It contains causal paths for only the single exogenous variable, education. The coefficients for each of these three paths, which represent the total effects of education, are estimated with the bivariate regression equations shown in Table 8.2. The covariances between the error terms for income, subjective income, and self-esteem are estimated by saving the residuals from the three bivariate regression equations and computing the covariances between these residuals (footnote 1 gives the SPSS commands for correlating the residuals of the two endogenous variables in the reduced form of the two-equation model). Figure 8.14 shows the semi-reduced model. In addition to causal paths from the exogenous variable education, the semi-reduced model shows the total effects of the first endogenous

FIGURE 8.12 A Three-Equation Model for Self-Esteem

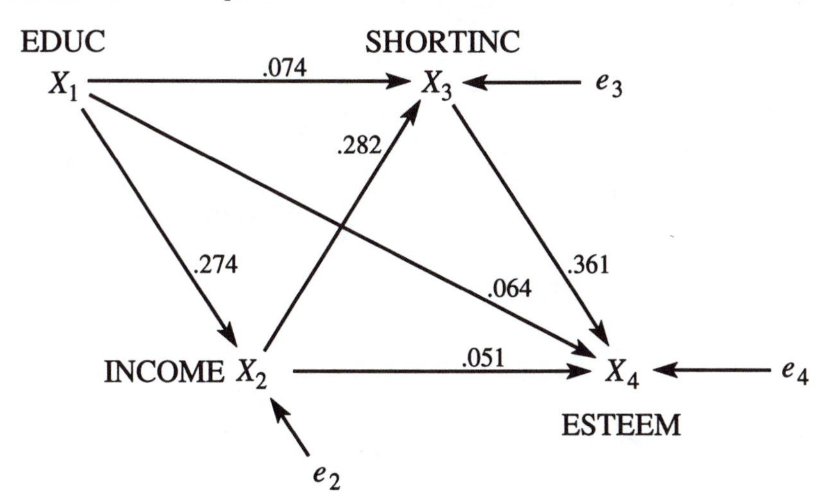

FIGURE 8.13 Reduced Form of the Three-Equation Model for Self-Esteem

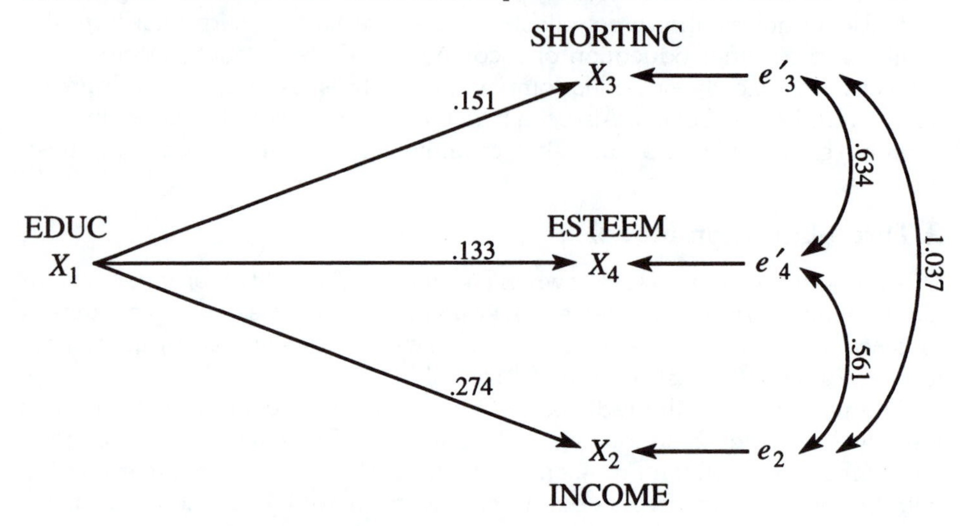

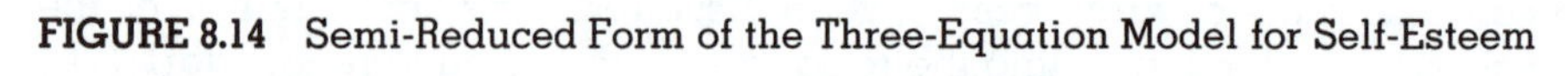

FIGURE 8.14 Semi-Reduced Form of the Three-Equation Model for Self-Esteem

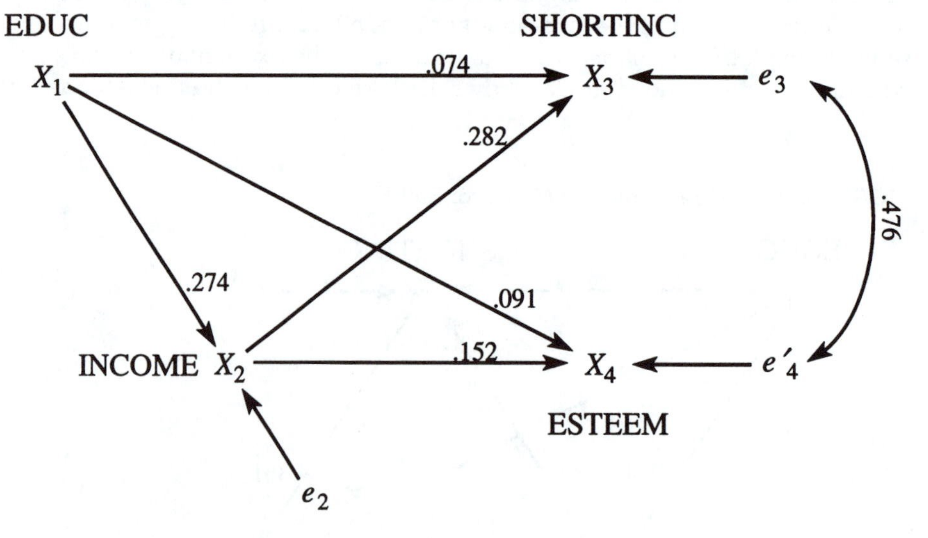

TABLE 8.4 Direct, Indirect, Total, and Spurious Effects (DITS) of Education in the Three-Equation Self-Esteem Model

Dep. Var.	*Indep. Var.*	*DITS Formulas*	*Effects*	
ESTEEM	EDUC	Path-Diagram Method		
(X_4)	(X_1)	$D_{41} = b_{41.23} =$		.064
		$I_{421} = b_{21}b_{42.13} = (.274)(.051) =$	.014	
		$I_{4321} = b_{21}b_{32.1}b_{43.12} = (.274)(.282)(.361) =$	.028	
		$I_{431} = b_{31.2}b_{43.12} = (.074)(.361) =$	.027	
		$\sum I_{4...1} =$		.069
		$T_{41} = D_{41} + \sum I_{4...1} =$		.133
		$S_{41} = b_{41} - T_{41} = .133 - .133 =$		.000
		Hierarchical-Equations Method		
		$D_{41} = b_{41.23} =$		.064
		$I_{421} + I_{4321} = b_{41} - b_{41.2} = .133 - .091 =$	.042	
		$I_{431} = b_{41.2} - b_{41.23} = .091 - .064 =$	.027	
		$\sum I_{4...1} = b_{41} - b_{41.23} = .133 - .064 =$		.069
		$T_{41} = b_{41} =$		.133
		$S_{41} = b_{41} - b_{41} =$		.000
SHORTINC	EDUC	Path-Diagram Method		
(X_3)	(X_1)	$D_{31} = b_{31.2} =$		.074
		$I_{321} = b_{21}b_{32.1} = (.274)(.282) =$		.077
		$T_{31} = D_{31} + I_{321} =$		.151
		$S_{31} = b_{31} - T_{31} = .151 - .151 =$		.000
		Hierarchical-Equations Method		
		$D_{31} = b_{31.2} =$		.074
		$I_{321} = b_{31} - b_{31.2} = .151 - .074 =$		.077
		$T_{31} = b_{31} =$		.151
		$S_{31} = b_{31} - b_{31} = .151 - .151 =$		.000
INCOME	EDUC	Path-Diagram Method		
(X_2)	(X_1)	$D_{21} = b_{21} =$		.274
		$I_{2..1} =$ none		—
		$T_{21} = D_{21} + I_{2..1} =$		.274
		$S_{21} = b_{21} - T_{21} = .274 - .274 =$		.000

variable, income, on the other two endogenous variables. Notice that except for the path from education to income, the semi-reduced form in Figure 8.14 is the same as the reduced form in Figure 8.11, including the covariance between the error terms.

The calculation of all new direct, indirect, total, and spurious effects for the three-equation model are shown in Table 8.4.

The three-equation model differs from the two-equation model only with respect to the total effects and indirect effects of education. By making income dependent upon education, several new indirect effects of education appeared. Table 8.4 gives all of the DITS effects for education. There are now two additional indirect effects of education on ESTEEM, both of which pass through INCOME; education increases income, which in turn has a slight effect on self-esteem (I_{421}), and education increases income, which increases subjective income, which increases self-esteem (I_{4321}). Notice that with the Alwin-Hauser method the difference between the reduced form and the semi-reduced form slope of self-esteem on education equals the sum of these two indirect effects. In many models, the hierarchical changes in slopes equal the sum of several indirect effects. Thus, although the Alwin-Hauser method is quick and accurate, if you want to know the values of each indirect effect, you may have to compute them by multiplying coefficients along the compound paths that define each indirect effect. But if you are mainly interested in total effects and total indirect effects, the Alwin-Hauser method is ideal.

The two new indirect effects of education on self-esteem increase the total indirect effect enough that it is now slightly greater than the direct effect of education. There is also an indirect effect of education on SHORTINC that is as large as its direct effect. Thus, not only does the three-equation model show the effect of education on income, it consequently opens up new indirect paths to subjective income and self-esteem. As a consequence, education assumes a more powerful explanatory role in the elaborated model. And since education is now the single exogenous variable in the model, none of its bivariate slope is spurious.

Nonrecursive Models

The models that we have examined are called **recursive** models. Recursive models do not have any causal loops; the causal flow is all in one direction. In a recursive model, the path effects leaving any particular variable will never return to that variable. As a consequence of the absence of any causal loops, *the parameters of recursive models can be estimated with ordinary least-squares regression.*

Nonrecursive models, however, are characterized by the presence of causal loops. Figure 8.15 provides an example. Figure 8.15 is the same as Figure 8.12 except that it includes a path from X_4 to X_3. This creates a causal loop between X_4 and X_3. A change in X_4 will cause a change in X_3, which will in turn feed back and cause a change in X_4. The return effect on X_4 would start another cycle around the loop, and so on. Although there are mathematical rules that can be used under certain circumstances to determine the total effect of the loop, they will not concern us here. The loop effect, however, is like an indirect effect, one that returns to cause a change in the original source of the effect. The loop effect could start with X_3 as well as with X_4.

FIGURE 8.15 A Nonrecursive Model

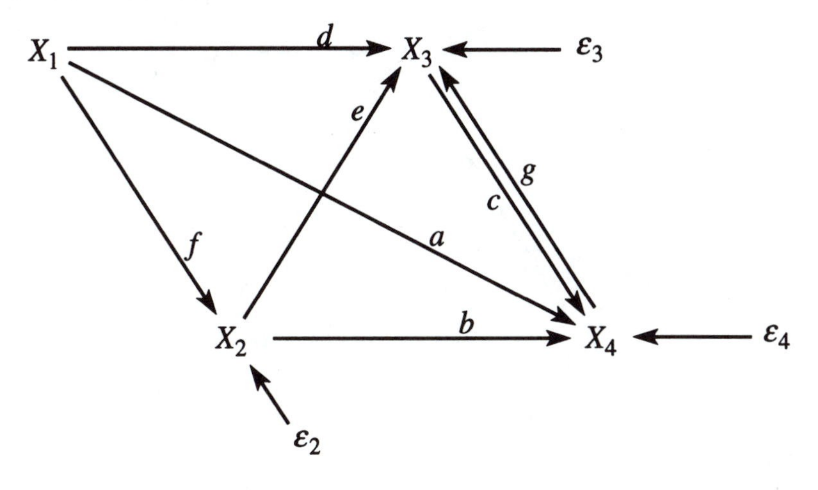

Figure 8.15 represents a three-equation model, one equation for each endogenous variable. Since the model is nonrecursive, not all of the parameters can be estimated with ordinary least-squares regression (for reasons to be demonstrated below). Therefore, we will not write these equations with regression notation, but instead, we will use the letters in the diagram as the coefficients for the variables in the equations. In order to avoid confusion between the effect *a* and the *a* that we have used as a constant in the regression equations, we will simply omit the intercept from the following equations (alternatively, we may assume that the *X*'s are deviation scores, in which case the intercept will equal zero).

$$X_2 = fX_1 + \varepsilon_2 \quad \textbf{(8.16)}$$

$$X_3 = dX_1 + eX_2 + gX_4 + \varepsilon_3 \quad \textbf{(8.17)}$$

$$X_4 = aX_1 + bX_2 + cX_3 + \varepsilon_4 \quad \textbf{(8.18)}$$

Equation 8.16 is actually a recursive equation because X_2 is not in a loop with any other variables. Thus, not all equations in a nonrecursive model are nonrecursive equations. Because Equation 8.16 is recursive, the parameter *f* can be estimated with an ordinary least-squares regression equation. Notice that the equation for X_3 (Equation 8.17) now contains X_4 because Figure 8.15 shows a path from X_4 to X_3. The equation for X_4 in turn contains X_3. Thus, each variable is included in the equation for the other.

Figure 8.15 shows an indirect path running from the error term for X_4 to X_3, that is, $\varepsilon_4 \rightarrow X_4 \rightarrow X_3$. This indirect path means that ε_4 and X_3 will covary or be correlated. Looking at Equation 8.18, this means that one of the independent

FIGURE 8.16 Another Nonrecursive Model

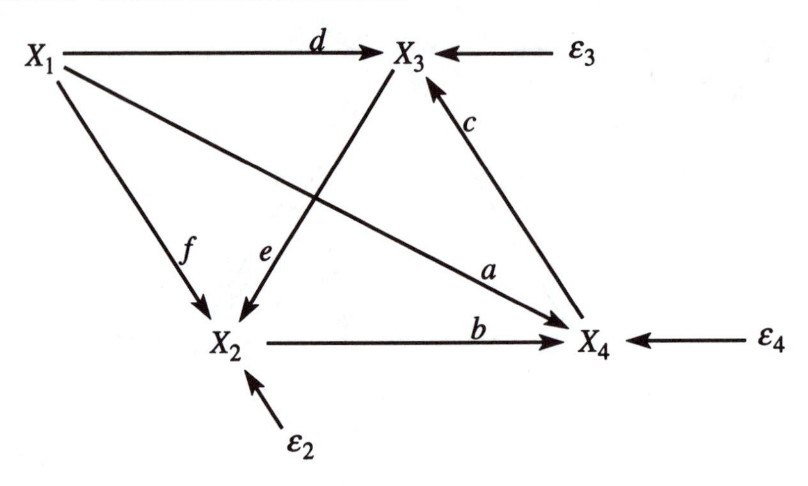

variables in the equation for X_4, namely X_3, is correlated with the error term for X_4. A basic assumption for using ordinary least-squares regression is that the error term must be uncorrelated with the independent variables. Since this assumption is clearly violated due to the loop in the model, least-squares regression should not be used to estimate the parameters in the equation for X_4. If we were to use the regression equation

$$\hat{X}_4 = b_{41 \cdot 23}X_1 + b_{42 \cdot 13}X_2 + b_{43 \cdot 12}X_3$$

to estimate the effects on X_4, $b_{43 \cdot 12}$ would be a biased estimate of c in Equation 8.18 because X_3 is correlated with the error term for X_4.

Figure 8.15 also shows the indirect path $\varepsilon_3 \rightarrow X_3 \rightarrow X_4$. This path will cause X_4 to be correlated with ε_3. Therefore, X_4 will be correlated with the error term for X_3 in Equation 8.17. If least-squares regression were used to estimate the effect of X_4 on X_3, $b_{34 \cdot 12}$ would be a biased estimate of parameter g.

To summarize, Figure 8.15 shows that the error term for each of the variables in the loop will be correlated with the other variable in the loop. This is the characteristic of nonrecursive models or equations that causes ordinary least-squares regression to give biased estimates of the effects of the variables in the loop.

Figure 8.16 shows another nonrecursive model. At a glance, this diagram looks like Figure 8.12. However, the direction of the paths between X_3 and X_4 and between X_3 and X_2 have been reversed in Figure 8.16. This creates a loop between X_3, X_4, and X_2.

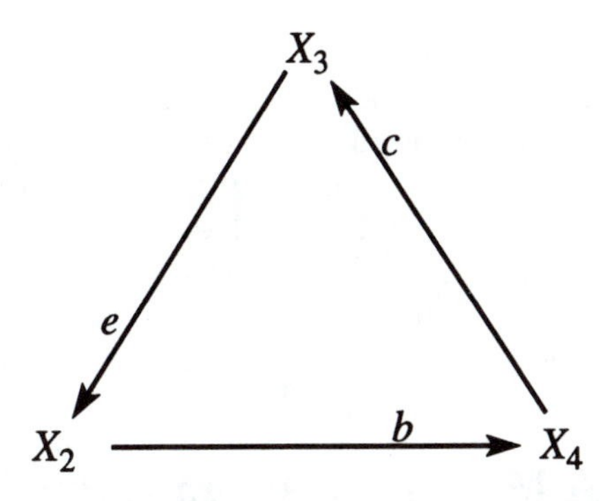

There are three equations in Figure 8.16, each containing two independent variables:

$$X_3 = dX_1 + cX_4 + \varepsilon_3$$

$$X_2 = fX_1 + eX_3 + \varepsilon_2$$

$$X_4 = aX_1 + bX_2 + \varepsilon_4$$

Because of the loop, each of the variables in the loop will be correlated with the error term in the equation in which they are one of the independent variables. This means that the least-squares regression equations will provide biased estimates of the effects of the variables in the loop, as shown:

$\varepsilon_4 \rightarrow X_4 \rightarrow X_3 \rightarrow X_2$, $\therefore r(\varepsilon_4 X_2) \neq 0$, $\therefore b_{42 \cdot 1}$ is a biased estimate of b

$\varepsilon_3 \rightarrow X_3 \rightarrow X_2 \rightarrow X_4$, $\therefore r(\varepsilon_3 X_4) \neq 0$, $\therefore b_{34 \cdot 1}$ is a biased estimate of c

$\varepsilon_2 \rightarrow X_2 \rightarrow X_4 \rightarrow X_3$, $\therefore r(\varepsilon_2 X_3) \neq 0$, $\therefore b_{23 \cdot 1}$ is a biased estimate of e

Bias is always a matter of degree. If one or more paths in a loop are weak compared to the others, the bias resulting from using least-squares regression may not be severe. Furthermore, there is undoubtedly always some bias in the regression estimates of recursive models because it is often the case that not all relevant independent variables are included in the regression equation, creating at least some covariance between the included variables and the error term.

Still, it is hard to justify using ordinary least-squares regression when we believe that causal loops are present in our models. There are other techniques that may be used to attempt to estimate the parameters of nonrecursive models, such as indirect least-squares and two-stage least-squares (see Duncan 1975; Heise 1975; Wonnacott and Wonnacott 1979). These techniques, however, do not provide easy solutions to the difficult problems of estimating nonrecursive models. In order to use these techniques, additional variables, called *instru-*

mental variables, must be found and added to the models. The instrumental variables must have strong statistical properties, and we must make strong theoretical assumptions about the absence of certain causal relationships between these variables and the variables in the loops. Further discussion, however, of two-stage least-squares and related techniques is beyond the scope of this book. For the present, we must be aware of the potential for causal loops in our models and recognize the inappropriateness of using ordinary least-squares regression when we believe that nonrecursiveness is present in our models.

A Model for Anomia and Self-Esteem

In this chapter we used a model for self-esteem that included education, income, and subjective income. In Chapter 4, anomia was used as a dependent variable in an equation that also used education, income, and subjective income as independent variables. Would it therefore be possible to add anomia as an endogenous variable to the three-equation model for self-esteem, thus converting it to a four-equation model? If so, would anomia be specified as a cause of self-esteem or vice versa? Some would feel it would be difficult to choose between these two alternatives and might want to specify a nonrecursive model in which anomia and self-esteem have reciprocal effects on each other. Such a model is shown in Figure 8.17.

As indicated in the previous discussion, estimating the reciprocal relationship between anomia and self-esteem would be difficult and beyond the scope of this book. However, there is a way out of this dilemma that allows us to salvage much of the model. We can choose not to attempt to estimate the reciprocal relationship but instead to estimate the remainder of the model with ordinary least-squares. If we eliminate the paths between anomia and self-esteem, we will be left with four equations, one each for INCOME, SHORTINC, ESTEEM, and ANOMIA, each of which can be estimated with ordinary least-squares. In fact, all of these regression equations have already been computed. The equation for ANOMIA was reported as Equation 1 in Table 4.3. The equations for INCOME, SHORTINC, and ESTEEM were given in Table 8.2. If we redraw Figure 8.17 and enter the coefficients from these tables, we get Figure 8.18.

Figure 8.18 is actually a semi-reduced form of Figure 8.17 produced by eliminating the causal paths between the last two endogenous variables, ESTEEM and ANOMIA. As such, the structural coefficients on the arrows leading to ESTEEM and ANOMIA are equal to a direct effect plus an indirect effect that passes through the arrows between ANOMIA and ESTEEM that are included in the full model (Figure 8.17). For example, $-.213$ on the path from SHORTINC to ANOMIA equals the direct effect of SHORTINC on ANOMIA (D_{53}) plus the

FIGURE 8.17 A Nonrecursive Model for Self-Esteem and Anomia

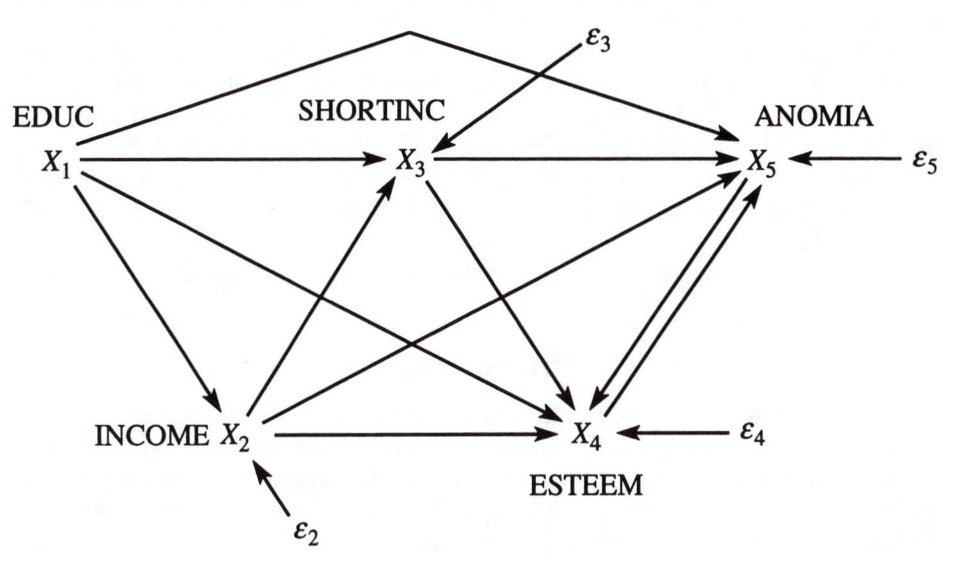

FIGURE 8.18 Semi-Reduced Model for Anomia and Self-Esteem

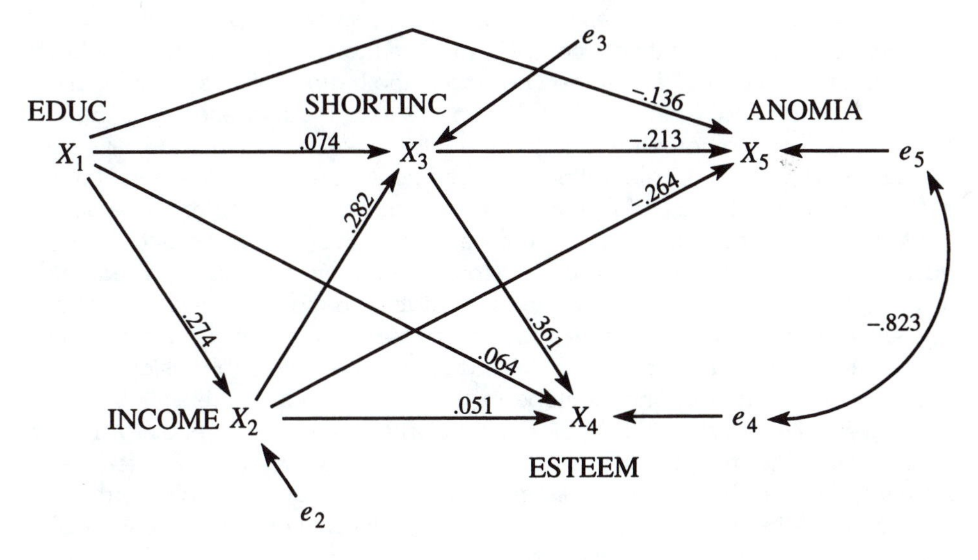

indirect effect that runs from SHORTINC to ESTEEM to ANOMIA (I_{543}).[3] Thus, from this semi-reduced model we can compute valid estimates of the total effects of education, income, and subjective income on anomia and self-esteem. Because we have chosen not to try to estimate the effects of anomia and self-esteem on one another, however, we cannot estimate the direct effects of EDUC, INCOME, and SHORTINC on ANOMIA and ESTEEM. Finally, the path between the error terms for ANOMIA and ESTEEM contains the covariance between these two errors. This covariance is produced by the causal effects of ANOMIA and ESTEEM on one another. The covariance may also be due to common causes outside the system that are independent of the education, income, and the subjective income variables.

Summary

In this chapter we have used multi-equation causal models to compute direct, indirect, total, and spurious (DITS) effects of one variable on another. The rules for constructing path diagrams to represent causal systems were described. If the model is recursive (i.e., no feedback loops), the structural coefficients or path coefficients for each path can be estimated by running an OLS regression equation for each endogenous variable. These coefficients represent the direct effects of one variable on another. Indirect effects between two variables (i.e., those that are mediated by at least one intervening variable) can then be calculated by multiplying the coefficients along the compound path that connects the two variables. In relatively complex models there may be several indirect paths between a pair of variables. The sum of the indirect effects for all of these paths gives the total indirect effect. When the total indirect effect is added to the direct effect, we get what is called the total effect.

A second method of calculating indirect and total effects (the Alwin-Hauser method) involves the use of reduced-form equations and semi-reduced equations. Reduced-form equations are created by omitting all of the endogenous variables that are included among the independent variables (i.e., a reduced-form equation contains only exogenous variables as independent variables). The slopes from a reduced-form equation represent estimates of the total effects of the exogenous variables. The difference between a variable's slope in the reduced-form equation and its slope in the full equation represents the total indirect effect of the variable. Adding the first intervening endogenous variable to the reduced-form equation produces a semi-reduced equation. The reduced-form slope minus the semi-reduced slope of an exogenous variable represents the sum of all indirect effects that pass through the first endogenous variable, and the slope for the first endogenous variable in the semi-reduced equation represents its total effect. The logic of this hierarchical approach can be used

3. Special algebraic rules that we have not covered are needed to compute indirect effects that pass through variables that are in a loop.

to calculate additional indirect and total effects by adding successive endogenous variables to the semi-reduced equation.

When feedback loops exist (nonrecursive models), ordinary least-squares cannot, in general, be used to estimate the structural or path coefficients, although some paths in a nonrecursive model may be recursive and are thus estimable with OLS regression. The reason that OLS regression gives biased estimates of nonrecursive paths is that if an independent variable is involved in a feedback loop that also involves the dependent variable, the independent variable will be correlated with the error term for the dependent variable, which is a violation of one of the regression assumptions.

This chapter considered the calculation of indirect effects for systems of equations involving only linear and additive effects. Methods for computing indirect effects in nonlinear and nonadditive models are described and illustrated in Stolzenberg (1979). Tests of statistical significance for indirect effects also were not covered here. Such tests are not readily available from regression programs. However, methods for estimating the standard errors of indirect effects are given by Sobel (1982).

References

Alwin, Duane F., and Robert M. Hauser. 1975. "The Decomposition of Effects in Path Analysis." *American Sociological Review* 40:37–47.

Duncan, Otis Dudley. 1975. *Introduction to Structural Equation Models.* New York: Academic Press.

Heise, David. 1975. *Causal Analysis.* New York: Wiley.

Jöreskog, Karl G., and Dag Sörbom. 1989. *LISREL 7 User's Reference Guide.* Mooresville, Indiana: Scientific Software, Inc.

Rosenberg, Morris. 1965. *Society and the Adolescent Self-Image.* Princeton, New Jersey: Princeton University Press.

Sobel, Michael. 1982. "Asymptotic Confidence Intervals for Indirect Effects in Structural Equation Models." Pp. 290–312 in *Sociological Methodology 1982*, edited by S. Leinhardt. San Francisco: Jossey-Bass.

Stolzenberg, Ross M. 1979. "The Measurement and Decomposition of Causal Effects in Nonlinear and Nonadditive Models." Pp. 459–488 in *Sociological Methodology 1980*, edited by Karl Schuessler. San Francisco: Jossey-Bass.

Wonnacott, Ronald J., and Thomas H. Wonnacott. 1979. *Econometrics.* New York: Wiley.

Wright, Sewell. 1921. "Correlation and Causation." *Journal of Agricultural Research* 20:557–585.

9 Causal Analysis II

We have seen how multiple regression can be used to estimate causal parameters. These parameters, however, are not directly observable or measurable. Instead, by using a causal model and the observed variances and covariances for a sample of cases, we are able to make inferences about the values of the causal parameters of the system. That is, we use statistical summaries of observable phenomena to make inferences about unobservable phenomena. For example, in the simplest causal model, we divide the covariance between X and Y by the variance of X to get the least-squares estimate of the structural coefficient.

In this chapter, we will look at this observable–unobservable dichotomy from the opposite direction. That is, we will see how the causal parameters of a system, along with the variances and covariances of exogenous variables (which are taken as givens), create all of the additional observed variances and covariances in a system. Once we have estimated the parameters of the model, we can use these estimated parameters to partition the observed variances and covariances into components that result from the various causal processes included in the model. To begin, we will learn rules for reading covariance equations directly from a path diagram. These equations express the covariance between two variables in terms of components that are due to such processes as direct causes, indirect causes, correlated causes, and common causes. We will then learn similar rules for reading variance equations from a path diagram, equations that also involve direct, indirect, and correlated effects. These equations will show, however, that generally it is not possible to allocate unambiguously the variance of an endogenous variable among the various variables that are causes of it. Furthermore, in general it is also not possible to allocate explained variance between direct effects and indirect effects. However, reduced-form equations that ascribe all of the variances and covariances to exogenous variables may help to clarify these issues.

The Simplest Causal Model

Let us start with the estimated parameters of a simple bivariate model,

$$X \xrightarrow{b} Y \leftarrow e$$

The slope b of this linear causal system represents the change in Y caused by a unit increase in X. Since changes in X are not restricted to unity but instead can take on any value ΔX, the change in Y caused by ΔX will be $\Delta Y = b\Delta X$. The origin of a change in X equal to ΔX can occur at many different points in the range of values of X. Let us now think of measuring changes in X and Y as deviations from their respective means, that is, $\Delta X = X - \overline{X}$ and $\Delta Y = Y - \overline{Y}$. When we think of the mean as the origin of changes in X and Y, the effect of X on Y is the amount that Y will be caused to change or deviate from its mean when X changes or deviates from its mean by $\Delta X = X - \overline{X}$. This can be seen when we write the linear equation for the model in terms of deviation scores, in which case the intercept will equal zero:

$$Y - \overline{Y} = b(X - \overline{X}) + e \tag{9.1}$$

If X changes or deviates from its mean by $X - \overline{X}$, it will cause a change or deviation in Y of $\hat{Y} - \overline{Y} = b(X - \overline{X})$. Since deviation scores are used to define the variance of variables, we can use Equation 9.1 to determine how much variance in Y is caused by X. The sample variance of Y equals

$$s_Y^2 = \frac{\sum (Y - \overline{Y})^2}{n}$$

Substituting Equation 9.1 for $Y - \overline{Y}$, we get

$$s_Y^2 = \frac{\sum [b(X - \overline{X}) + e]^2}{n}$$

Squaring the expression in the numerator to the right of the summation operator gives

$$s_Y^2 = \frac{\sum [b^2(X - \overline{X})^2 + 2b(X - \overline{X})e + e^2]}{n}$$

$$= \frac{\sum b^2(X - \overline{X})^2 + \sum 2b(X - \overline{X})e + \sum e^2}{n}$$

$$= \frac{b^2\sum (X - \overline{X})^2}{n} + \frac{2b\sum (X - \overline{X})e}{n} + \frac{\sum e^2}{n}$$

Dividing the sums of squares and sums of products by n gives the following variances and covariances:

$$s_Y^2 = b^2 s_X^2 + 2bs_{Xe} + s_e^2$$

Since the covariance between X and e, s_{Xe}, is equal to zero for the sample regression results and is assumed to be zero in the population model, the formula for the variance of Y reduces to

$$s_Y^2 = b^2 s_X^2 + s_e^2 \tag{9.2}$$

Equation 9.2 shows that the variance of Y has been decomposed into two components, the variance caused by the observed variable X and the variance caused by all the unobserved variables e (which may also include measurement error). The first component in Equation 9.2 is the "explained" variance, and the second component is the "unexplained" variance. Although we already knew that the variance could be divided into explained and unexplained components, this new expression for the explained variance is very informative. It shows that the variance in Y caused by X, $b^2 s_X^2$, equals the squared estimate of the structural coefficient times the variance of X. Thus, the more X varies, the greater will be the variance in Y caused by X. Also, the greater the absolute value of the structural coefficient, the greater will be the variance in Y caused by X. The derivation of Equation 9.2 shows that the structural coefficient is squared because the variance in Y consists of squared changes or deviations around its mean.

The variances of X and e are variances of exogenous variables and thus are created by causes outside of this system. If any outside sources caused the variances of either X or e either to increase or decrease, this would cause an increase or a decrease in the variance of the endogenous variable Y, even if the structural coefficient b were to remain the same. This fact has important implications for the coefficient of determination, which is the explained variance in Y divided by the total variance in Y, or

$$r_{XY}^2 = \frac{b^2 s_X^2}{b^2 s_X^2 + s_e^2}$$

If the variance of X increased, the numerator would increase proportionately more than the denominator; thus, the proportion of variance in Y that is explained by X would increase *even though the effect of X (i.e., b) did not change.* If our sample of observations were taken over a restricted range of X (i.e., we did not have a representative sample of X), the variance of X would decrease and the coefficient of determination would also decrease; if the effect of X were truly linear throughout its range, however, b would still be the same in the restricted range of observations. These two possibilities show how the measure of the strength of association is dependent on a factor, the variance of X, that is independent of the causal effect of X on Y. In an analogous situation, outside sources might cause the variance of e to increase without any changes occurring in the structural coefficient or the variance of X. This would increase the denominator of the coefficient of determination (it would increase the variance of Y) and thus reduce the value of the strength of association, even though the actual effect of X and its variance have not changed. Thus, the *proportion* of variance explained by X (r^2) might decrease even though the absolute amount of variance that is explained or caused by X does not change. In sum, Equation 9.2 makes it clear how changes in the variances of exogenous variables (X and e) may create changes in r^2 even though the effect of X on Y remains the same.

It is also possible to derive a formula that shows how changes in X (i.e., $\Delta X = X - \overline{X}$) create covariance between X and Y. The covariance is

$$s_{XY} = \frac{\sum (X - \overline{X})(Y - \overline{Y})}{n}$$

Substituting Equation 9.1 for $Y - \overline{Y}$,

$$s_{XY} = \frac{\sum (X - \overline{X})[b(X - \overline{X}) + e]}{n}$$

$$= \frac{\sum [b(X - \overline{X})(X - \overline{X}) + e(X - \overline{X})]}{n}$$

$$= \frac{b\sum (X - \overline{X})^2 + \sum e(X - \overline{X})}{n}$$

$$= bs_X^2 + s_{eX}$$

Since the covariance between the error term and X equals zero,

$$s_{XY} = bs_X^2 \qquad \textbf{(9.3)}$$

Thus, the covariance equals the effect of X times the variance of X. Since $Y - \overline{Y}$ is not squared in the formula for a covariance, b is not squared in Equation 9.3. The greater the effect of X and the greater the variance of X, the greater will be the covariance between X and Y. In this case, all of the covariance between X and Y is created by X; there is no spurious component to the covariance. Notice that Equation 9.3 is just a rearrangement of the terms in the formula for the bivariate regression slope $b = s_{XY}/s_X^2$.

A Two-Equation Model

Covariances

We have derived formulas that show how the variance of a dependent variable and the covariance between the dependent variable and an independent variable are created in a two-variable causal model. When more than one independent variable is included in the equation or when we have a multi-equation model, the algebra for deriving such formulas becomes rather complex and tedious. The procedure is the same as for a two-variable model, however. We first write an equation for each endogenous variable in the model in terms of deviation scores, such as Equation 9.1. We then follow the same procedures used above to derive an equation for the variance of each endogenous variable and an equation for each of the covariances that involve one or more endogenous variables. We will not go through these algebraic derivations but instead will use some relatively simple rules for "reading" these equations directly from the path diagram for the causal system.

Figure 9.1 presents the same two-equation model previously shown in Figure 8.3. The roman letter e for the error terms indicates that we will be examining formulas in terms of the sample statistics rather than the population parameters.

The regression equations for each endogenous variable are

$$\hat{X}_3 = a_{3 \cdot 12} + b_{31}X_1 + b_{32}X_2$$

$$\hat{X}_4 = a_{4 \cdot 123} + b_{41}X_1 + b_{42}X_2 + b_{43}X_3$$

The regression coefficients, which are the estimates of the structural coefficients, could be placed on the appropriate paths to make Figure 9.1 complete.

Chain Rule for Covariances. Sewell Wright (1921) formulated a multiplication rule for reading each correlation from a diagram containing path (standardized) coefficients. The same rule may be used, with a slight modification, to read covariances from a diagram containing structural coefficients or unstandardized regression coefficients. The rule for determining the covariance s_{ij} involves finding each distinct chain that links X_i and X_j. Each chain has X_i at one end and X_j at the other end. To find each chain, read back from X_j to X_i, where X_j appears "later" in the model, along each path or compound path that connects the two variables. A chain may reverse directions from backward to forward, if necessary, but only one reversal is permitted. It may also pass through a covariance between two exogenous variables, but only one covariance is allowed in each chain. Each path in a chain may be traversed only once. Each chain has an origin or source, which is the "earliest" link in a chain. If a chain has a reversal of direction, the variable at which the reversal occurs is the origin. If there is a covariance between two exogenous variables in the chain, that covariance is the origin (it is the point at which the reversal occurs).

FIGURE 9.1 **A Two-Equation Model for the Sample Observations**

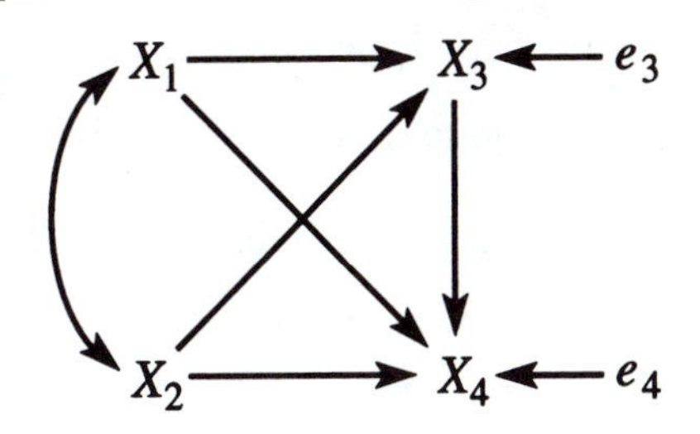

If there is no reversal, X_i is the origin (it is the earliest point in the chain). To find out how much each chain contributes to the covariance, form the product of all the coefficients along the chain, including either the variance of the origin variable or the covariance at the origin. Finally, the covariance between X_i and X_j equals the sum of the products obtained for all of the distinct chains linking X_i and X_j.

Now, let us apply the above rule to compute the covariances for the variables in the first equation specified by Figure 9.1, the equation for X_3. The chains and the products for each chain that contribute to the covariance between X_3 and X_1 are shown below.

Cov (X_1, X_3)

$X_1 \longrightarrow X_3$

Direct Effect: $b_{31}s_1^2$

X_1 X_3

X_2

Correlated Cause: $b_{32}s_{12}$

The first chain is formed by the direct effect of X_1 on X_3. X_1 is the origin of this chain, and its contribution equals the product of the variance of X_1 and the estimated structural coefficient. This contribution is analogous to that derived in Equation 9.3. The second chain is formed by the covariance between X_1 and X_2 and the direct effect of X_2 on X_3. The covariance is the origin. The contribution of the second chain is spurious because it is produced by a correlated cause rather than a direct or indirect effect of X_1 on X_3. We are assuming that there is no causal relationship between the two exogenous variables; they are merely correlated, as the diagram specifies.

The covariance between X_2 and X_3 is analogous to that shown above for X_1 and X_3. The covariance s_{23} is due to a nonspurious direct effect and a spurious correlated cause.

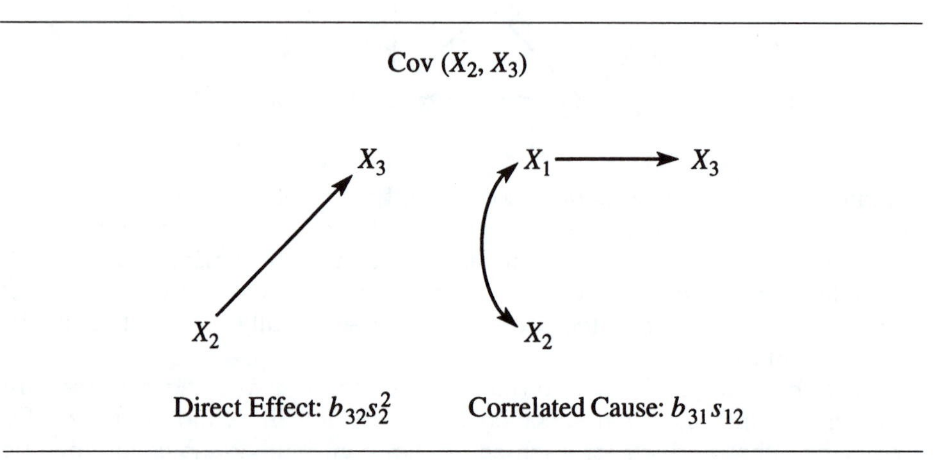

For both of the above covariances, notice that if the spurious component has the same sign as the causal component (e.g., the direct effects are both positive, and the covariance between the exogenous variables is also positive), we will have redundancy, and the covariance will be inflated. If the spurious component is of opposite sign, however, we will have suppression, and the covariance will be deflated or of opposite sign to that component created by the causal relationship.

Next, we will use the chain rule to determine the covariances between X_4 and the variables in its equation. The components of the covariance between X_4 and X_1 are more complex because there are three variables that are antecedent to X_4. There are now two causal sources of covariance and two spurious sources.

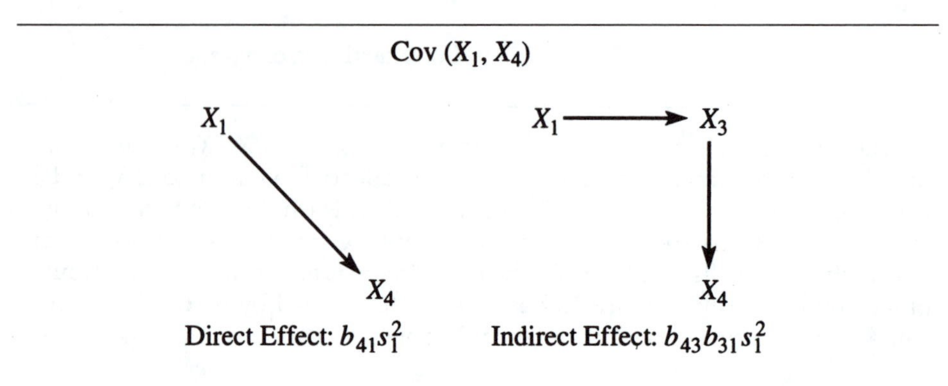

Cov (X_1, X_4) (continued)

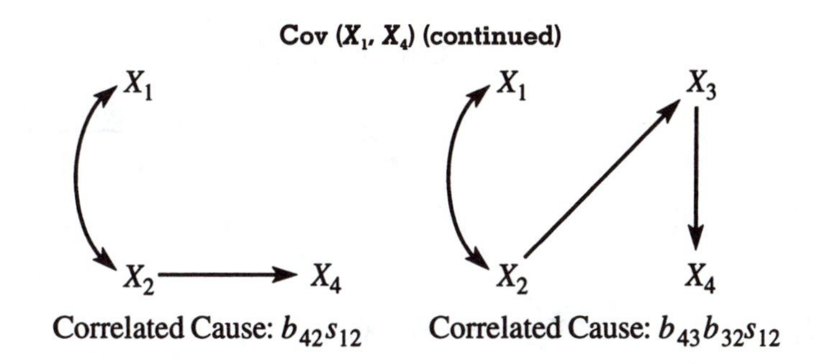

The chains for the covariance for X_2 and X_4 are shown below. The sources of the covariance between X_2 and X_4 are completely analogous to those for X_1 and X_4: two causal components and two spurious components.

Cov (X_2, X_4)

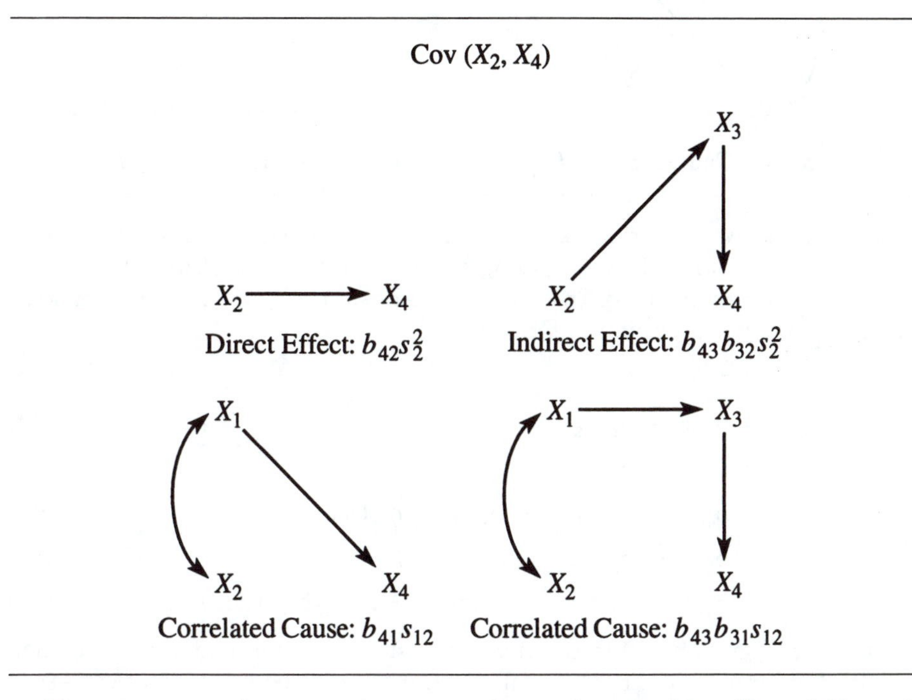

The covariance between the two endogenous variables, X_3 and X_4, is obtained from the five chains shown below. Since X_3 is dependent upon X_1 and X_2, the possibilities for spurious covariance between X_3 and X_4 are enhanced. There is only one nonspurious source, the direct effect of X_3 on X_4. Depending upon the signs of the various structural coefficients and the covariance between the exogenous variables, the covariance might be either enhanced or reduced due to the various sources of spuriousness.

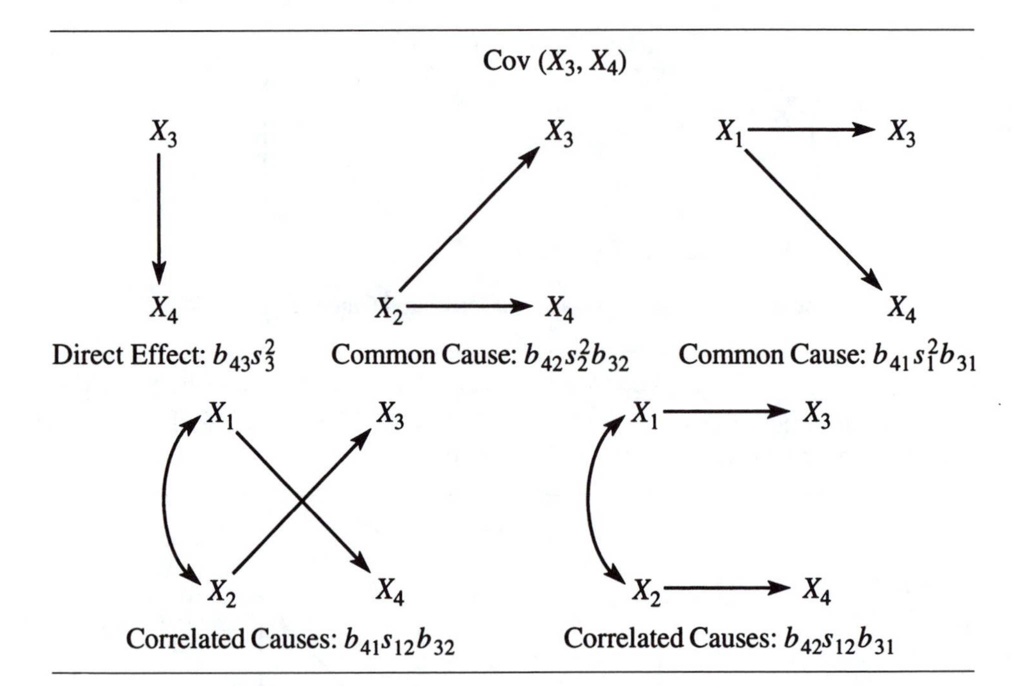

Each of the five covariances whose components have been diagrammed above may be obtained by summing the products for all of the chains that contribute to the covariance. The covariance between X_1 and X_2 does not have to be computed; it is a given in Figure 9.1.

$$s_{12} = s_{12}$$

$$s_{13} = b_{31}s_1^2 + b_{32}s_{12} \tag{9.4}$$

$$s_{23} = b_{32}s_2^2 + b_{31}s_{12} \tag{9.5}$$

$$s_{14} = b_{41}s_1^2 + b_{43}b_{31}s_1^2 + b_{42}s_{12} + b_{43}b_{32}s_{12} \tag{9.6}$$

$$s_{24} = b_{42}s_2^2 + b_{43}b_{32}s_2^2 + b_{41}s_{12} + b_{43}b_{31}s_{12} \tag{9.7}$$

$$s_{34} = b_{43}s_3^2 + b_{42}s_2^2b_{32} + b_{41}s_1^2b_{31} + b_{42}s_{12}b_{31} + b_{41}s_{12}b_{32} \tag{9.8}$$

The terms in Equations 9.4 through 9.8 consist of the estimated structural coefficients of the model (the b_{ji}'s) and the variances and covariance of the exogenous variables, with one exception. The single exception is in the last equation, which contains the variance of an endogenous variable, X_3. Since X_3 is dependent upon the exogenous variables, however, its variance is created by the exogenous variables. Once we have learned how to write an equation for the variance of X_3, we can enter it into the last equation above to obtain an expression containing only the variances and covariances of the exogenous variables and the structural coefficients.

TABLE 9.1 Variance-Covariance Matrix

	X_1 *EDUC*	X_2 *INCOME*	X_3 *SHORTINC*	X_4 *ESTEEM*
EDUC	7.4475	2.0411	1.1248	.9897
INCOME	2.0411	4.2322	1.3428	.8312
SHORTINC	1.1248	1.3428	1.7767	.7823
ESTEEM	.9897	.8312	.7823	4.8185

TABLE 9.2 Decomposition of Covariances

s_{13}:	$b_{31}s_1^2 = (.0738)(7.4475) =$	.5496
	$b_{32}s_{12} = (.2817)(2.0411) =$	.5750
	$s_{13} =$	1.1246
s_{23}:	$b_{32}s_2^2 = (.2817)(4.2322) =$	1.1922
	$b_{31}s_{12} = (.0738)(2.0411) =$	.1506
	$s_{23} =$	1.3428
s_{14}:	$b_{41}s_1^2 = (.0644)(7.4475) =$	.4796
	$b_{43}b_{31}s_1^2 = (.3612)(.0738)(7.4475) =$	.1985
	$b_{42}s_{12} = (.0507)(2.0411) =$	.1035
	$b_{43}b_{32}s_{12} = (.3612)(.2817)(2.0411) =$	.2077
	$s_{14} =$	.9893
s_{24}:	$b_{42}s_2^2 = (.0507)(4.2322) =$	.2146
	$b_{43}b_{32}s_2^2 = (.3612)(.2817)(4.2322) =$	.4306
	$b_{41}s_{12} = (.0644)(2.0411) =$	.1314
	$b_{43}b_{31}s_{12} = (.3612)(.0738)(2.0411) =$	.0544
	$s_{24} =$	.8310
s_{34}:	$b_{43}s_3^2 = (.3612)(1.7767) =$	.6417
	$b_{42}s_2^2b_{32} = (.0507)(4.2322)(.2817) =$	.0604
	$b_{41}s_1^2b_{31} = (.0644)(7.4475)(.0738) =$	.0354
	$b_{42}s_{12}b_{31} = (.0507)(2.0411)(.0738) =$	.0076
	$b_{41}s_{12}b_{32} = (.0644)(2.0411)(.2817) =$	.0370
	$s_{34} =$	.7821

Covariances for Self-Esteem Model. The self-esteem model (Figure 8.10) and data described in Chapter 8 will be used to illustrate the computation of the components of covariance. The covariances and variances for the four variables in the self-esteem model are shown in Table 9.1.

Equations 9.4 through 9.8 are used to compute the components of the five covariances shown in Table 9.2. Whereas about half of the covariance between

X_1 (EDUC) and X_3 (SHORTINC) is spuriously created by a correlated cause ($b_{32}s_{12}$ = .5750), very little of the covariance between X_2 (INCOME) and X_3 is spurious. With respect to X_1 and X_4 (ESTEEM), however, the covariance that is due to the direct and indirect effects of education (.4796 + .1985 = .6781) is about twice as much as the spurious component created by the correlation of education with income (.1035 + .2077 = .3112). The covariance between income and self-esteem has an even greater nonspurious component (.2146 + .4306 = .6452), relative to its spurious component (.1314 + .0544 = .1850). Finally, even though there are four spurious components to the covariance between X_3 (SHORTINC) and X_4, in comparison to only one causal component ($b_{43}s_3^2$), the great majority of the covariance is nonspurious (.6417/.7821 = .82).

Correlations. Before turning to the equations for the variances of the endogenous variables, we should note what the covariance equations would look like for the path coefficients of a standardized model. The variances of all variables in a standardized model equal unity; thus, the variances in the above equations can all be omitted. Also, the covariance between two standardized variables equals their correlation; thus, the covariances can all be replaced with correlations. Using B_{ji} for a standardized regression coefficient, the equation for the covariance between X_3 and X_4, for example, becomes

$$r_{43} = B_{43} + B_{42}B_{32} + B_{41}B_{31} + B_{42}r_{12}B_{31} + B_{41}r_{12}B_{32}$$

Variances

Chain Rule for Variances. The rules for reading the variance of an endogenous variable from a causal diagram are analogous to those for covariances. The variance of a variable can be thought of as the covariance of a variable with itself (see Chapter 2). Therefore, the chains that define contributions to the variance of X_j have X_j at both ends of the chain; the chain starts and ends with X_j. To find each distinct chain, trace backward from X_j to an origin, as defined by the covariance rules, and then trace forward along another path or compound path to get back to X_j. Thus, the chain may be thought of as a loop (but not a causal loop, as in a nonrecursive model) that returns to X_j. There is one difference, however, between a variance chain and a covariance chain. A variance chain passes over the same path twice when the origin is a variable that directly affects X_j; we trace back to a variable that directly affects X_j and then move back to X_j over the same path. The value of a chain's contribution to the variance of X_j is obtained by taking the product of all the coefficients along the chain and the variance, or covariance, at the origin. For chains that do not consist of a direct path between X_j and an antecedent variable, the product must also be multiplied by two, a rule that did not apply to covariance chains.

The rules will be illustrated by reading the variances of X_3 and X_4 from Figure 9.1. The variance for X_3 is shown below.

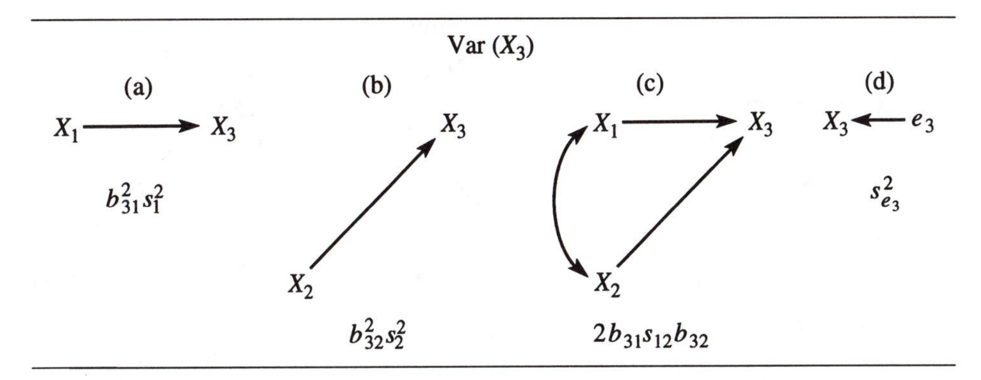

The contributions due to the direct effects of the exogenous variables, (a) and (b), are identical in form to Equation 9.2. Since the variance chain rule indicates that we must trace over the direct path twice, the structural coefficient is squared in (a) and (b). Chain (c) is a new variance component that arises when there are two or more independent variables. Since this component contains the effects of both exogenous variables and their covariance, (c) is a portion of the variance that cannot be allocated to either variable. Since (a) and (b) must be positive because of squaring, the sign of (c) determines whether the covariance between the exogenous variables increases or decreases the variance in X_3. If the direct effects are both positive and there is a positive covariance, for example, the variance in X_3 will be greater than the sum of the direct contributions. If both direct effects are positive but the covariance is negative, the variance will be less than the sum of the direct contributions. If the two exogenous variables are not correlated, however, the variance will equal the sum of the two direct contributions. Thus, correlated input variables may either increase or decrease the diversity/variance of outcome variables relative to inputs that are not correlated. Finally, the contribution of the error term equals its variance since its structural coefficient equals unity by definition.

The variance chains and components for X_4 are as follows:

Var (X_4)

(a) X_1 → X_4 $\quad b_{41}^2 s_1^2$

(b) X_2 → X_4 $\quad b_{42}^2 s_2^2$

(c) X_3 → X_4 $\quad b_{43}^2 s_3^2$

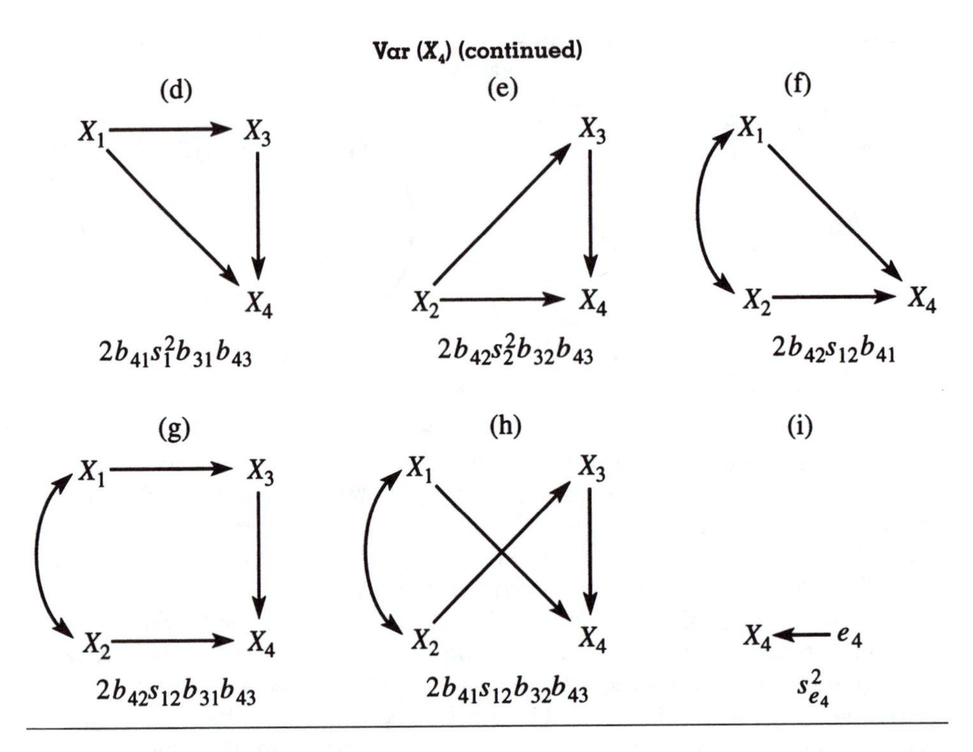

Chains (a), (b), and (c) define direct contributions analogous to those given before. Chains (d) and (e) each consist of the products of direct and indirect effects of the exogenous variable, times its variance. For example, (d) includes the product of the direct effect of X_1 (b_{41}), the indirect effect of X_1 ($b_{31}b_{43}$), and the variance of X_1. The contribution due to chain (f) is identical in form to chain (c) for the variance of X_3; it involves the direct effects of the exogenous variables and the covariance between them. Chains (g) and (h) also include this covariance; each of these cases, however, includes the product of the direct effect of one exogenous variable and the indirect effect of the other.

Summing the values of each distinct variance chain for X_3 and each distinct chain for X_4 gives the following variance equations:

$$s_3^2 = b_{31}^2 s_1^2 + b_{32}^2 s_2^2 + 2b_{31} s_{12} b_{32} + s_{e_3}^2 \tag{9.9}$$

$$s_4^2 = b_{41}^2 s_1^2 + b_{42}^2 s_2^2 + b_{43}^2 s_3^2 + 2b_{41} s_1^2 b_{31} b_{43} + 2b_{42} s_2^2 b_{32} b_{43}$$
$$+ 2b_{41} s_{12} b_{42} + 2b_{43} b_{31} s_{12} b_{42} + 2b_{43} b_{32} s_{12} b_{41} + s_{e_4}^2 \tag{9.10}$$

Equation 9.10 is rather lengthy. It demonstrates that the variance of a dependent variable can be a rather complex function of the direct effects, indirect effects, variances, and covariances of the independent variables, even in a relatively simple model containing only three independent variables. Some of these terms, however, might be zero or nearly zero when empirically estimated; thus, the equation might turn out to be less complicated than it appears.

It is tempting to try to use Equations 9.9 and 9.10 to divide the explained variance into components that can be allocated to each independent variable and to divide the variance that each variable explains into various direct and indirect components. Notice that each term in each equation contains either a variance of an independent variable X_i or a covariance between two exogenous variables. We might sum the terms containing s_i^2 and allocate this portion of the explained variance to X_i. The desire to allocate the explained variance among the independent variables runs into problems, however, because of the terms that contain covariances. These terms represent amounts by which variance is *amplified* or *dampened* due to the correlated causes. Therefore, these portions cannot be allocated to a specific variable. If the covariance is small, or if one of the variables has a small effect, then the value of the term containing the covariance might be quite small and of no importance. If most of the terms containing covariances are small, then most of the variance may be allocated among the various independent variables. If there is a great deal of multicollinearity, however, then large portions of the explained variance will go unallocated. Thus, the degree to which Equations 9.9 and 9.10 can be used to allocate variance among the independent variables will have to be determined on a case-by-case basis.

There is also a problem in dividing explained variance into direct and indirect components: some of the terms containing s_i^2 also contain products of the direct and indirect effects of X_i, such as $2b_{41}s_1^2b_{31}b_{43}$. The explained variance represented by this term cannot be allocated to either the direct effect or the indirect effect. Notice that this term may be either positive or negative; if the direct effect (b_{41}) and the indirect effect ($b_{31}b_{43}$) have the same signs, the term will be positive, and if they have opposite signs, it will be negative. If the term is positive, the variance explained by direct and indirect effects combined will be greater than would be expected by summing the variances that each would explain in the absence of the other. If they have opposite signs, the combined explained variance will be less than the sum of the variances that each would explain in the absence of the other. Thus, the terms that contain the products of direct and indirect effects prevent us from dividing the variance explained by an independent variable into direct and indirect components.

Variances for Self-Esteem Model. Equations 9.9 and 9.10 are used to decompose the variances of X_3 (SHORTINC) and X_4 (ESTEEM) in Table 9.3. Over 70 percent of the explained variance in X_4 is allocated to the direct effect of INCOME (.3358/.4613 = .728). The positive covariance between the two exogenous variables X_1 and X_2, however, does increase the variance in SHORTINC some-

TABLE 9.3 Decomposition of Variances

s_3^2:	$b_{31}^2 s_1^2 = (.0738)^2(7.4475) =$	.0406
	$b_{32}^2 s_2^2 = (.2817)^2(4.2322) =$	.3358
	$2b_{31}s_{12}b_{32} = 2(.0738)(2.0411)(.2817) =$	.0849
	$s_{e_3}^2 = (1 - R_3^2)s_3^2 = (1 - .2596)(1.7767) =$	1.3155
	$s_3^2 =$	1.7768
s_4^2:	$b_{41}^2 s_1^2 = (.0644)^2(7.4475) =$	.0309
	$b_{42}^2 s_2^2 = (.0507)^2(4.2322) =$	.0109
	$b_{43}^2 s_3^2 = (.3612)^2(1.7767) =$	.2318
	$2b_{41}s_1^2 b_{31}b_{43} = 2(.0644)(7.4475)(.0738)(.3612) =$	.0256
	$2b_{42}s_2^2 b_{32}b_{43} = 2(.0507)(4.2322)(.2817)(.3612) =$	.0437
	$2b_{41}s_{12}b_{42} = 2(.0544)(2.0411)(.0507) =$	.0133
	$2b_{43}b_{31}s_{12}b_{42} = 2(.3612)(.0738)(2.0411)(.0644) =$	.0055
	$2b_{43}b_{32}s_{12}b_{41} = 2(.3612)(.2817)(2.0411)(.0644) =$	.0267
	$s_{e_4}^2 = (1 - R_4^2)s_4^2 = (1 - .0806)(4.8185) =$	4.4301
	$s_4^2 =$	4.8185

what (.0849). With respect to ESTEEM, the majority of the explained variance is allocated to the direct effect of SHORTINC [.2318/(4.8185 – 4.4301) = .60]. Even though there are three components containing s_{12}, the positive covariance between the two exogenous variables does very little to increase the variance in self-esteem (.0133 + .0055 + .0267 = .0455).

Reduced-Form Variances and Covariances

It is also important to note that Equation 9.10 also contains a term that includes the variance of X_3, namely, $b_{43}^2 s_3^2$. This represents variance in X_4 that is directly caused by X_3, which is an endogenous variable. There is nothing wrong with including a term that represents the variance in one endogenous variable that is explained by another endogenous variable. However, the variance of X_3 is partly caused by the two exogenous variables. Thus, this term represents variance in X_4 that is partly caused by the indirect effects of X_1 and X_2 that pass through X_3. If we substitute Equation 9.9 (the equation for the variance of X_3) for s_3^2 in Equation 9.10, we can derive an equation that contains terms for all of the indirect effects of the exogenous variables:

$$s_4^2 = b_{41}^2 s_1^2 + b_{42}^2 s_2^2 + b_{43}^2(b_{31}^2 s_1^2 + b_{32}^2 s_2^2 + 2b_{31}s_{12}b_{32} + s_{e3}^2) + 2b_{41}s_1^2 b_{31}b_{43}$$
$$+ 2b_{42}s_2^2 b_{32}b_{43} + 2b_{41}s_{12}b_{42} + 2b_{43}b_{31}s_{12}b_{42} + 2b_{43}b_{32}s_{12}b_{41} + s_{e_4}^2$$

The expression in parentheses is Equation 9.9. If we multiply the parenthetical expression by b_{43}^2 and rearrange slightly, we get

$$
\begin{aligned}
s_4^2 = {} & b_{41}^2 s_1^2 + b_{42}^2 s_2^2 \\
& + b_{43}^2 b_{31}^2 s_1^2 + b_{43}^2 b_{32}^2 s_2^2 \\
& + 2b_{41} s_1^2 b_{31} b_{43} + 2b_{42} s_2^2 b_{32} b_{43} \\
& + 2b_{41} s_{12} b_{42} + 2b_{43} b_{31} s_{12} b_{42} + 2b_{43} b_{32} s_{12} b_{41} + 2b_{43} b_{31} s_{12} b_{32} b_{43} \\
& + b_{43}^2 s_{e_3}^2 + s_{e_4}^2
\end{aligned} \qquad \textbf{(9.11)}
$$

At first glance, Equation 9.11 appears to be quite imposing. However, let us note the structure in the equation. First, Equation 9.11 is a *reduced-form equation* because X_3 has been eliminated. As such, each term contains either a variance of an exogenous variable or a covariance between two exogenous variables, with the error terms rightfully defined as exogenous variables. Therefore, Equation 9.11 shows that the variance of an endogenous variable can be accounted for entirely by the structural coefficients of the model plus the variances and covariances of the exogenous variables. The first row contains terms for the variance caused by direct effects. The second row contains terms for the variance caused by indirect effects. The terms in the third row include the products of the direct and indirect effects for each exogenous variable. Thus, Row 3 shows that the variance cannot be divided between direct and indirect effects. The fourth row contains correlated effects (both direct and indirect) between variables. Row 4 shows that the explained variance cannot be divided up between the exogenous variables when they are correlated. Row 5 contains the variance contributed by the error terms.

Note the next-to-last term in the equation, $b_{43}^2 s_{e_3}^2$. This is the variance contributed by the error term for X_3. It is the variance in X_4 caused by variance in X_3 that is not explained by X_1 and X_2. Thus, it is the unique contribution that X_3 makes to the variance of X_4. It is *analogous to a squared semipartial correlation*, except that it is the amount of variance, rather than the proportion of variance, accounted for by X_3.

Equation 9.11 can be simplified considerably by writing it in terms of total effects (T_{ji}'s) instead of structural coefficients (b_{ji}'s), as follows:

$$
\begin{aligned}
s_4^2 &= (b_{41} + b_{31} b_{43})^2 s_1^2 + (b_{42} + b_{32} b_{43})^2 s_2^2 \\
&\quad + 2(b_{41} + b_{31} b_{43}) s_{12} (b_{42} + b_{32} b_{43}) + b_{43}^2 s_{e_3}^2 + s_{e_4}^2 \\
&= T_{41}^2 s_1^2 + T_{42}^2 s_2^2 + 2T_{41} s_{12} T_{42} + T_{43}^2 s_{e_3}^2 + s_{e_4}^2
\end{aligned} \qquad \textbf{(9.12)}
$$

The terms $(b_{41} + b_{31} b_{43})$ and $(b_{42} + b_{32} b_{43})$ in the first row of Equation 9.12 are the total effects of X_1 and X_2, respectively. Equation 9.12 expresses the variance of X_4 in terms of the variance of each exogenous variable times its squared total effect (the total effect of e_4 equals unity), with one exception. The presence of the term $2T_{41} s_{12} T_{42}$ shows that part of the variance of X_4 is due to the correlated total effects of X_1 and X_2. Thus, there is still a component of the variance of X_4 that cannot be attributed uniquely to X_1 or X_2.

Now we return to Equation 9.8 for the covariance between X_3 and X_4, an equation that also contained the variance of the endogenous variable X_3. If we substitute Equation 9.9 for s_3^2 in Equation 9.8, we get

$$s_{34} = b_{41}s_1^2b_{31} + b_{42}s_2^2b_{32} + b_{31}s_{12}b_{42} + b_{32}s_{12}b_{41} + b_{43}b_{31}^2s_1^2 + b_{43}b_{32}^2s_2^2 + 2b_{43}b_{31}s_{12}b_{32} + b_{43}s_{e_3}^2 \tag{9.13}$$

This is a reduced-form equation containing only variances and covariances of exogenous variables. The terms in the second row that contain the covariance between the exogenous variables show that we cannot allocate the covariance of X_3 and X_4 to unique contributions made by X_1 and X_2, due to the covariance between the latter two variables.

Self-Esteem Example. Equation 9.12 is used to allocate the variance of self-esteem (X_4) among the exogenous variables (Table 9.4). Education and income are now allocated more variance (.0617 + .0984 + .0568 = .2169) than is SHORTINC ($T_{43}^2s_{e_3}^2$ = .1716). This is because the variance in self-esteem caused by the variance in SHORTINC that is explained by education and income is now allocated to education and income. Only the variance in self-esteem caused by the unique variance in SHORTINC ($s_{e_3}^2$) is allocated to SHORTINC.

Equation 9.13 is used to allocate the covariance between X_3 and X_4 among the exogenous variables (Table 9.4). In Table 9.2, eighty-two percent of the covariance was attributed to the direct effect of X_3. The reduced-form equation now allocates some of that component to X_1 and X_2 (.0147 + .1213 + .0307

TABLE 9.4 Reduced-Form Components of Variance and Covariance

s_4^2:	$T_{41}^2s_1^2$ = (.0911)²(7.4975) =	.0617
	$T_{42}^2s_2^2$ = (.1525)²(4.2322) =	.0984
	$2T_{41}^2s_{12}T_{42}$ = 2(.0911)(2.0411)(.1525) =	.0568
	$T_{43}^2s_{e_3}^2$ = (.3612)²(1.3154) =	.1716
	$s_{e_4}^2 = (1 - R_4^2)s_4^2$ = (1 − .0806)(4.8185) =	4.4301
	s_4^2 =	4.8186
s_{43}:	$b_{42}^2s_2^2b_{32}$ = (.0507)(4.2322)(.2817) =	.0604
	$b_{41}s_1^2b_{31}$ = (.0644)(7.4475)(.0738) =	.0354
	$b_{42}s_{12}b_{31}$ = (.0507)(2.0411)(.0738) =	.0076
	$b_{41}s_{12}b_{32}$ = (.0644)(2.0411)(.2817) =	.0370
	$b_{43}b_{31}^2s_1^2$ = (.3612)(.0738)²(7.4475) =	.0147
	$b_{43}b_{32}^2s_2^2$ = (.3612)(.2817)²(4.2322) =	.1213
	$2b_{43}b_{31}s_{12}b_{32}$ = 2(.3612)(.0738)(2.0411)(.2817) =	.0307
	$b_{43}s_{e_3}^2$ = (.3612)(1.3154)	.4751
	s_{43} =	.7821

FIGURE 9.2 Three-Equation Model

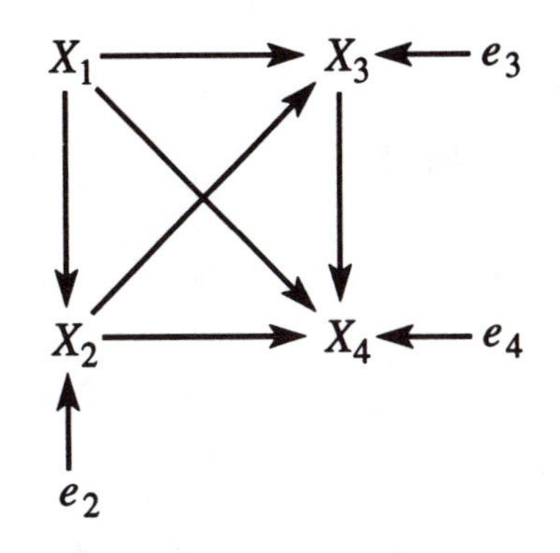

= .1667). Most of that component, however, is attributed to the unique variance of X_3 ($b_{43}s^2_{e_3}$ = .4751).

A Three-Equation Model

Covariances

We will now modify Figure 9.1 by specifying a causal path from X_1 to X_2 (Figure 9.2). We now have an equation for X_2. We will use the chain rules for reading covariances from the diagram. All of the chains that previously included a covariance between X_1 and X_2 will change; the other chains will remain the same. The covariance between X_1 and X_2 is now read as

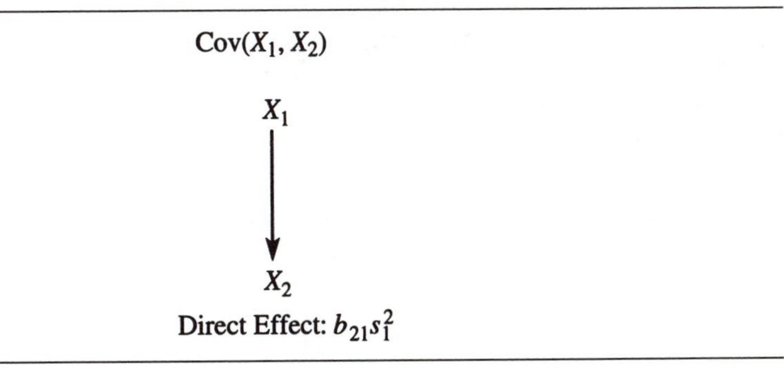

The fact that this covariance can now be expressed as a structural coefficient times the variance of the exogenous variable will have major ramifications for the remainder of the covariances.

The covariance between X_1 and X_3 is

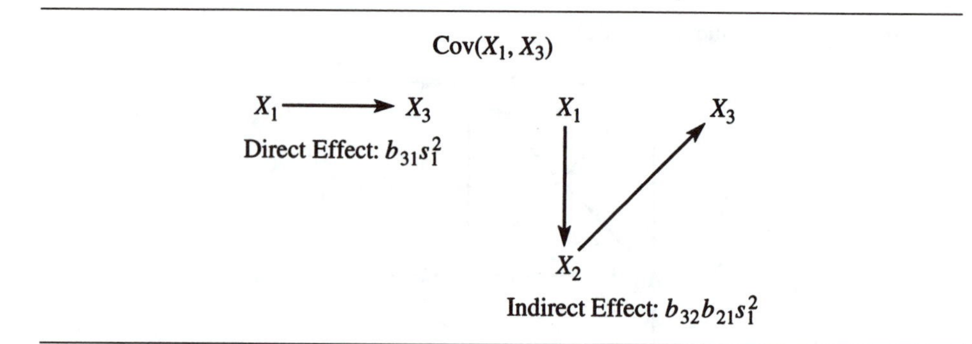

The second chain was a correlated effect in the two-equation model, but now it is an indirect-effect contribution to the covariance; it is no longer a spurious component of covariance. The change in the expression and its interpretation is due to the newly specified causal path between X_1 and X_2.

The covariance between X_2 and X_3 is given by

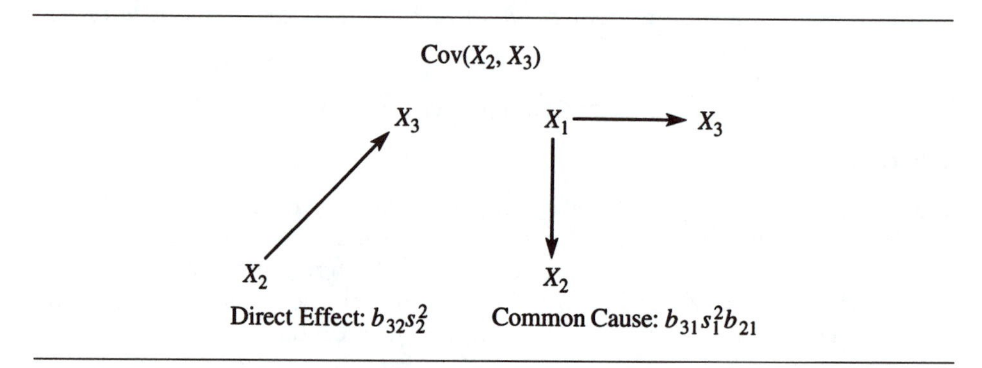

The second chain was previously due to a correlated cause but now it is due to a common cause. In this case, the change in specification does not alter the conclusion that the second component is a spurious contribution to the covariance.

The covariance between X_1 and X_4 is read as follows:

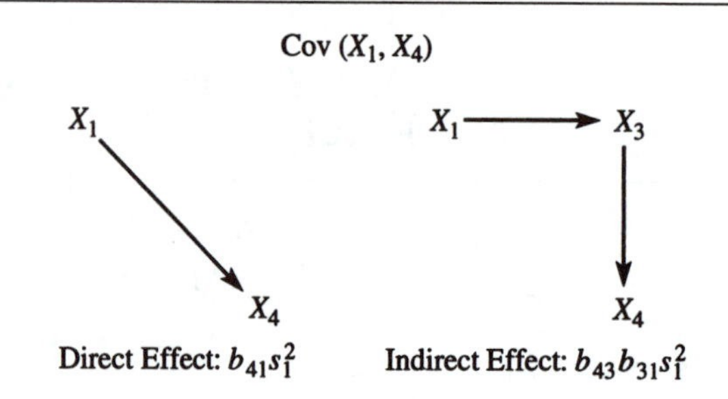

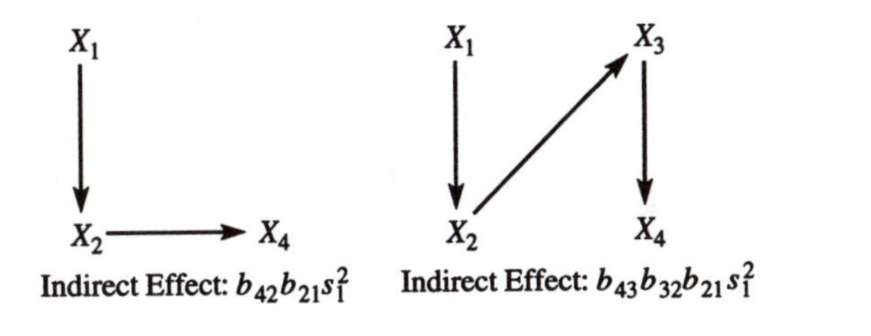

Whereas in the two-equation model the bottom two chains were spurious contributions, in the three-equation model we are using now, they are due to indirect effects and thus are nonspurious. All of the covariance between these two variables is now a valid causal relationship (nonspurious).

The covariance between X_2 and X_4 is diagrammed next. The bottom two chains were previously read as correlated causes and thus were spurious; they are now read as common causes, but they are still spurious.

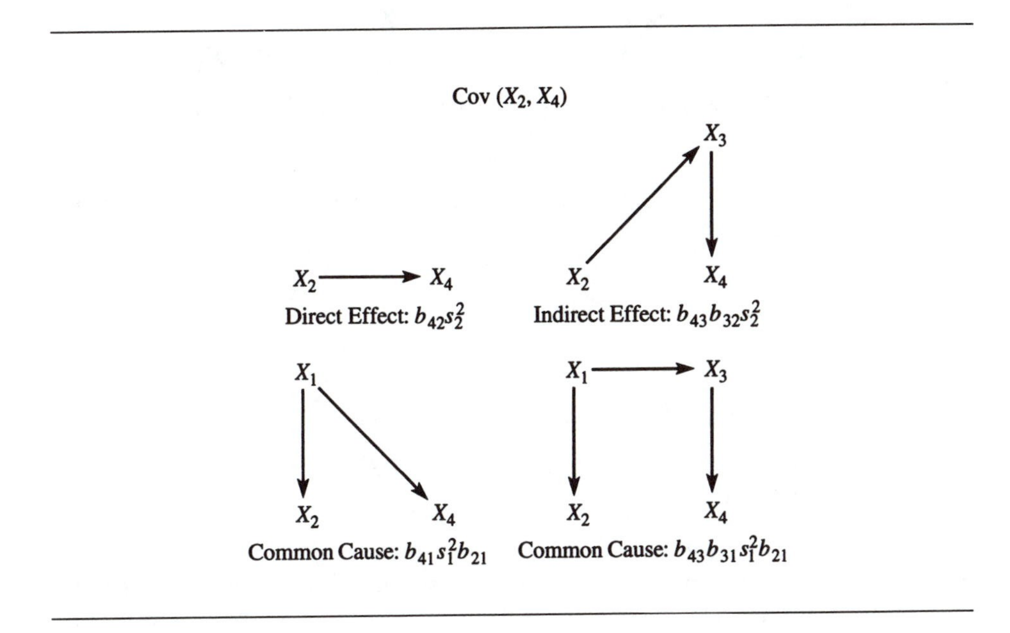

The covariance between X_3 and X_4 is shown below. The bottom two chains were due to correlated causes in the two-equation model. They are now due to a common cause, since X_1 is now the origin of the chain. They both are still spurious sources of variance. All of the sources are spurious except for the covariance created by the direct effect.

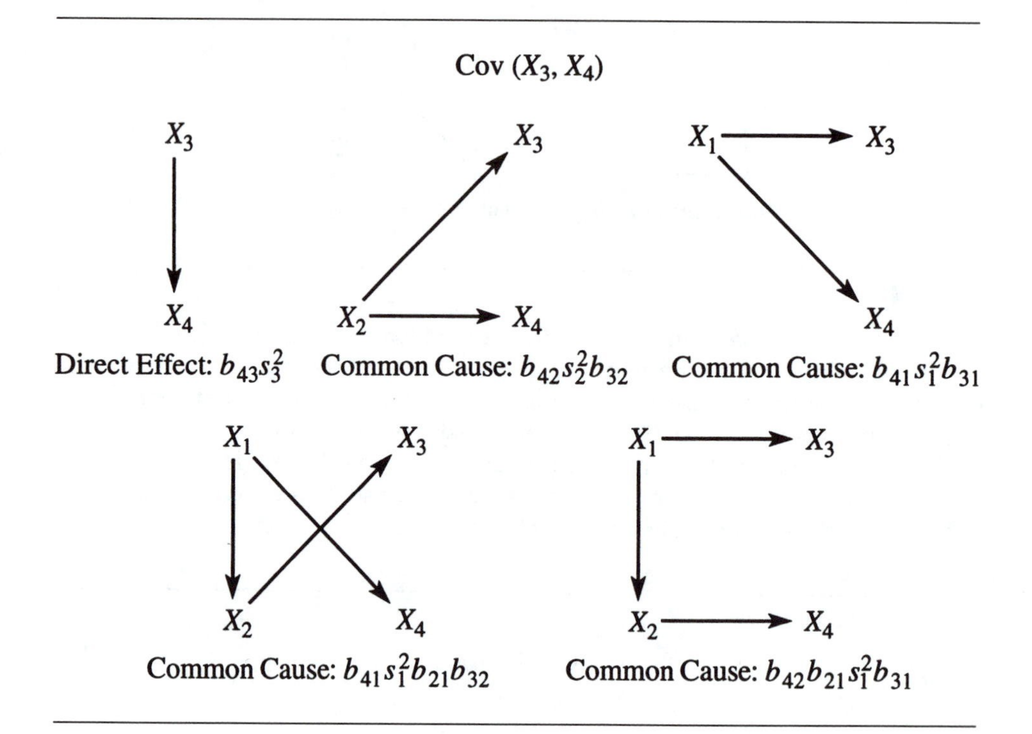

Summing the products obtained for each distinct covariance chain gives the following covariance equations:

$$s_{12} = b_{21}s_1^2$$
$$s_{13} = b_{31}s_1^2 + b_{32}b_{21}s_1^2$$
$$s_{23} = b_{32}s_2^2 + b_{31}s_1^2b_{21}$$
$$s_{14} = b_{41}s_1^2 + b_{43}b_{31}s_1^2 + b_{42}b_{21}s_1^2 + b_{43}b_{32}b_{21}s_1^2$$
$$s_{24} = b_{42}s_2^2 + b_{43}b_{32}s_2^2 + b_{41}s_1^2b_{21} + b_{43}b_{31}s_1^2b_{21}$$
$$s_{34} = b_{43}s_3^2 + b_{42}s_2^2b_{32} + b_{41}s_1^2b_{31} + b_{31}s_1^2b_{21}b_{42} + b_{32}b_{21}s_1^2b_{41}$$

Each term in the equation for each s_{ij} includes a variance. If that variance is s_i^2, then that term represents a nonspurious source of covariance. If the variance is not s_i^2, then the term represents a spurious source of covariance. There are no correlated exogenous variables in the three-equation model, since there is only one measured exogenous variable in the model. Consequently, there are no covariances in the above equations. Where s_{12} appeared in Equations 9.4 through 9.8, it has been replaced by $b_{21}s_1^2$ in the above equations. Therefore, each term in the above equations will have the same value it had in Equations

9.4 through 9.8. For example, the values of the covariance components of the three-equation self-esteem model are the same as those given in Table 9.2. Since a variance appears in each term of the above equations, each chain's contribution to the covariance s_{ij} can be assigned to a specific variable. This was not always possible with the two-equation model because that model did not specify a causal relationship between X_1 and X_2.

The decomposition of covariances is similar to the decomposition of the bivariate slope that was covered earlier. The sum of the nonspurious covariance components is analogous to the total effect of an independent variable on a dependent variable. The sum of the spurious covariance components is analogous to the difference between the bivariate slope and the total effect.

Some of the above covariance equations contain variances of endogenous variables. After reading equations for the variances of the endogenous variables from the diagram of the three-equation model (Figure 9.2), we could substitute them in the above equations to obtain equations containing only variances of the exogenous variables. These reduced-form equations would be analogous to Equation 9.13 for the covariance between X_3 and X_4. It is recommended that you derive these equations in order to demonstrate that all of the covariances can be accounted for by the structural coefficients and the variances of the exogenous variables.

Variances

We will now read the equations for the variances of the endogenous variables from Figure 9.2. The variance of X_2 was not accounted for by the two-equation model.

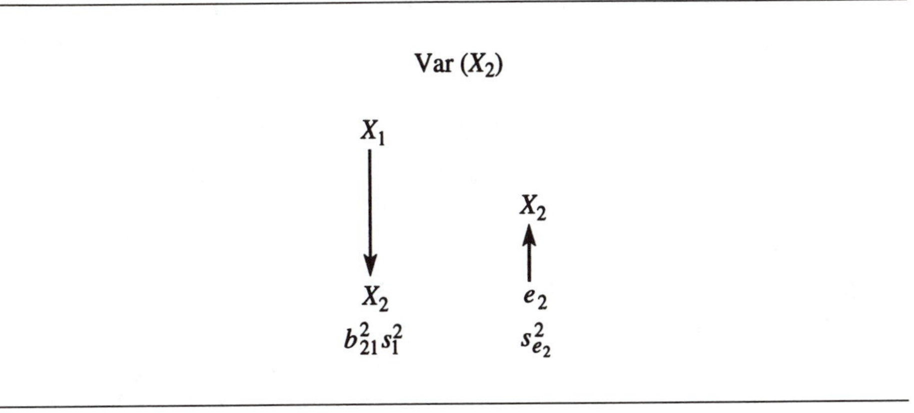

The variance of X_3 is given by

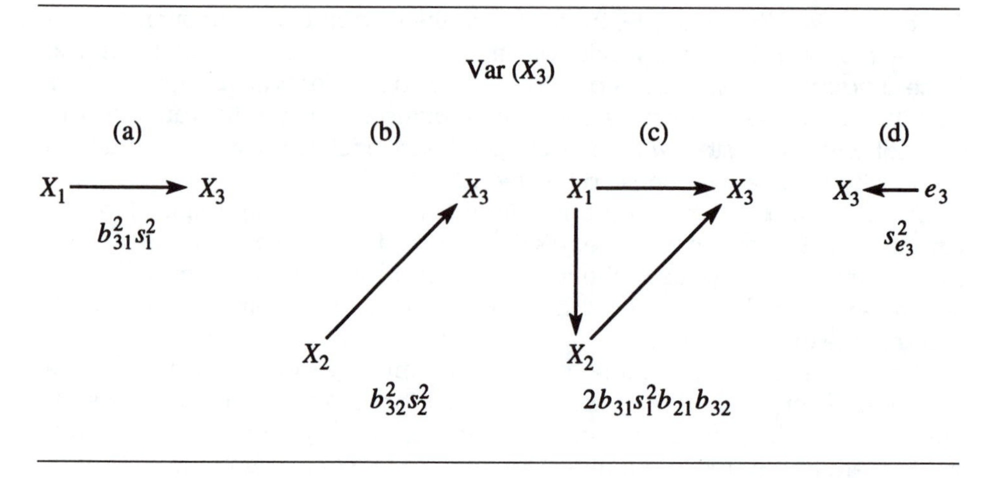

Component (c) is the only one that changes in the three-equation model. It was previously a portion of variance due to correlated causes. Now (c) is attributed to X_1.

The sources of variance in X_4 are shown below.

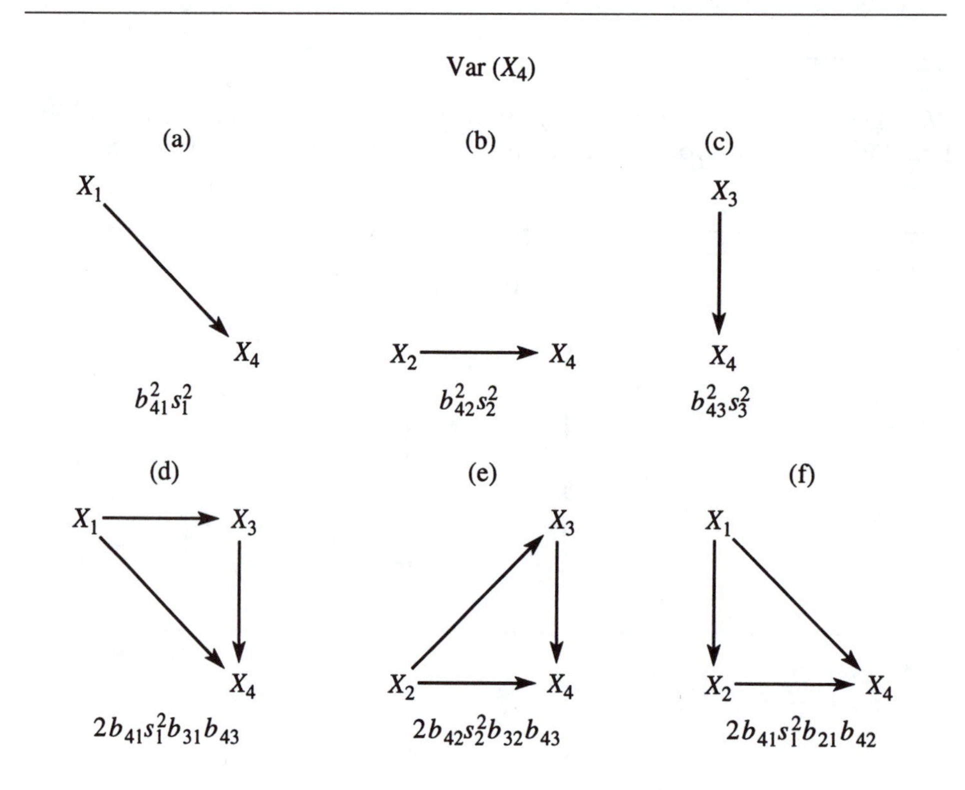

Var (X_4) (continued)

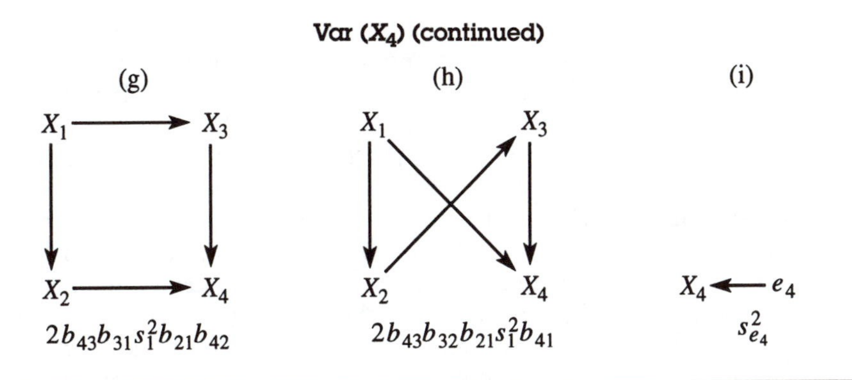

Components (f), (g), and (h) are the three components that have changed from the first model. They were sources of variance due to correlated causes in the first model.

The equations for the variances of the endogenous variables are

$$s_2^2 = b_{21}^2 s_1^2 + s_{e_2}^2 \tag{9.14}$$

$$s_3^2 = b_{31}^2 s_1^2 + b_{32}^2 s_2^2 + 2b_{31} s_1^2 b_{21} b_{32} + s_{e_3}^2 \tag{9.15}$$

$$s_4^2 = b_{41}^2 s_1^2 + b_{42}^2 s_2^2 + b_{43}^2 s_3^2 + 2b_{41} s_1^2 b_{31} b_{43} + 2b_{42} s_2^2 b_{32} b_{43}$$
$$+ 2b_{41} s_1^2 b_{21} b_{42} + 2b_{43} b_{31} s_1^2 b_{21} b_{42} + 2b_{43} b_{32} b_{21} s_1^2 b_{41} + s_{e_4}^2 \tag{9.16}$$

Equation 9.14 is new for this model. Each term in Equations 9.15 and 9.16 contains a variance of one of the variables; there are no longer any covariances in the equations. Where s_{12} appeared in Equations 9.9 and 9.10, it has been replaced by $b_{21} s_1^2$ in Equations 9.15 and 9.16, respectively. Thus, the components of Equations 9.15 and 9.16 would have the same numeric values as those given in Table 9.3 for the self-esteem example.

Some of the variances in Equations 9.15 and 9.16, however, are the variances of the endogenous variables X_2 and X_3. Since s_2^2 and s_3^2 are partly determined by X_1 and by X_1 and X_2, respectively, these terms include some of the variance that is indirectly caused by X_1 and X_2. The only way to achieve an unambiguous allocation of the variance of X_3 and X_4 is to remove the variance of X_2 from Equation 9.15 and to remove the variances of X_2 and X_3 from Equation 9.16. This may be accomplished by substituting Equation 9.14 for s_2^2 in Equation 9.15 and by substituting Equations 9.14 and 9.15 for s_2^2 and s_3^2, respectively, in Equation 9.16. The result of this tedious operation, after much rearranging and simplifying, is

$$s_3^2 = (b_{31} + b_{21} b_{32})^2 s_1^2 + b_{32}^2 s_{e_2}^2 + s_{e_3}^2 \tag{9.17}$$

$$s_4^2 = (b_{41} + b_{21} b_{42} + b_{31} b_{43} + b_{21} b_{32} b_{43})^2 s_1^2$$
$$+ (b_{42} + b_{32} b_{43})^2 s_{e_2}^2 + b_{43}^2 s_{e_3}^2 + s_{e_4}^2 \tag{9.18}$$

Let us examine carefully the terms in these equations. The terms in parentheses in Equation 9.17 are the direct and indirect effects of X_1 on X_3. The sum of these effects is the total effect of X_1 on X_3. When this total effect is squared and multiplied times the variance of X_1, we get the total variance caused by X_1. The next term in Equation 9.17, $b_{32}^2 s_{e_2}^2$, is the square of the direct effect of X_2 on X_3 multiplied by the error variance of X_2. This error variance is the portion that is not explained by X_1, i.e., the unique variance of X_2 relative to X_1. Thus, the second term in Equation 9.17 is the amount of variance in X_3 that is uniquely caused by X_2 or the total variance explained by X_2. The last term, of course, is the residual variance in X_3 that cannot be explained by X_1 and X_2. Thus, Equation 9.17 equals the total variance explained by X_1, plus the total variance explained by X_2, plus the unexplained variance of X_3. Thus, we have arrived at an unambiguous allocation of the variance among the variables.

Equation 9.18, although more complex, has the same interpretation. The terms in parentheses are the total effects of X_1 and X_2 on X_4. The equation also contains the square of the direct effect of X_3 on X_4, which is the total effect of X_3. When these terms are multiplied by the appropriate variances, we get the total variance explained by each independent variable.

We should note that when we square these expressions that represent the total effects, we will get some terms that consist of the product of direct and indirect effects. These products cannot be allocated to either the direct or the indirect effect. Thus, it still is not possible to say how much of the total explained variance is due to the direct effect and how much is due to each of the indirect effects. We must be content to know the total variance explained by the sum of the direct and indirect effects.

Since the equations contain total effects, we may simplify them as follows:

$$s_3^2 = T_{31}^2 s_1^2 + T_{32}^2 s_{e_2}^2 + s_{e_3}^2$$

$$s_4^2 = T_{41}^2 s_1^2 + T_{42}^2 s_{e_2}^2 + T_{43}^2 s_{e_3}^2 + s_{e_4}^2$$

It is important to remember how we were able to accomplish this unambiguous decomposition of the variance. It was made possible by the fact that we were able to specify a causal order for *all* of the variables in the model. If we were not able to do this for some set of variables being investigated, we would have two or more exogenous variables that are merely correlated but that are not believed to be involved in a causal relationship. In that case, Equations 9.17 and 9.18 would include some terms that contain a covariance between the exogenous variables. This would result in some portion of the variance that could not be allocated to one variable or the other.

Summary

A chain rule for reading covariance equations from a path diagram was presented. Each component of the equation for s_{ij} consists of a chain of variables with X_j at one end and X_i at the other end. The contribution of the chain equals the product of all the structural coefficients along the chain times either the variance of the "earliest" variable or the covariance between two exogenous variables. Each component represents either a nonspurious contribution to the covariance, resulting from either a direct or indirect effect, or a spurious contribution, resulting from a common cause or correlated cause. Elaborating a model by specifying a causal link between some of the exogenous variables will change some of the spurious components to nonspurious indirect effects. The reduced-form version of the covariance equation allocates all of the covariance to the exogenous variables, both measured and unmeasured.

The chain rule for reading variance equations from a path diagram is similar to that for covariances, except that each chain begins and ends with X_j. The variance components in models with two or more exogenous variables consist of contributions due to direct effects, combinations of direct and indirect effects, and correlated causes. It is not possible to allocate the explained variance between direct–effect components and indirect–effect components. It is also not possible to allocate unambiguously the variance of an endogenous variable into components due to each of the independent variables. If, however, the model can be respecified to make endogenous variables out of all but one of the exogenous variables (the three-equation example), then the explained variance can be allocated into distinct components due to each of the independent variables. Finally, the reduced forms of the variance equations allocate the variance of each endogenous variable into unique components that represent the total effect of the single measured exogenous variable and the total effect of each unmeasured exogenous variable.

Reference

Wright, Sewell. 1921. "Correlation and Causation." *Journal of Agricultural Research* 20:557–585.

Appendix:
Statistical Tables

TABLE A.1 Areas Under the Standard Normal Distribution Between Z and +∞

Z	.00	.01	.02	.03	.04	.05	.06	.07	.08	.09
.0	.50000	.49601	.49202	.48803	.48405	.48006	.47608	.47210	.46812	.46414
.1	.46017	.45620	.45224	.44828	.44433	.44038	.43644	.43251	.42858	.42465
.2	.42074	.41683	.41294	.40905	.40517	.40129	.39743	.39358	.38974	.38591
.3	.38209	.37828	.37448	.37070	.36693	.36317	.35942	.35569	.35197	.34827
.4	.34458	.34090	.33724	.33360	.32997	.32636	.32276	.31918	.31561	.31207
.5	.30854	.30503	.30153	.29806	.29460	.29116	.28774	.28434	.28096	.27760
.6	.27425	.27093	.26763	.26435	.26109	.25785	.25463	.25143	.24825	.24510
.7	.24196	.23885	.23576	.23270	.22965	.22663	.22363	.22065	.21770	.21476
.8	.21186	.20897	.20611	.20327	.20045	.19766	.19489	.19215	.18943	.18673
.9	.18406	.18141	.17879	.17619	.17361	.17106	.16853	.16602	.16354	.16109
1.0	.15866	.15625	.15386	.15151	.14917	.14686	.14457	.14231	.14007	.13786
1.1	.13567	.13350	.13136	.12924	.12714	.12507	.12302	.12100	.11900	.11702
1.2	.11507	.11314	.11123	.10935	.10749	.10565	.10383	.10204	.10027	.09853
1.3	.09680	.09510	.09342	.09176	.09012	.08851	.08691	.08534	.08379	.08226
1.4	.08076	.07927	.07780	.07636	.07493	.07353	.07215	.07078	.06944	.06811
1.5	.06681	.06552	.06426	.06301	.06178	.06057	.05938	.05821	.05705	.05592
1.6	.05480	.05370	.05262	.05155	.05050	.04947	.04846	.04746	.04648	.04551
1.7	.04457	.04363	.04272	.04182	.04093	.04006	.03920	.03836	.03754	.03673
1.8	.03593	.03515	.03438	.03362	.03288	.03216	.03144	.03074	.03005	.02938
1.9	.02872	.02807	.02743	.02680	.02619	.02559	.02500	.02442	.02385	.02330
2.0	.02275	.02222	.02169	.02118	.02068	.02018	.01970	.01923	.01876	.01831
2.1	.01786	.01743	.01700	.01659	.01618	.01578	.01539	.01500	.01463	.01426
2.2	.01390	.01355	.01321	.01287	.01255	.01222	.01191	.01160	.01130	.01101
2.3	.01072	.01044	.01017	.00990	.00964	.00939	.00914	.00889	.00866	.00842
2.4	.00820	.00798	.00776	.00755	.00734	.00714	.00695	.00676	.00657	.00639
2.5	.00621	.00604	.00587	.00570	.00554	.00539	.00523	.00508	.00494	.00480
2.6	.00466	.00453	.00440	.00427	.00415	.00402	.00391	.00379	.00368	.00357
2.7	.00347	.00336	.00326	.00317	.00307	.00298	.00289	.00280	.00272	.00264
2.8	.00256	.00248	.00240	.00233	.00226	.00219	.00212	.00205	.00199	.00193
2.9	.00187	.00181	.00175	.00169	.00164	.00159	.00154	.00149	.00144	.00139
3.0	.00135	.00131	.00126	.00122	.00118	.00114	.00111	.00107	.00104	.00100
3.1	.00097	.00094	.00090	.00087	.00084	.00082	.00079	.00076	.00074	.00071
3.2	.00069	.00066	.00064	.00062	.00060	.00058	.00056	.00054	.00052	.00050
3.3	.00048	.00047	.00045	.00043	.00042	.00040	.00039	.00038	.00036	.00035
3.4	.00034	.00032	.00031	.00030	.00029	.00028	.00027	.00026	.00025	.00024
3.5	.00023	.00022	.00022	.00021	.00020	.00019	.00019	.00018	.00017	.00017
3.6	.00016	.00015	.00015	.00014	.00014	.00013	.00013	.00012	.00012	.00011
3.7	.00011	.00010	.00010	.00010	.00009	.00009	.00008	.00008	.00008	.00008
3.8	.00007	.00007	.00007	.00006	.00006	.00006	.00006	.00005	.00005	.00005
3.9	.00005	.00005	.00004	.00004	.00004	.00004	.00004	.00004	.00003	.00003

Entries were computed with SYSTAT's cumulative normal distribution function ACF(z).

TABLE A.2 Percentiles of the *t* Distribution

	Level of Significance for One-Tailed Test						
	.10	.05	.025	.01	.005	.0005	.00005
	Level of Significance for Two-Tailed Test						
df	.20	.10	.05	.02	.01	.001	.0001
1	3.078	6.314	12.706	31.821	63.657	636.619	6366.198
2	1.886	2.920	4.303	6.965	9.925	31.599	99.992
3	1.638	2.353	3.182	4.541	5.841	12.924	28.000
4	1.533	2.132	2.776	3.747	4.604	8.610	15.544
5	1.476	2.015	2.571	3.365	4.032	6.869	11.178
6	1.440	1.943	2.447	3.143	3.707	5.959	9.082
7	1.415	1.895	2.365	2.998	3.499	5.408	7.885
8	1.397	1.860	2.306	2.896	3.355	5.041	7.120
9	1.383	1.833	2.262	2.821	3.250	4.781	6.594
10	1.372	1.812	2.228	2.764	3.169	4.587	6.211
11	1.363	1.796	2.201	2.718	3.106	4.437	5.921
12	1.356	1.782	2.179	2.681	3.055	4.318	5.694
13	1.350	1.771	2.160	2.650	3.012	4.221	5.513
14	1.345	1.761	2.145	2.624	2.977	4.140	5.363
15	1.341	1.753	2.131	2.602	2.947	4.073	5.239
16	1.337	1.746	2.120	2.583	2.921	4.015	5.134
17	1.333	1.740	2.110	2.567	2.898	3.965	5.044
18	1.330	1.734	2.101	2.552	2.878	3.922	4.966
19	1.328	1.729	2.093	2.539	2.861	3.883	4.897
20	1.325	1.725	2.086	2.528	2.845	3.850	4.837
22	1.321	1.717	2.074	2.508	2.819	3.792	4.736
24	1.318	1.711	2.064	2.492	2.797	3.745	4.654
26	1.315	1.706	2.056	2.479	2.779	3.707	4.587
28	1.313	1.701	2.048	2.467	2.763	3.674	4.530
30	1.310	1.697	2.042	2.457	2.750	3.646	4.482
35	1.306	1.690	2.030	2.438	2.724	3.591	4.389
40	1.303	1.684	2.021	2.423	2.704	3.551	4.321
50	1.299	1.676	2.009	2.403	2.678	3.496	4.228
60	1.296	1.671	2.000	2.390	2.660	3.460	4.169
80	1.292	1.664	1.990	2.374	2.639	3.416	4.096
100	1.290	1.660	1.984	2.364	2.626	3.390	4.053
150	1.287	1.655	1.976	2.351	2.609	3.357	3.998
200	1.286	1.653	1.972	2.345	2.601	3.340	3.970
300	1.284	1.650	1.968	2.339	2.592	3.323	3.944
400	1.284	1.649	1.966	2.336	2.588	3.315	3.930
500	1.283	1.648	1.965	2.334	2.586	3.310	3.922
750	1.283	1.647	1.963	2.331	2.582	3.304	3.912
1000	1.282	1.646	1.962	2.330	2.581	3.300	3.906
1500	1.282	1.646	1.962	2.329	2.579	3.297	3.901
∞	1.282	1.645	1.960	2.326	2.576	3.291	3.891

Entries were computed with SYSTAT's inverse *t* distribution function TIF (α , df).

TABLE A.3a Percentiles of the *F* Distribution, $p = .05$

	Degrees of Freedom for Numerator (df_1)														
df_2	1	2	3	4	5	6	7	8	9	10	12	16	20	40	∞
1	161.4	199.5	215.7	224.5	230.1	233.9	236.7	238.8	240.5	241.8	243.9	246.4	248.0	251.1	254.3
2	18.51	19.00	19.16	19.25	19.30	19.33	19.35	19.37	19.38	19.40	19.41	19.43	19.45	19.47	19.50
3	10.13	9.55	9.28	9.12	9.01	8.94	8.89	8.85	8.81	8.79	8.74	8.69	8.66	8.59	8.53
4	7.71	6.94	6.59	6.39	6.26	6.16	6.09	6.04	6.00	5.96	5.91	5.84	5.80	5.72	5.63
5	6.61	5.79	5.41	5.19	5.05	4.95	4.88	4.82	4.77	4.74	4.68	4.60	4.56	4.46	4.37
6	5.99	5.14	4.76	4.53	4.39	4.28	4.21	4.15	4.10	4.06	4.00	3.92	3.87	3.77	3.67
7	5.59	4.74	4.35	4.12	3.97	3.87	3.79	3.73	3.68	3.64	3.57	3.49	3.44	3.34	3.23
8	5.32	4.46	4.07	3.84	3.69	3.58	3.50	3.44	3.39	3.35	3.28	3.20	3.15	3.04	2.93
9	5.12	4.26	3.86	3.63	3.48	3.37	3.29	3.23	3.18	3.14	3.07	2.99	2.94	2.83	2.71
10	4.96	4.10	3.71	3.48	3.33	3.22	3.14	3.07	3.02	2.98	2.91	2.83	2.77	2.66	2.54
11	4.84	3.98	3.59	3.36	3.20	3.09	3.01	2.95	2.90	2.85	2.79	2.70	2.65	2.53	2.41
12	4.75	3.89	3.49	3.26	3.11	3.00	2.91	2.85	2.80	2.75	2.69	2.60	2.54	2.43	2.30
13	4.67	3.81	3.41	3.18	3.03	2.92	2.83	2.77	2.71	2.67	2.60	2.51	2.46	2.34	2.21
14	4.60	3.74	3.34	3.11	2.96	2.85	2.76	2.70	2.65	2.60	2.53	2.44	2.39	2.27	2.13
15	4.54	3.68	3.29	3.06	2.90	2.79	2.71	2.64	2.59	2.54	2.48	2.38	2.33	2.20	2.07
16	4.49	3.63	3.24	3.01	2.85	2.74	2.66	2.59	2.54	2.49	2.42	2.33	2.28	2.15	2.01
17	4.45	3.59	3.20	2.96	2.81	2.70	2.61	2.55	2.49	2.45	2.38	2.29	2.23	2.10	1.96
18	4.41	3.55	3.16	2.93	2.77	2.66	2.58	2.51	2.46	2.41	2.34	2.25	2.19	2.06	1.92
19	4.38	3.52	3.13	2.90	2.74	2.63	2.54	2.48	2.42	2.38	2.31	2.21	2.16	2.03	1.88
20	4.35	3.49	3.10	2.87	2.71	2.60	2.51	2.45	2.39	2.35	2.28	2.18	2.12	1.99	1.84
22	4.30	3.44	3.05	2.82	2.66	2.55	2.46	2.40	2.34	2.30	2.23	2.13	2.07	1.94	1.78
24	4.26	3.40	3.01	2.78	2.62	2.51	2.42	2.36	2.30	2.25	2.18	2.09	2.03	1.89	1.73
26	4.23	3.37	2.98	2.74	2.59	2.47	2.39	2.32	2.27	2.22	2.15	2.05	1.99	1.85	1.69
28	4.20	3.34	2.95	2.71	2.56	2.45	2.36	2.29	2.24	2.19	2.12	2.02	1.96	1.82	1.65
30	4.17	3.32	2.92	2.69	2.53	2.42	2.33	2.27	2.21	2.16	2.09	1.99	1.93	1.79	1.62
35	4.12	3.27	2.87	2.64	2.49	2.37	2.29	2.22	2.16	2.11	2.04	1.94	1.88	1.74	1.56
40	4.08	3.23	2.84	2.61	2.45	2.34	2.25	2.18	2.12	2.08	2.00	1.90	1.84	1.69	1.51
50	4.03	3.18	2.79	2.56	2.40	2.29	2.20	2.13	2.07	2.03	1.95	1.85	1.78	1.63	1.44
60	4.00	3.15	2.76	2.53	2.37	2.25	2.17	2.10	2.04	1.99	1.92	1.82	1.75	1.59	1.39
80	3.96	3.11	2.72	2.49	2.33	2.21	2.13	2.06	2.00	1.95	1.88	1.77	1.70	1.54	1.33
100	3.94	3.09	2.70	2.46	2.31	2.19	2.10	2.03	1.97	1.93	1.85	1.75	1.68	1.52	1.28
150	3.90	3.06	2.66	2.43	2.27	2.16	2.07	2.00	1.94	1.89	1.82	1.71	1.64	1.48	1.22
200	3.89	3.04	2.65	2.42	2.26	2.14	2.06	1.98	1.93	1.88	1.80	1.69	1.62	1.46	1.19
300	3.87	3.03	2.63	2.40	2.24	2.13	2.04	1.97	1.91	1.86	1.78	1.68	1.61	1.43	1.15
400	3.86	3.02	2.63	2.39	2.24	2.12	2.03	1.96	1.90	1.85	1.78	1.67	1.60	1.42	1.13
500	3.86	3.01	2.62	2.39	2.23	2.12	2.03	1.96	1.90	1.85	1.77	1.66	1.59	1.42	1.12
750	3.85	3.01	2.62	2.38	2.23	2.11	2.02	1.95	1.89	1.84	1.77	1.66	1.58	1.41	1.09
1000	3.85	3.00	2.61	2.38	2.22	2.11	2.02	1.95	1.89	1.84	1.76	1.65	1.58	1.41	1.08
1500	3.85	3.00	2.61	2.38	2.22	2.10	2.02	1.94	1.89	1.84	1.76	1.65	1.58	1.40	1.07
∞	3.84	3.00	2.61	2.37	2.21	2.10	2.01	1.94	1.88	1.83	1.75	1.64	1.57	1.40	1.00

Entries were computed with SYSTAT's inverse *F* distribution function FIF (α, *df*1, *df*2).

TABLE A.3b Percentiles of the F Distribution, $p = .01$

	Degrees of Freedom for Numerator (df_1)														
df_2	1	2	3	4	5	6	7	8	9	10	12	16	20	40	∞
1	.	.	.	.	.	.	.	.	.	.	.	.	.	.	.
2	98.50	99.00	99.17	99.25	99.30	99.33	99.36	99.37	99.39	99.40	99.42	99.44	99.45	99.47	99.50
3	34.12	30.82	29.46	28.71	28.24	27.91	27.67	27.49	27.35	27.23	27.05	26.83	26.69	26.41	26.13
4	21.20	18.00	16.69	15.98	15.52	15.21	14.98	14.80	14.66	14.55	14.37	14.15	14.02	13.75	13.46
5	16.26	13.27	12.06	11.39	10.97	10.67	10.46	10.29	10.16	10.05	9.89	9.68	9.55	9.29	9.02
6	13.75	10.92	9.78	9.15	8.75	8.47	8.26	8.10	7.98	7.87	7.72	7.52	7.40	7.14	6.88
7	12.25	9.55	8.45	7.85	7.46	7.19	6.99	6.84	6.72	6.62	6.47	6.28	6.16	5.91	5.65
8	11.26	8.65	7.59	7.01	6.63	6.37	6.18	6.03	5.91	5.81	5.67	5.48	5.36	5.12	4.86
9	10.56	8.02	6.99	6.42	6.06	5.80	5.61	5.47	5.35	5.26	5.11	4.92	4.81	4.57	4.31
10	10.04	7.56	6.55	5.99	5.64	5.39	5.20	5.06	4.94	4.85	4.71	4.52	4.41	4.17	3.91
11	9.65	7.21	6.22	5.67	5.32	5.07	4.89	4.74	4.63	4.54	4.40	4.21	4.10	3.86	3.60
12	9.33	6.93	5.95	5.41	5.06	4.82	4.64	4.50	4.39	4.30	4.16	3.97	3.86	3.62	3.36
13	9.07	6.70	5.74	5.21	4.86	4.62	4.44	4.30	4.19	4.10	3.96	3.78	3.66	3.43	3.17
14	8.86	6.51	5.56	5.04	4.69	4.46	4.28	4.14	4.03	3.94	3.80	3.62	3.51	3.27	3.00
15	8.68	6.36	5.42	4.89	4.56	4.32	4.14	4.00	3.89	3.80	3.67	3.49	3.37	3.13	2.87
16	8.53	6.23	5.29	4.77	4.44	4.20	4.03	3.89	3.78	3.69	3.55	3.37	3.26	3.02	2.75
17	8.40	6.11	5.18	4.67	4.34	4.10	3.93	3.79	3.68	3.59	3.46	3.27	3.16	2.92	2.65
18	8.29	6.01	5.09	4.58	4.25	4.01	3.84	3.71	3.60	3.51	3.37	3.19	3.08	2.84	2.57
19	8.18	5.93	5.01	4.50	4.17	3.94	3.77	3.63	3.52	3.43	3.30	3.12	3.00	2.76	2.49
20	8.10	5.85	4.94	4.43	4.10	3.87	3.70	3.56	3.46	3.37	3.23	3.05	2.94	2.69	2.42
22	7.95	5.72	4.82	4.31	3.99	3.76	3.59	3.45	3.35	3.26	3.12	2.94	2.83	2.58	2.31
24	7.82	5.61	4.72	4.22	3.90	3.67	3.50	3.36	3.26	3.17	3.03	2.85	2.74	2.49	2.21
26	7.72	5.53	4.64	4.14	3.82	3.59	3.42	3.29	3.18	3.09	2.96	2.78	2.66	2.42	2.13
28	7.64	5.45	4.57	4.07	3.75	3.53	3.36	3.23	3.12	3.03	2.90	2.72	2.60	2.35	2.06
30	7.56	5.39	4.51	4.02	3.70	3.47	3.30	3.17	3.07	2.98	2.84	2.66	2.55	2.30	2.01
35	7.42	5.27	4.40	3.91	3.59	3.37	3.20	3.07	2.96	2.88	2.74	2.56	2.44	2.19	1.89
40	7.31	5.18	4.31	3.83	3.51	3.29	3.12	2.99	2.89	2.80	2.66	2.48	2.37	2.11	1.80
50	7.17	5.06	4.20	3.72	3.41	3.19	3.02	2.89	2.78	2.70	2.56	2.38	2.27	2.01	1.68
60	7.08	4.98	4.13	3.65	3.34	3.12	2.95	2.82	2.72	2.63	2.50	2.31	2.20	1.94	1.60
80	6.96	4.88	4.04	3.56	3.26	3.04	2.87	2.74	2.64	2.55	2.42	2.23	2.12	1.85	1.49
100	6.90	4.82	3.98	3.51	3.21	2.99	2.82	2.69	2.59	2.50	2.37	2.19	2.07	1.80	1.43
150	6.81	4.75	3.91	3.45	3.14	2.92	2.76	2.63	2.53	2.44	2.31	2.12	2.00	1.73	1.33
200	6.76	4.71	3.88	3.41	3.11	2.89	2.73	2.60	2.50	2.41	2.27	2.09	1.97	1.69	1.28
300	6.72	4.68	3.85	3.38	3.08	2.86	2.70	2.57	2.47	2.38	2.24	2.06	1.94	1.66	1.22
400	6.70	4.66	3.83	3.37	3.06	2.85	2.68	2.56	2.45	2.37	2.23	2.05	1.92	1.64	1.19
500	6.69	4.65	3.82	3.36	3.05	2.84	2.68	2.55	2.44	2.36	2.22	2.04	1.92	1.63	1.17
750	6.67	4.63	3.81	3.34	3.04	2.83	2.66	2.53	2.43	2.34	2.21	2.02	1.90	1.62	1.13
1000	6.66	4.63	3.80	3.34	3.04	2.82	2.66	2.53	2.43	2.34	2.20	2.02	1.90	1.61	1.11
1500	6.65	4.62	3.79	3.33	3.03	2.81	2.65	2.52	2.42	2.33	2.20	2.01	1.89	1.61	1.09
∞	6.63	4.61	3.78	3.32	3.02	2.80	2.64	2.51	2.41	2.32	2.18	2.00	1.88	1.59	1.00

Entries were computed with SYSTAT's inverse F distribution function FIF (α, $df1$, $df2$).

TABLE A.3c Percentiles of the F Distribution, $p = .001$

	Degrees of Freedom for Numerator (df_1)														
df_2	1	2	3	4	5	6	7	8	9	10	12	16	20	40	∞
1	.	.	.	.	.	.	.	.	.	.	.	.	.	.	.
2	998.	999.	999.	999.	999.	999.	999.	999.	999.	999.	999.	999.	999.	999.	999.
3	167.0	148.5	141.1	137.1	134.5	132.8	131.5	130.6	129.8	129.2	128.3	127.1	126.4	124.9	123.4
4	74.14	61.25	56.18	53.44	51.71	50.53	49.66	49.00	48.47	48.05	47.41	46.60	46.10	45.09	44.05
5	47.18	37.12	33.20	31.09	29.75	28.83	28.16	27.65	27.24	26.92	26.42	25.78	25.39	24.60	23.79
6	35.51	27.00	23.70	21.92	20.80	20.03	19.46	19.03	18.69	18.41	17.99	17.45	17.12	16.44	15.75
7	29.25	21.69	18.77	17.20	16.21	15.52	15.02	14.63	14.33	14.08	13.71	13.23	12.93	12.33	11.70
8	25.41	18.49	15.83	14.39	13.48	12.86	12.40	12.05	11.77	11.54	11.19	10.75	10.48	9.92	9.33
9	22.86	16.39	13.90	12.56	11.71	11.13	10.70	10.37	10.11	9.89	9.57	9.15	8.90	8.37	7.81
10	21.04	14.91	12.55	11.28	10.48	9.93	9.52	9.20	8.96	8.75	8.45	8.05	7.80	7.30	6.76
11	19.69	13.81	11.56	10.35	9.58	9.05	8.66	8.35	8.12	7.92	7.63	7.24	7.01	6.52	6.00
12	18.64	12.97	10.80	9.63	8.89	8.38	8.00	7.71	7.48	7.29	7.00	6.63	6.40	5.93	5.42
13	17.82	12.31	10.21	9.07	8.35	7.86	7.49	7.21	6.98	6.80	6.52	6.16	5.93	5.47	4.97
14	17.14	11.78	9.73	8.62	7.92	7.44	7.08	6.80	6.58	6.40	6.13	5.78	5.56	5.10	4.60
15	16.59	11.34	9.34	8.25	7.57	7.09	6.74	6.47	6.26	6.08	5.81	5.46	5.25	4.80	4.31
16	16.12	10.97	9.01	7.94	7.27	6.80	6.46	6.19	5.98	5.81	5.55	5.20	4.99	4.54	4.06
17	15.72	10.66	8.73	7.68	7.02	6.56	6.22	5.96	5.75	5.58	5.32	4.99	4.78	4.33	3.85
18	15.38	10.39	8.49	7.46	6.81	6.35	6.02	5.76	5.56	5.39	5.13	4.80	4.59	4.15	3.67
19	15.08	10.16	8.28	7.27	6.62	6.18	5.85	5.59	5.39	5.22	4.97	4.64	4.43	3.99	3.51
20	14.82	9.95	8.10	7.10	6.46	6.02	5.69	5.44	5.24	5.08	4.82	4.49	4.29	3.86	3.38
22	14.38	9.61	7.80	6.81	6.19	5.76	5.44	5.19	4.99	4.83	4.58	4.26	4.06	3.63	3.15
24	14.03	9.34	7.55	6.59	5.98	5.55	5.23	4.99	4.80	4.64	4.39	4.07	3.87	3.45	2.97
26	13.74	9.12	7.36	6.41	5.80	5.38	5.07	4.83	4.64	4.48	4.24	3.92	3.72	3.30	2.82
28	13.50	8.93	7.19	6.25	5.66	5.24	4.93	4.69	4.50	4.35	4.11	3.80	3.60	3.18	2.70
30	13.29	8.77	7.05	6.12	5.53	5.12	4.82	4.58	4.39	4.24	4.00	3.69	3.49	3.07	2.59
35	12.90	8.47	6.79	5.88	5.30	4.89	4.59	4.36	4.18	4.03	3.79	3.48	3.29	2.87	2.38
40	12.61	8.25	6.59	5.70	5.13	4.73	4.44	4.21	4.02	3.87	3.64	3.34	3.14	2.73	2.23
50	12.22	7.96	6.34	5.46	4.90	4.51	4.22	4.00	3.82	3.67	3.44	3.14	2.95	2.53	2.03
60	11.97	7.77	6.17	5.31	4.76	4.37	4.09	3.86	3.69	3.54	3.32	3.02	2.83	2.41	1.89
80	11.67	7.54	5.97	5.12	4.58	4.20	3.92	3.70	3.53	3.39	3.16	2.87	2.68	2.26	1.72
100	11.50	7.41	5.86	5.02	4.48	4.11	3.83	3.61	3.44	3.30	3.07	2.78	2.59	2.17	1.62
150	11.27	7.24	5.71	4.88	4.35	3.98	3.71	3.49	3.32	3.18	2.96	2.67	2.48	2.06	1.47
200	11.15	7.15	5.63	4.81	4.29	3.92	3.65	3.43	3.26	3.12	2.90	2.61	2.42	2.00	1.39
300	11.04	7.07	5.56	4.75	4.22	3.86	3.59	3.38	3.21	3.07	2.85	2.56	2.37	1.94	1.31
400	10.99	7.03	5.53	4.71	4.19	3.83	3.56	3.35	3.18	3.04	2.82	2.53	2.34	1.92	1.26
500	10.96	7.00	5.51	4.69	4.18	3.81	3.54	3.33	3.16	3.02	2.81	2.52	2.33	1.90	1.23
750	10.91	6.97	5.48	4.67	4.15	3.79	3.52	3.31	3.14	3.00	2.78	2.49	2.31	1.88	1.18
1000	10.89	6.96	5.46	4.65	4.14	3.78	3.51	3.30	3.13	2.99	2.77	2.48	2.30	1.87	1.15
1500	10.87	6.94	5.45	4.64	4.13	3.77	3.50	3.29	3.12	2.98	2.76	2.47	2.29	1.86	1.12
∞	10.83	6.91	5.42	4.62	4.10	3.74	3.48	3.27	3.10	2.96	2.74	2.45	2.27	1.84	1.00

Entries were computed with SYSTAT's inverse F distribution function FIF (α, $df1$, $df2$).

Index